The Mocap Book

A Practical Guide to the Art of Motion Capture

Ricardo Tobón

To Alex

Acknowledgements

My profound thanks to Demian Gordon for his continued mentorship and guidance in everything related to the motion capture field.

I would also like to acknowledge and thank the Motion Capture Society and the Mocap Club for providing a platform for the advancement and strengthening of the motion capture community.

Thanks to Pedro Flores for designing the cover of this book.

Thanks to the Flight of Icarus Committee for creating the CG visuals that make part of the cover.

Thanks to Chris Maraffi for his insight on the difficult art of writing a book.

Many thanks to Full Sail University, my colleagues as well as the students for their daily support and inspiration.

Thanks to Motion Analysis for their help and support over the years.

Thanks to my family, to Olga and Gilberto for instilling in me the interest and love for the academia, to Eduardo for teaching me to never give up.

My gratitude towards Alejandra Restrepo for her patience and support.

And last but not least, I would like to thank Foris Force for publishing this book.

Contents

Book Files

Follow along files and bonus videos are available for download at:

www.MocapClub.com/TheMocapBook

Part 01
Motion Capture Basics

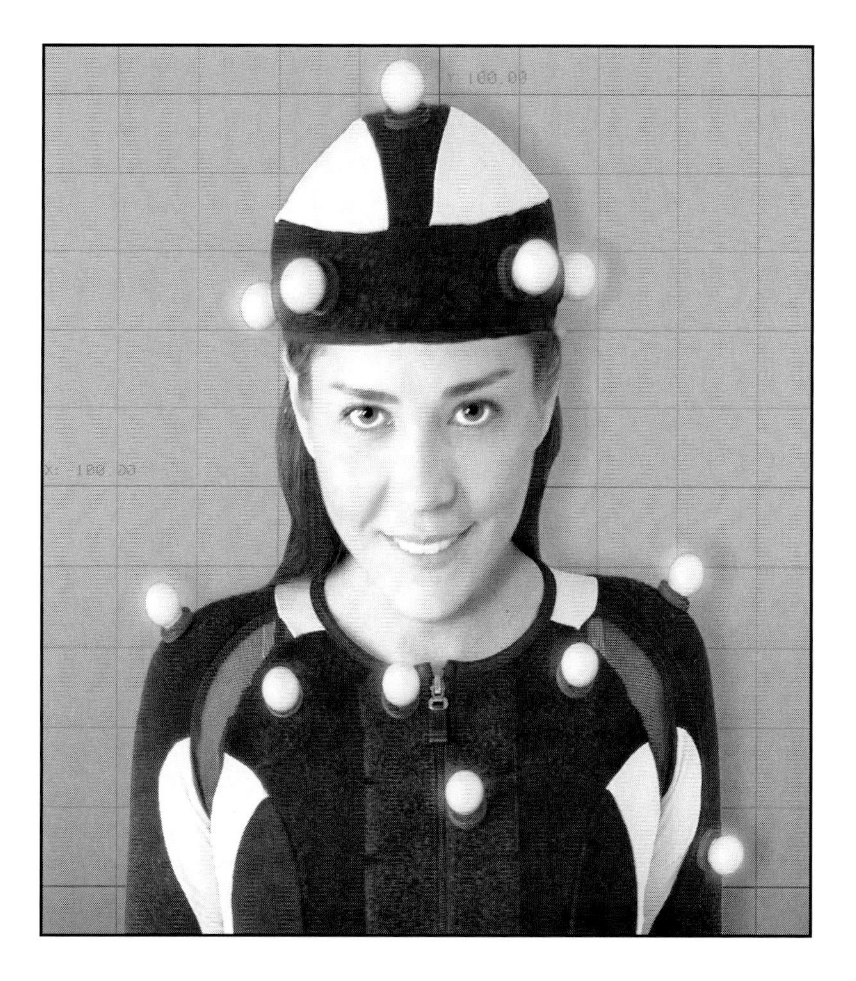

Chapter 01
Optical Motion Capture

Motion Capture is self-defined by its name.

Motion = the act of physically changing location.

Capture = take into possession, to seize, to acquire.

So motion capture is the acquisition of movement.

Because of the advances in computing capabilities in the last few decades modern motion capture in the practical sense pertains more to digitally acquiring motion or the digitization of motion (Figure 01_01).

Figure 01_01.

There are many ways to digitize motion; most of them combine the use of hardware and software to capture movement. In this book we will focus on the use of an optical system to capture motion (Figure 01_02).

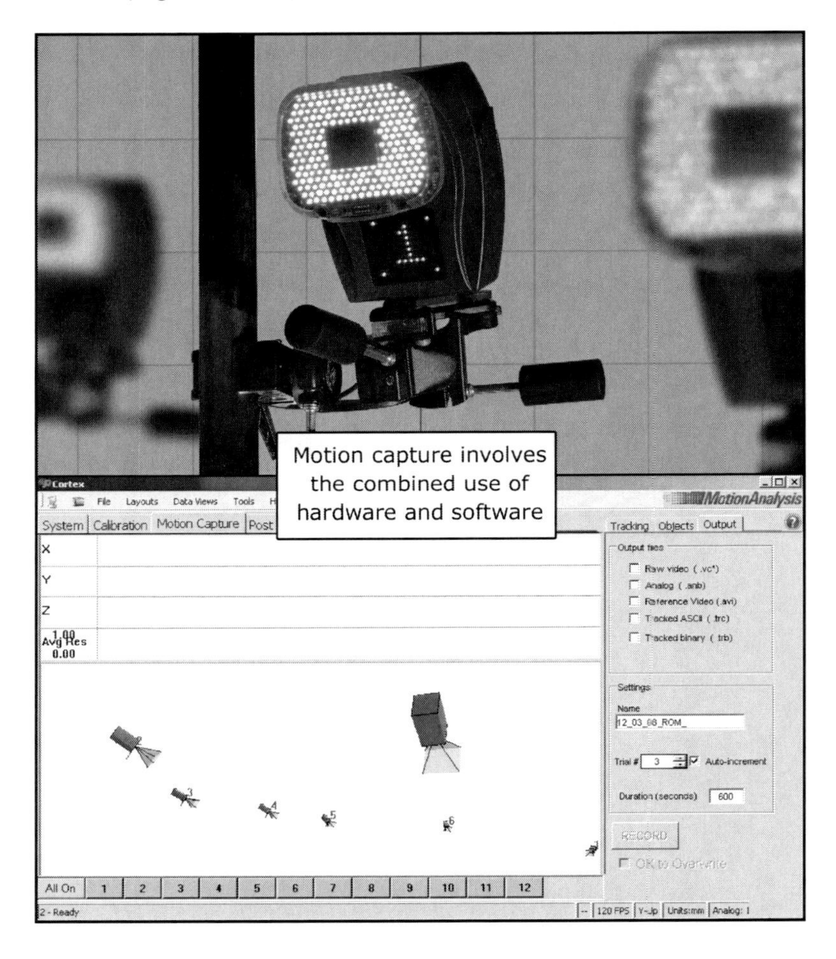

Figure 01_02.

An optical system uses an arrangement of cameras to track information. Most optical systems track reflective markers. The system then combines the information of these different views of the tracked marker to describe the 3D position of the object (Figure 01_03).

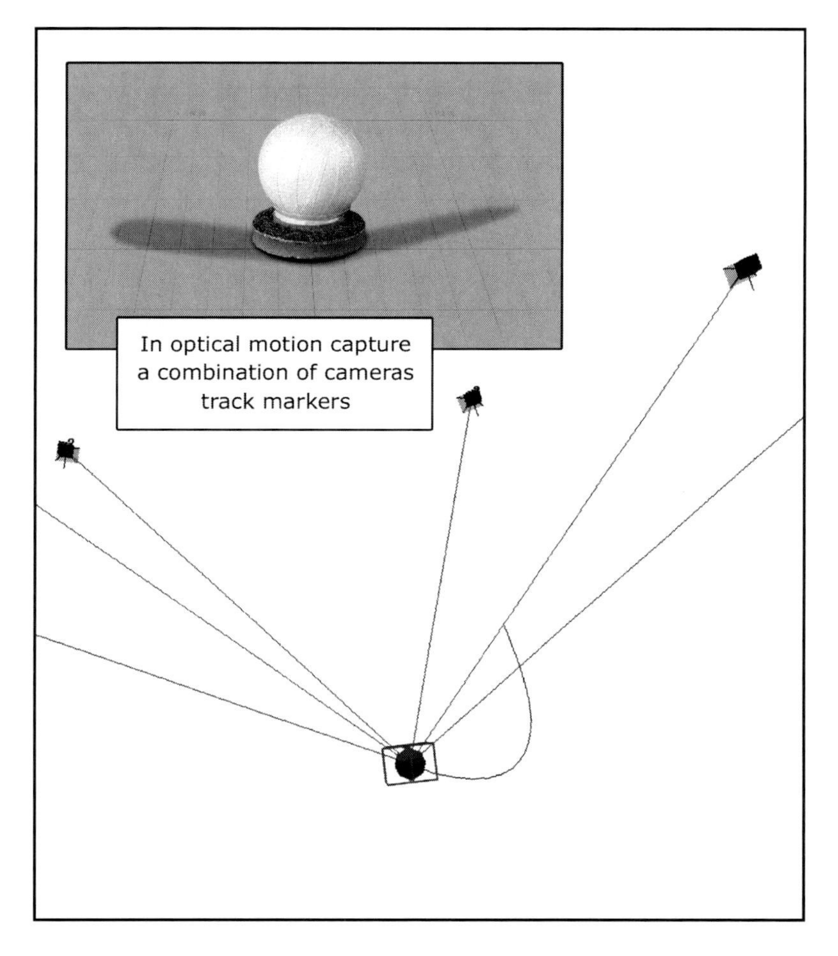

Figure 01_03.

This is quite similar to how a user can place an object in 3D with high accuracy by the use of the orthographic views of the animation software of their choice (Figure 01_04).

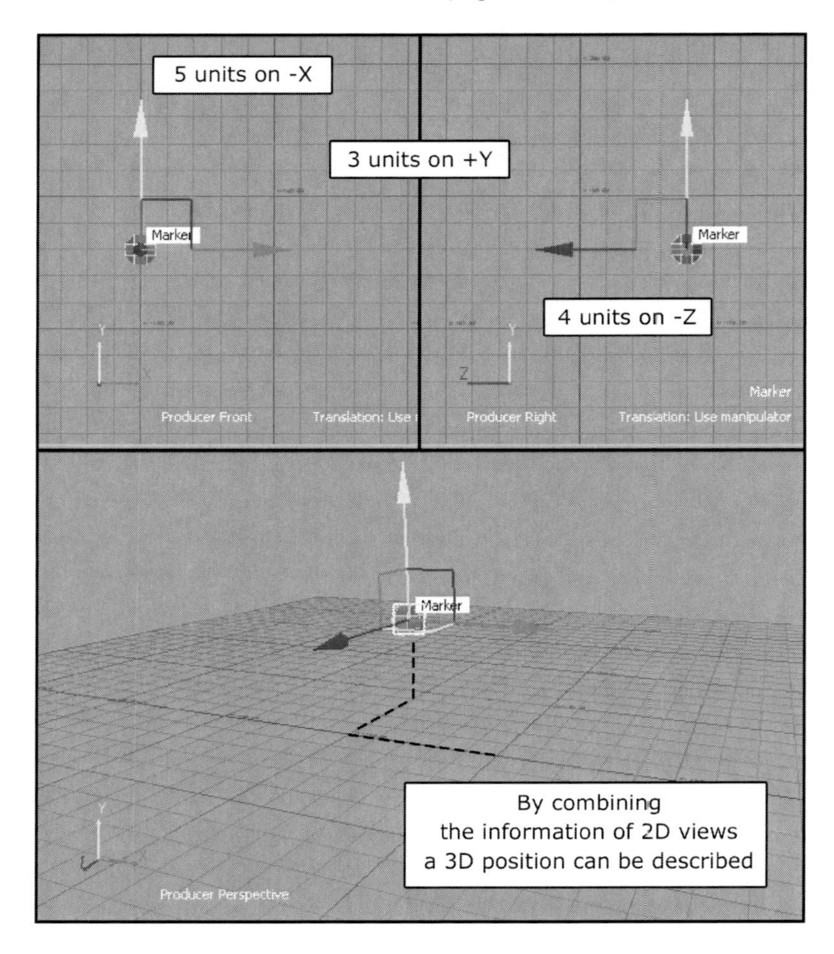

Figure 01_04.

By repeating this operation several times per second the system can provide us with the volumetric trajectory of the marker according to time and space. Most optical systems can capture anywhere from 30 to 120 fps (Figure 01_05).

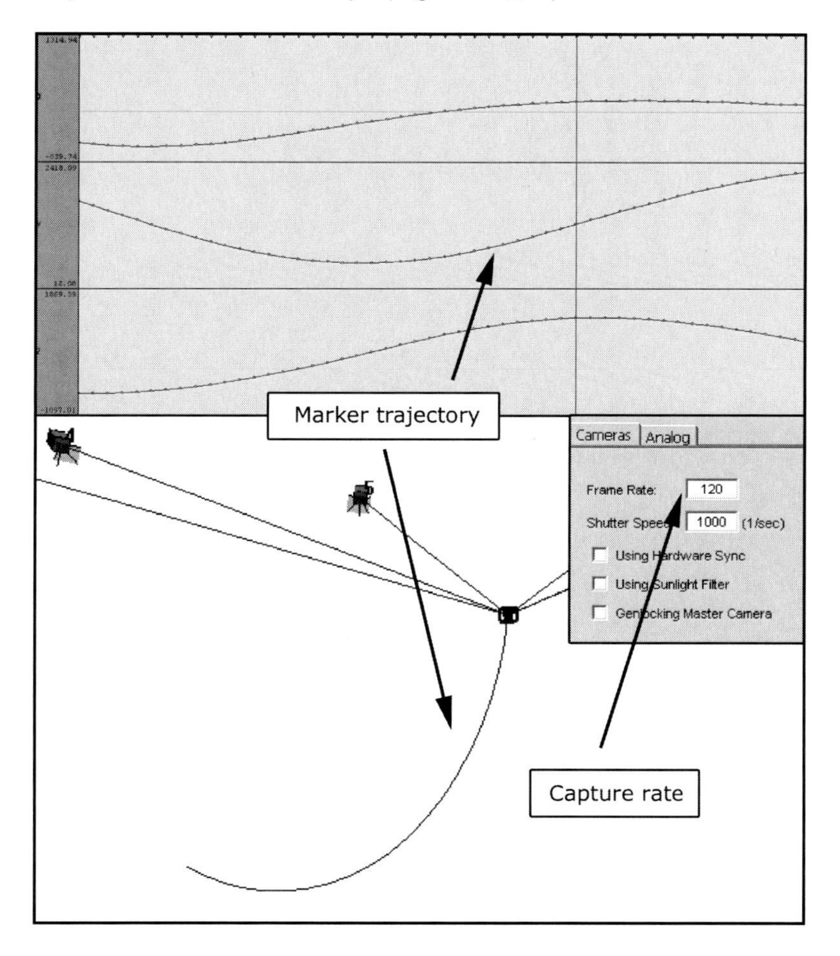

Figure 01_05.

The minimum amount of cameras needed to track an object is 2 (Figure 01_06).

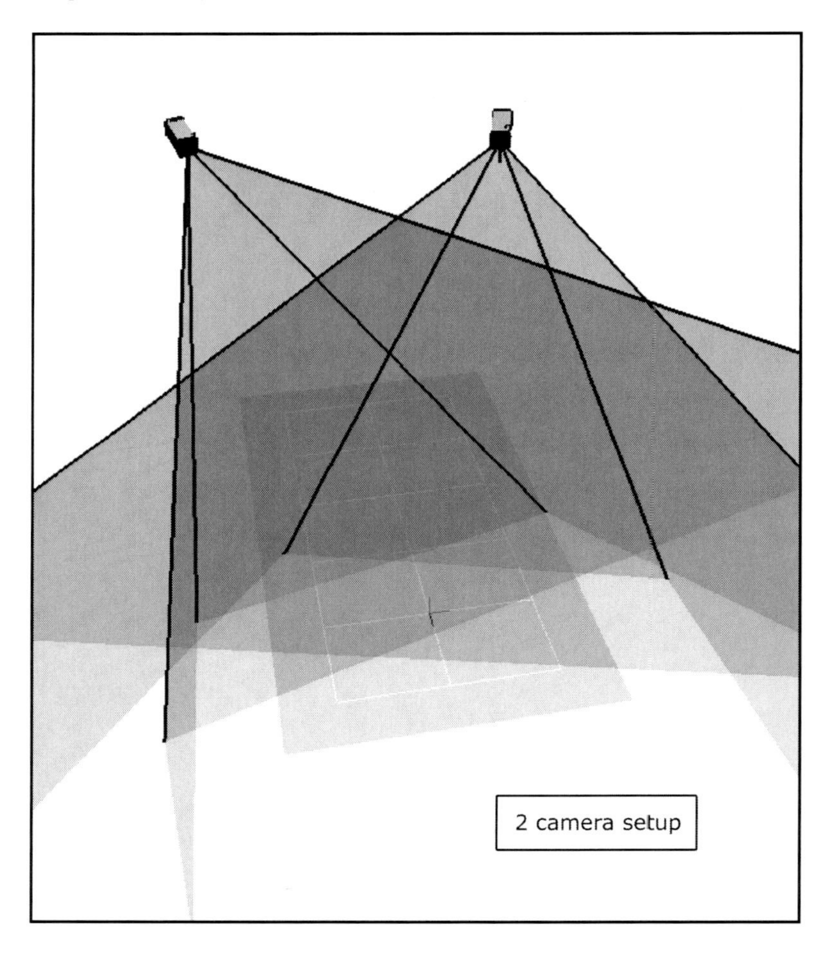

Figure 01_06.

When a non-transparent object is carrying markers a 2 camera setup is unable to see the ones that are pointing away from the cameras. A marker that is not seen by the cameras is called an *occluded marker* (Figure 01_07).

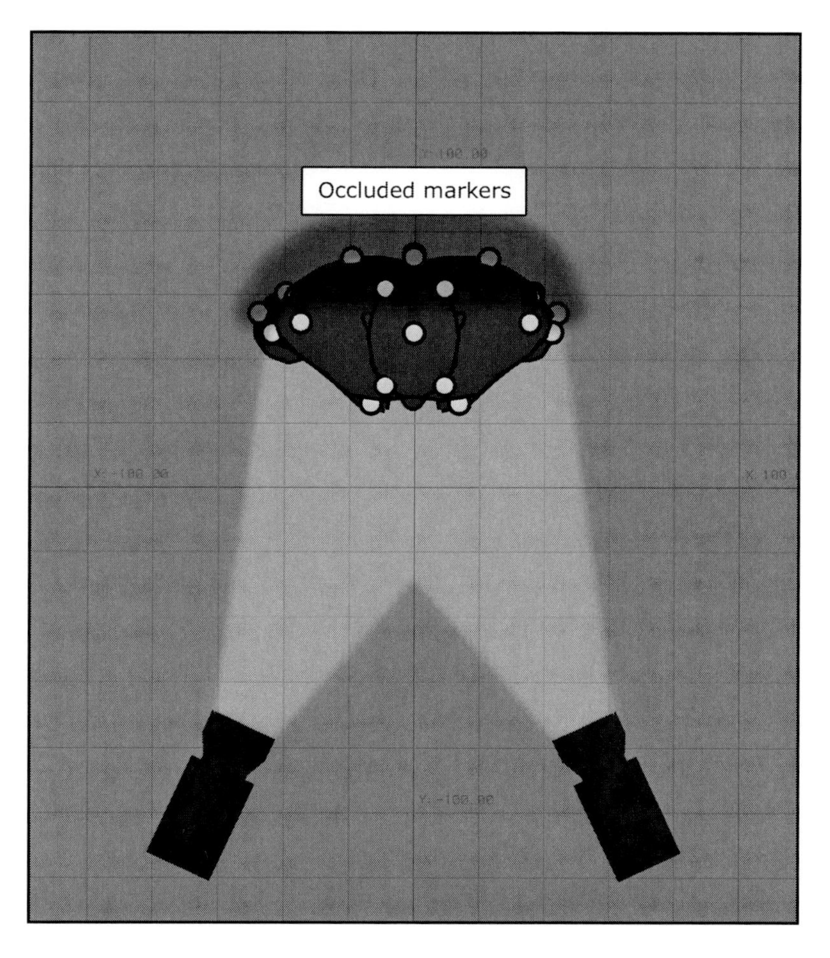

Figure 01_07.

A 4 camera setup will fix that problem (Figure 01_08)

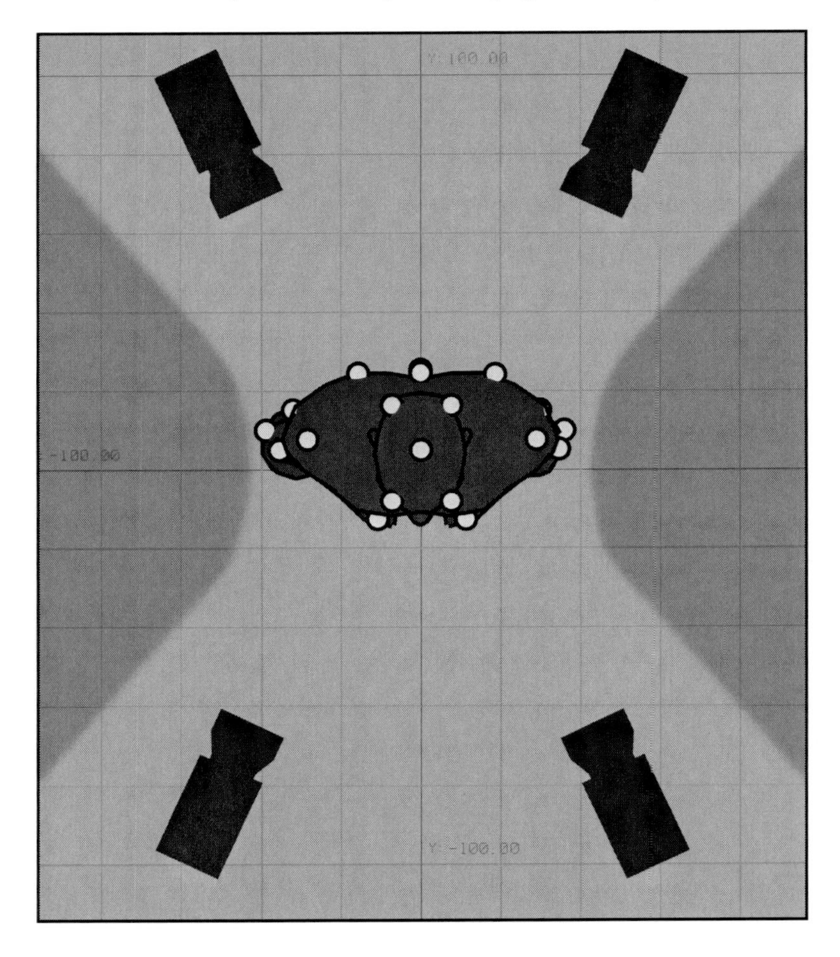

Figure 01_08.

The physical space where the cameras can combine their fields of view in order to describe the position of a marker is called the Capture Volume (Figure 01_09).

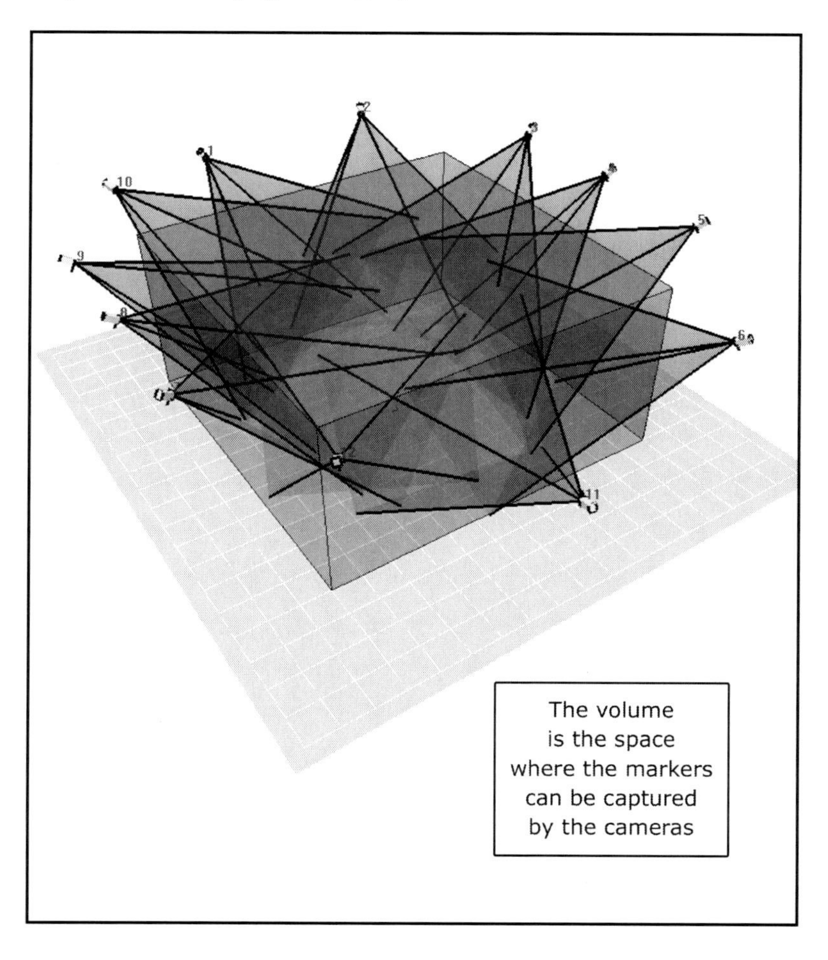

Figure 01_09.

As a general rule of thumb, the more cameras a system has, the more the chances that two or more cameras can track the markers, the bigger the volume can be. 8 to 16 cameras provide a good basic volume.

Most motion capture vendors will be able to help a costumer on figuring out a camera setup that will optimize the space to be used for motion capture. It is important to make sure that the camera setup that the client acquires fits their specific purposes. A 2 to 8 camera system will be very accurate for some medical applications, while a 16 to 32 camera system works better for a video game scenario. Motion Picture and VSFX setups can have over 200 cameras (Figure 01_10).

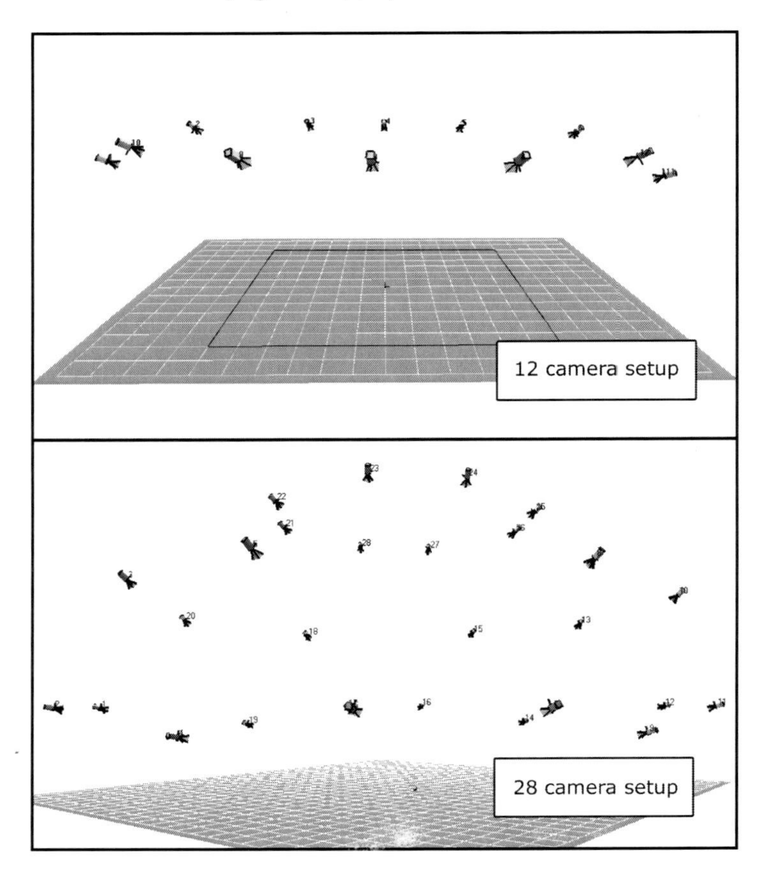

Figure 01_10.

Chapter 02
Marker Placement

One of the strengths of an optical system is the possibility to capture everything you can put markers on. You can capture, humans, dogs, horses, inanimate objects, etc. (Figure 02_02).

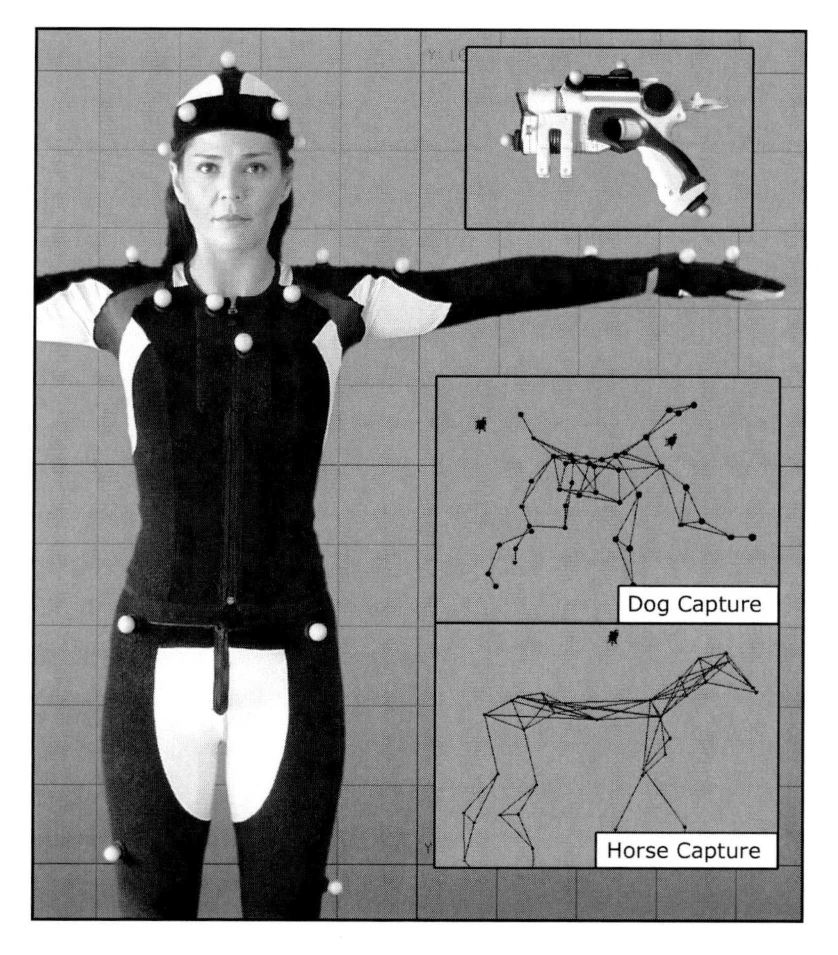

Figure 02_01.

It was explained in the previous chapter that an optical system uses cameras to track the position of a marker. The main set of data you can get out of a single reflective dot to transfer to a 3D animation package is X, Y and Z translation information (Figure 02_02).

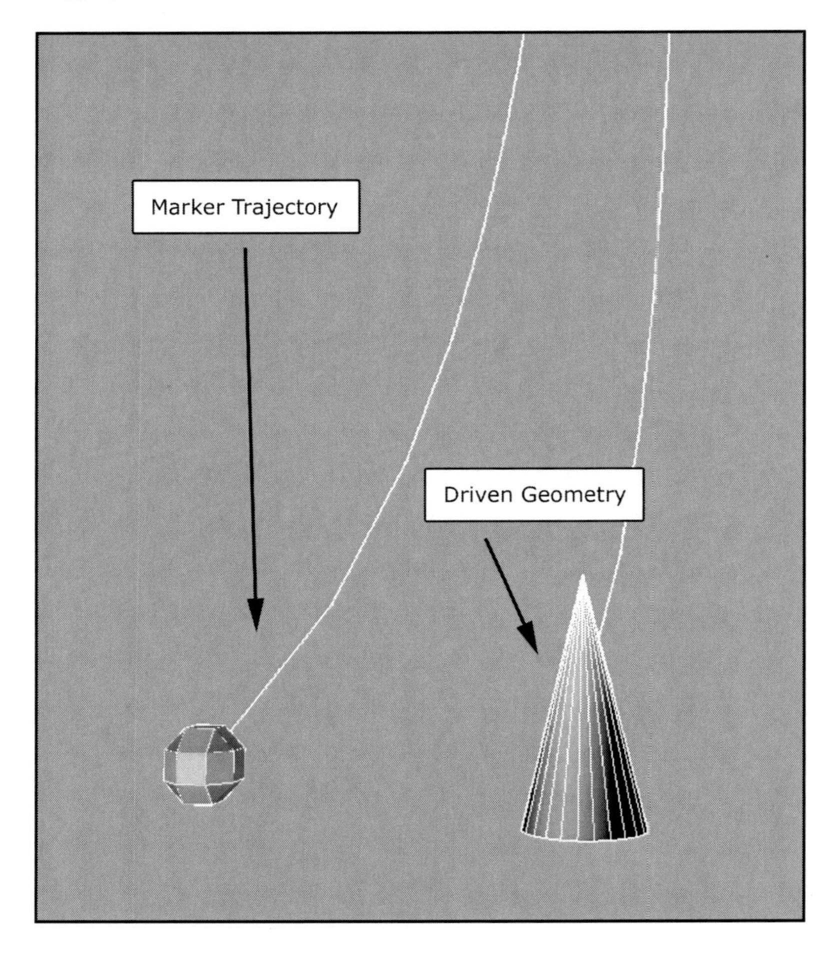

Figure 02_02.

If 2 markers are used then translation information from any of the 2 markers can be extracted, but also 2 rotation axis can be replicated based on the change of position of one marker against the other (Figure 02_03).

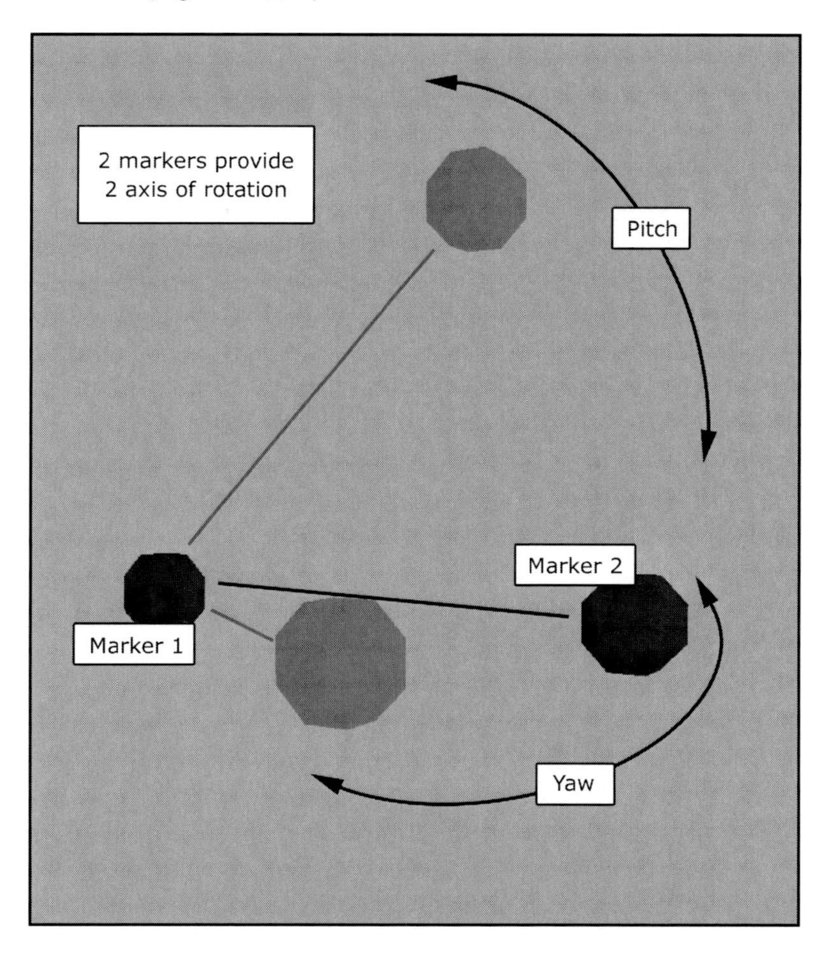

Figure 02_03.

3 markers can replicate any possible translation as well as any possible rotation (Figure 02_04).

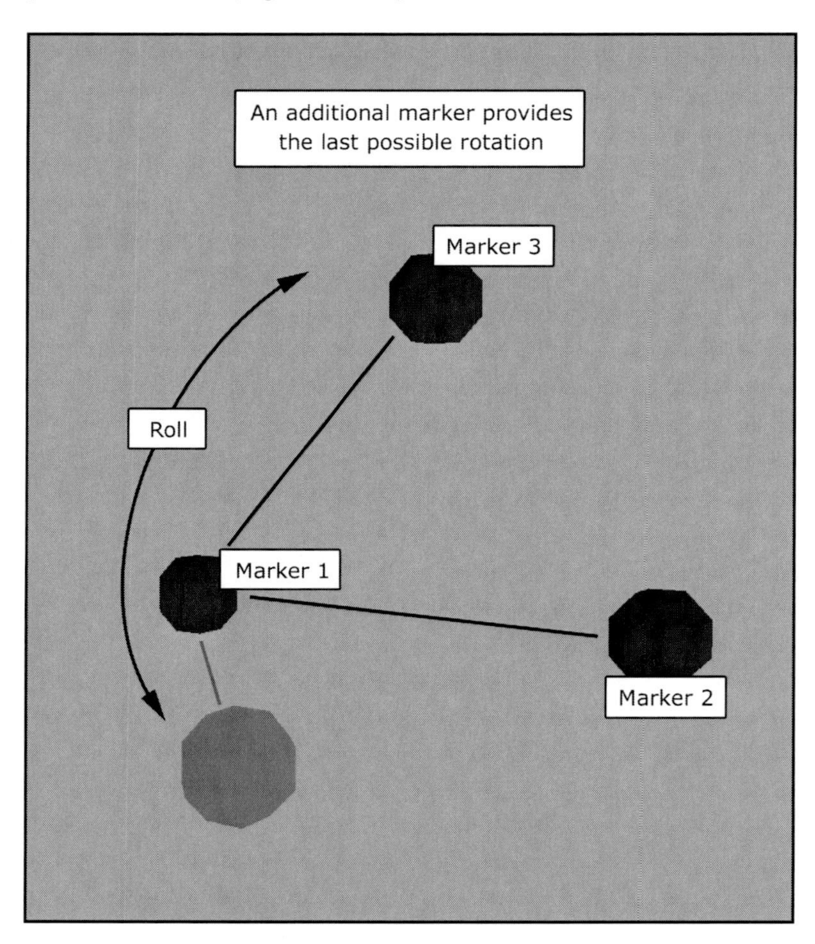

Figure 02_04.

In order to properly capture every possible translation and every possible rotation, it is recommended to place the markers as a close to a coordinate system as possible. 3 markers is the minimum amount of markers to be applied to an element that is to be captured. However, it is recommended to have 4 markers or more per object in case that the cameras cannot see a marker at a particular time (Figure 02_05).

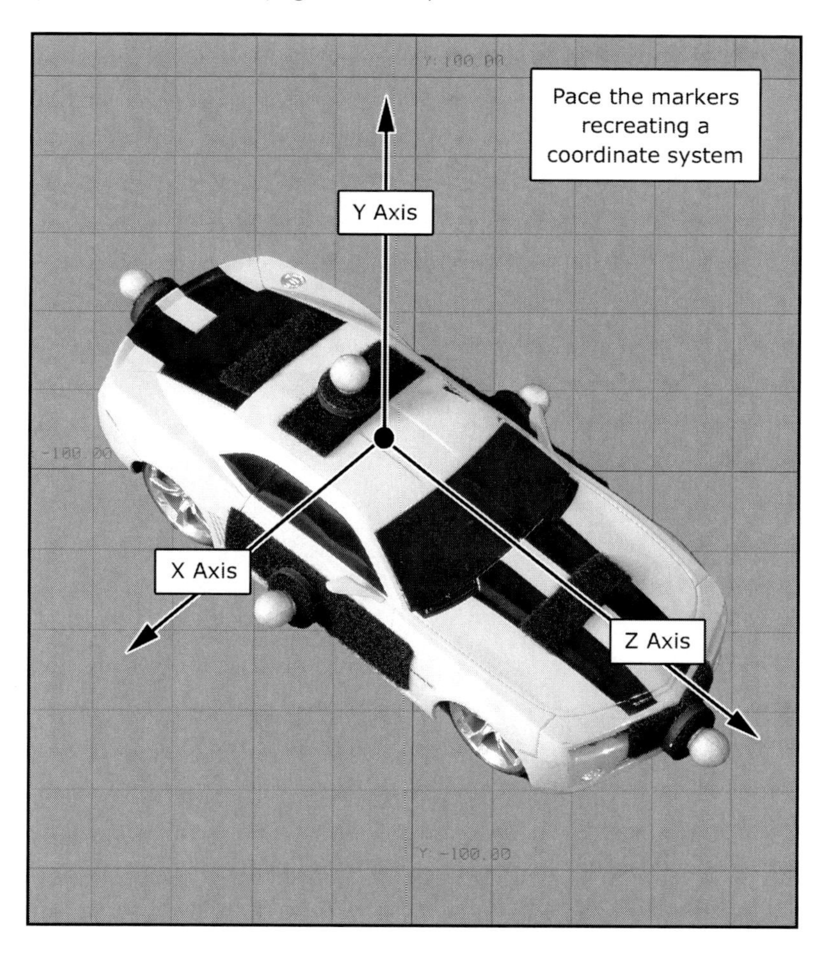

Figure 02_05.

When dealing with humans, you want to think of every articulated part of the performer as a separate object. 3 or more markers are to be applied per articulated section, whenever possible (Figure 02_06).

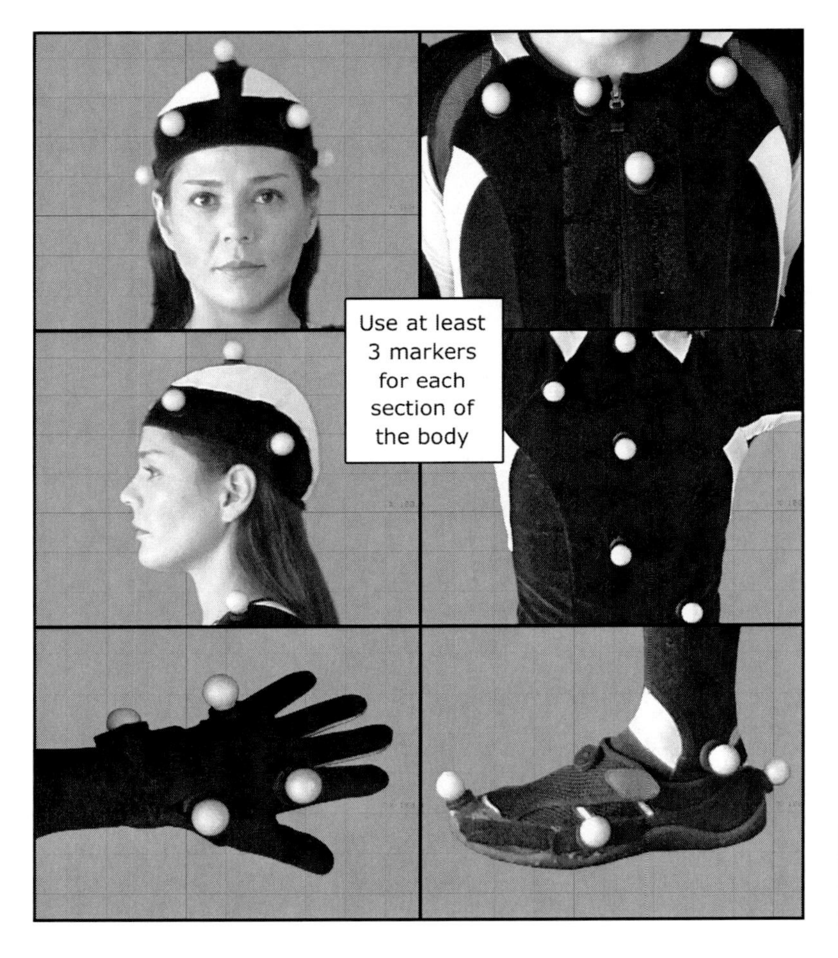

Figure 02_06.

A marker can make part of more that one section (Figure 02_07).

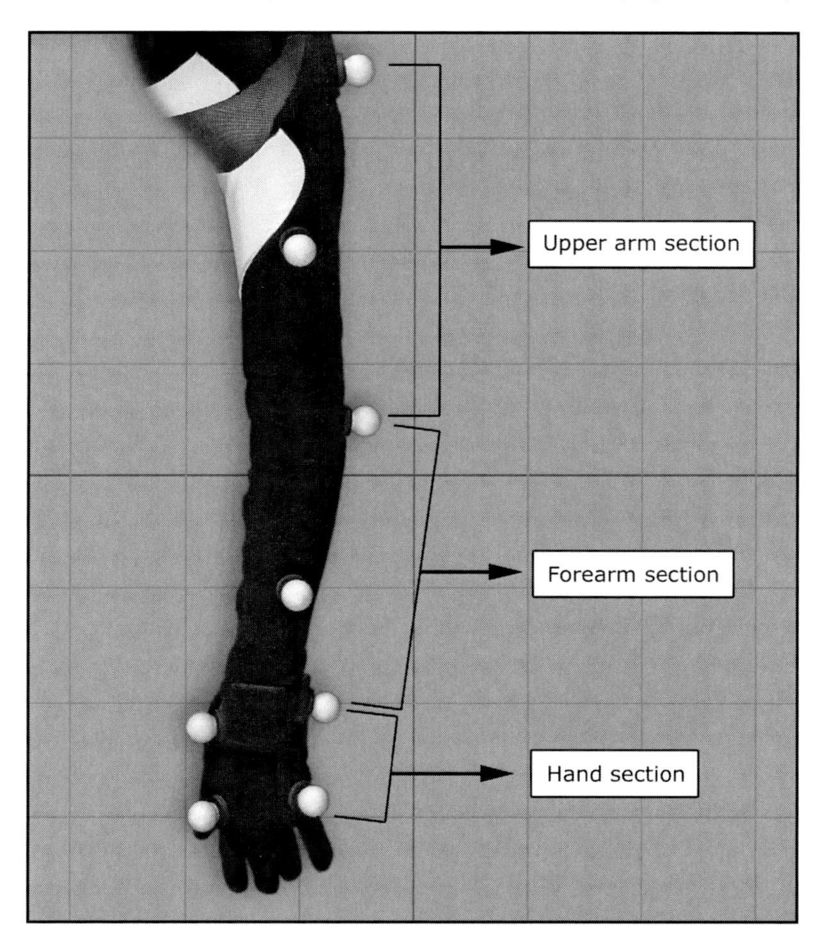

Figure 02_07.

The arrangement of markers on a performers body is called a *Marker Set*. It is important to note that there is no universal *Marker Set* that works for every possible motion capture situation. Here is an example of a basic 49 *Marker Set* (Figure 02_08).

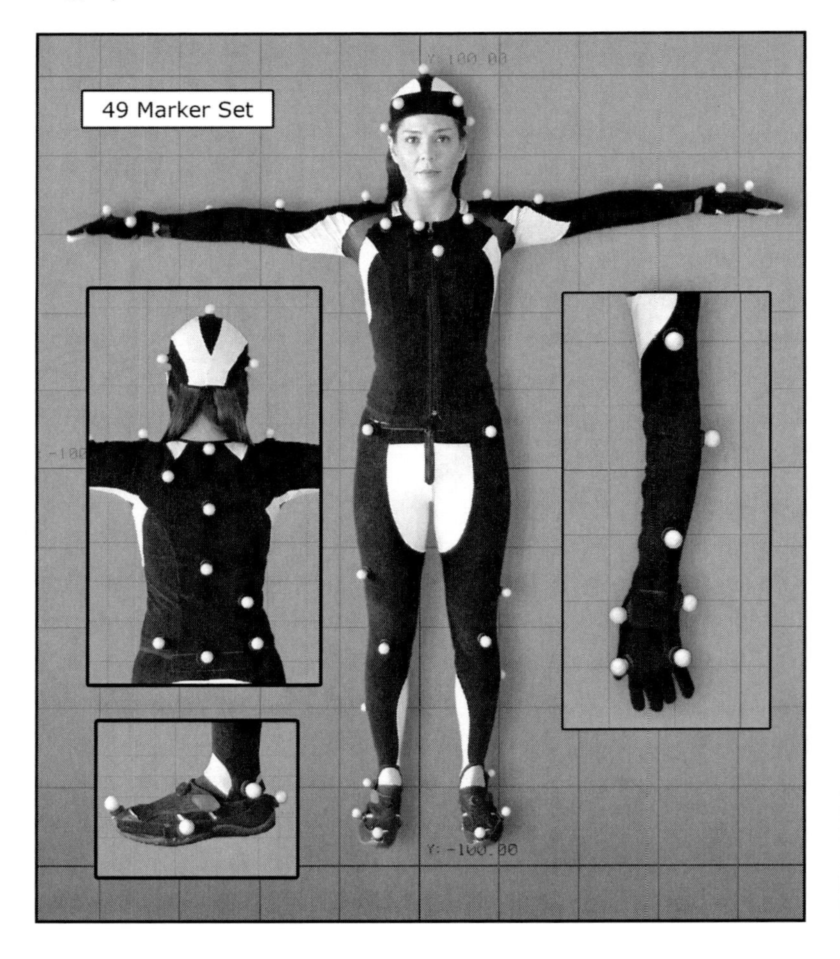

Figure 02_08.

This *Marker Set* works well for walking, talking, running and some acrobatic movement. Markers get *occluded* when the performer crouches, arcs their back or lays on the floor (Figure 02_09).

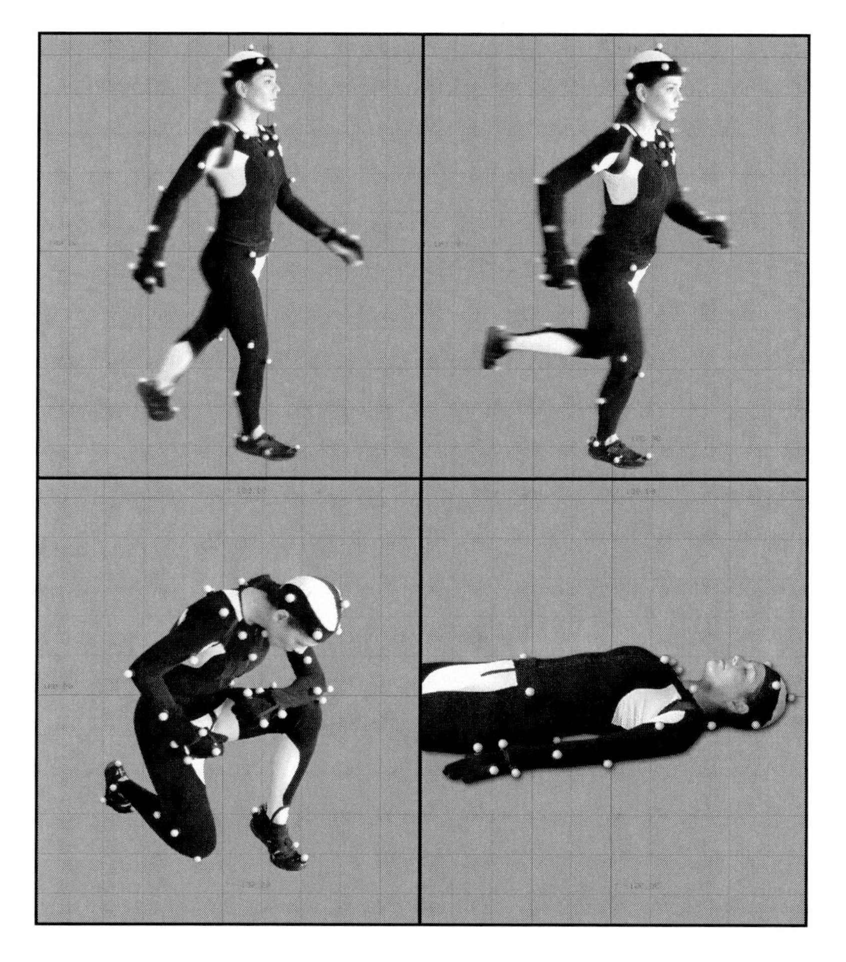

Figure 02_09.

Much like rigging in the 3D animation world where custom rigs can be built to accommodate the specific animation required for a shot; custom *Marker Sets* can be created that best fit the motions to be captured in a particular session (Figure 02_10).

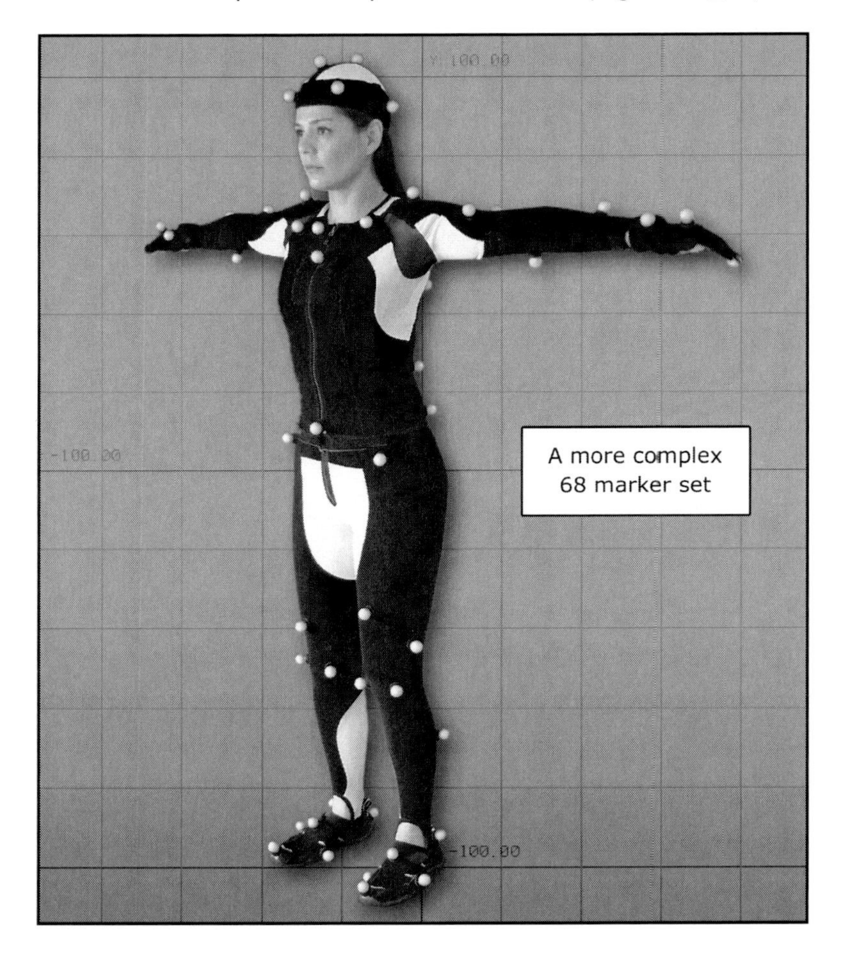

Figure 02_10.

Part 02
Motion Capture With Cortex

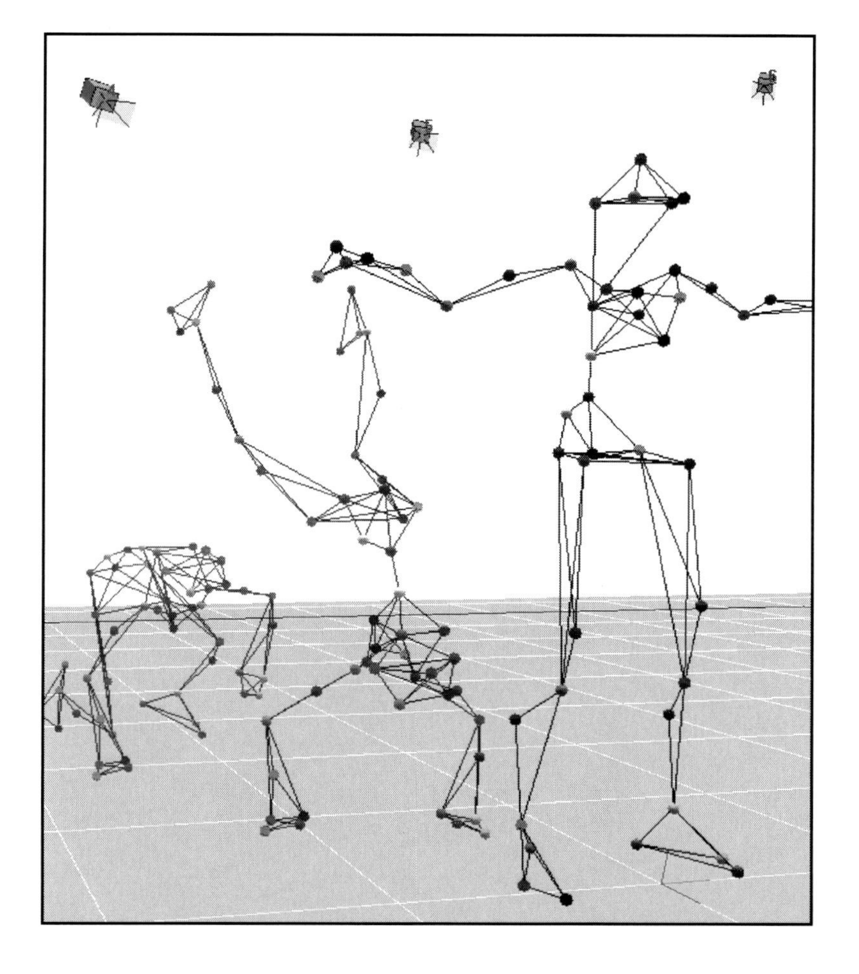

Chapter 03
System Calibration

In order for the cameras to be able to track the markers with precision, their position in physical space needs to match their position in the capturing software. The synchronization of the physical position and orientation of the camera with its digital counterpart is called *System Calibration*. There are three steps to the *System Calibration*; First comes the *Calibration with L-Frame*, then the *Calibration with Wand* and finally the *Floor Calibration* (Figure 03_01).

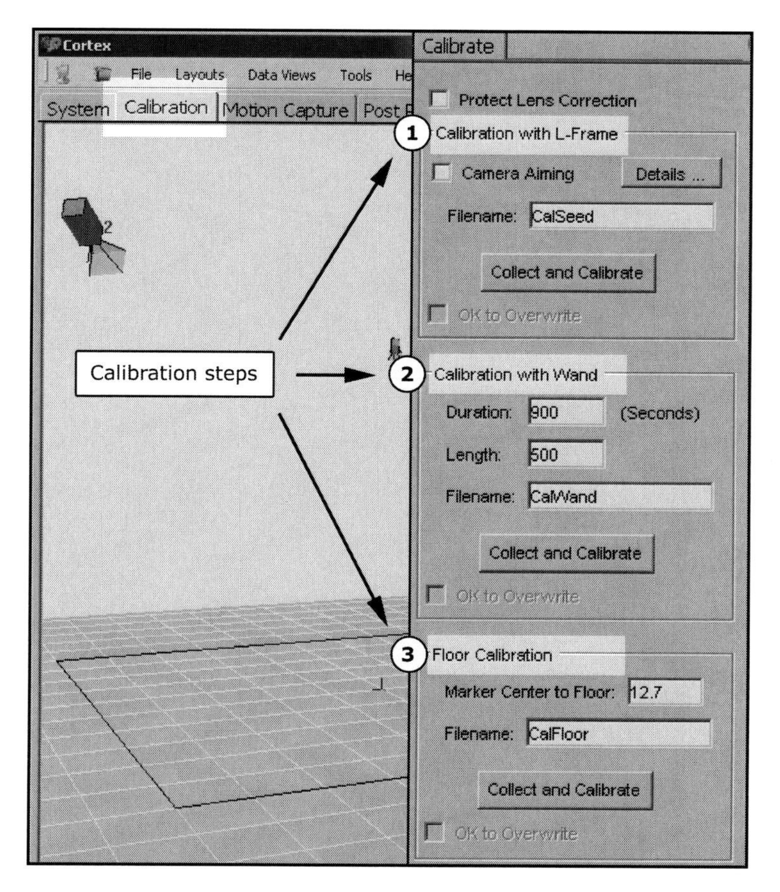

Figure 03_01.

First start by creating the folder you will use to manage the project. Let's call it "Chapter03". Save the **Cortex** project to this folder, we will call the project "Chapter03.prj". When you save a **Cortex** project in a specific folder, every aspect of the capture gets stored on that folder, calibration files, motion takes, etc. (Figure 03_02).

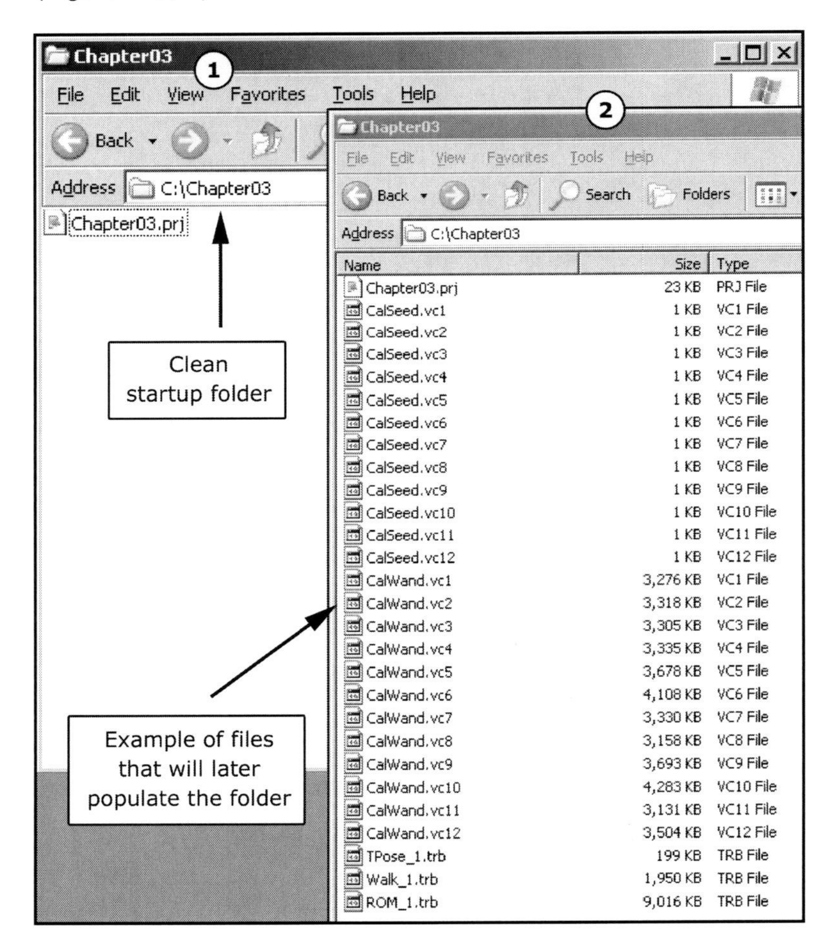

Figure 03_02.

Place the *L-Frame* in the center of your physical capture volume with its longest axis pointing towards what you want the +Z axis inside Cortex to be (Figure 03_03).

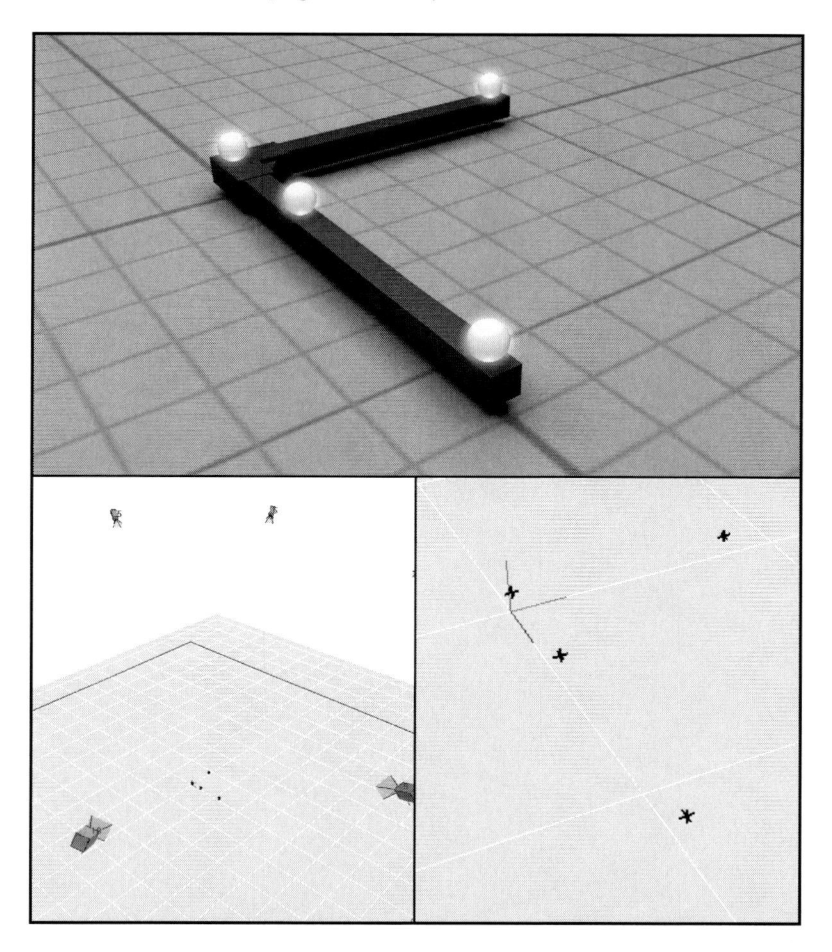

Figure 03_03.

Connect to cameras. The system will give you a message letting you know how many cameras its seeing. If the number fits the amount of physical cameras that you installed press the OK button. You will see the ring of the physical cameras turn red as they light up. It is recommended that you let the cameras warm up for 20 minutes before you start the calibration (Figure 03_04).

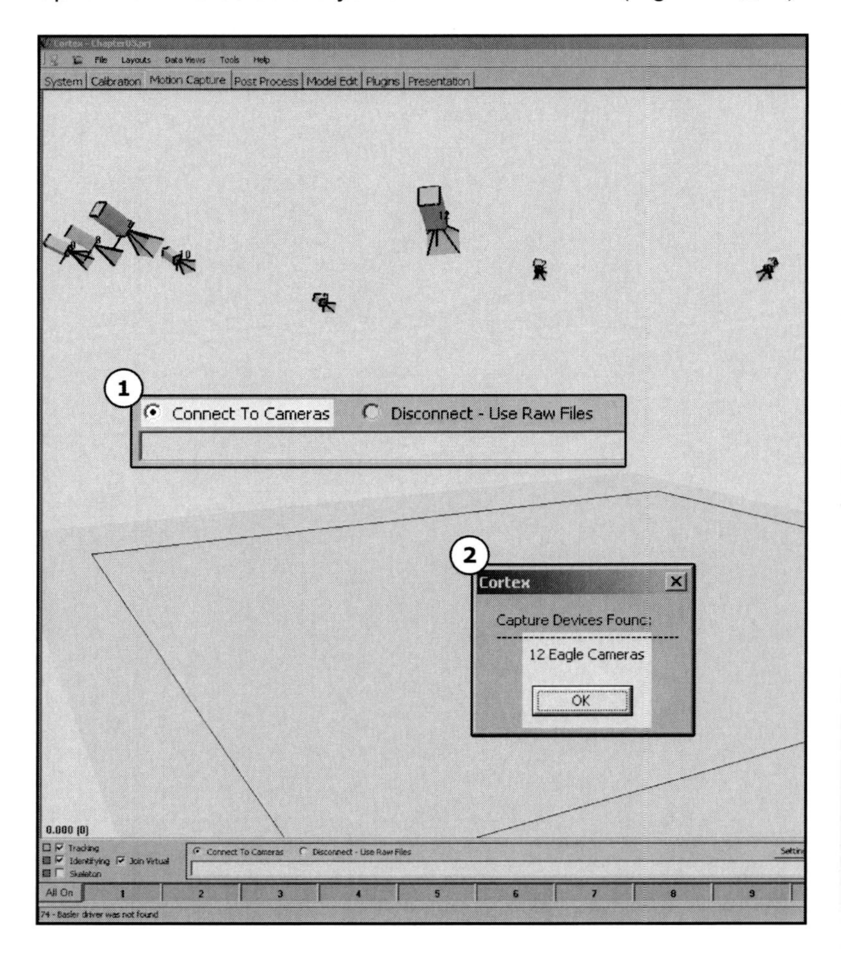

Figure 03_04.

Tip: If a warning appears saying that there is an amount of cameras set by the projects that does not fit the amount of physical cameras, cancel the operation and reboot all cameras. To Reboot All Cameras go to the Setup Panel, click on the Cameras sub panel and press the Reboot All Cameras Button. You need to calibrate again after rebooting (Figure 03_05).

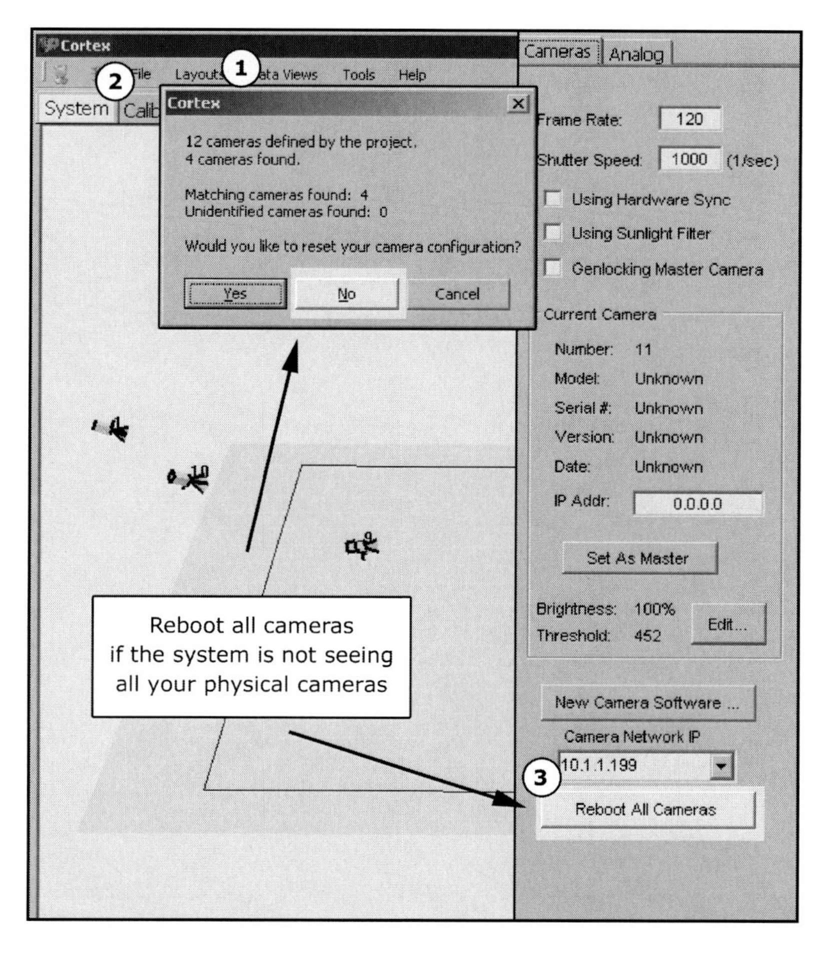

Figure 03_05.

Under the *Layouts* menu select *2 Panes: Top/Bottom* option. Click on the top pane and hit the *F2* key to turn the view into a *2D Display*. Click on the bottom pane and press the *F3* key to turn it into a *3D Display*. Drag on the panes division to make the 2D view larger (Figure 03_06).

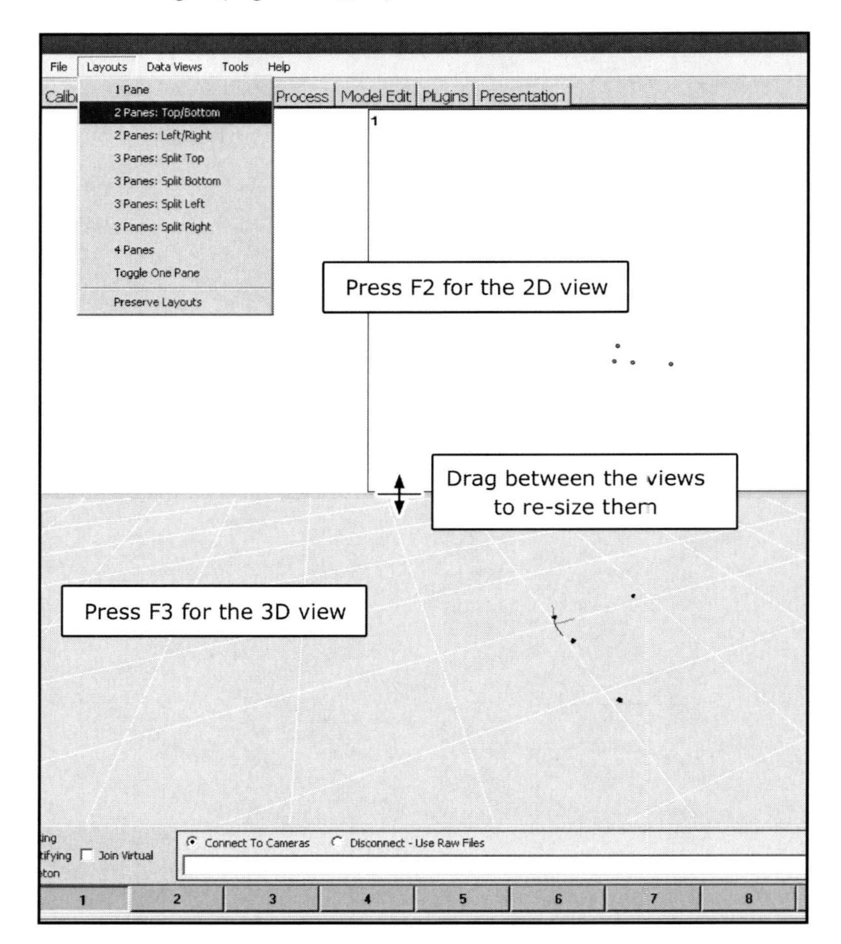

Figure 03_06

In the *Calibrate* panel inside the *Calibration* tab, check *Protect Lens Correction* off and *Camera Aiming* on. After you do a few calibrations the *Protect Lens Correction* can be turned on and the *Camera Aiming* off for faster calibrations (Figure 03_07).

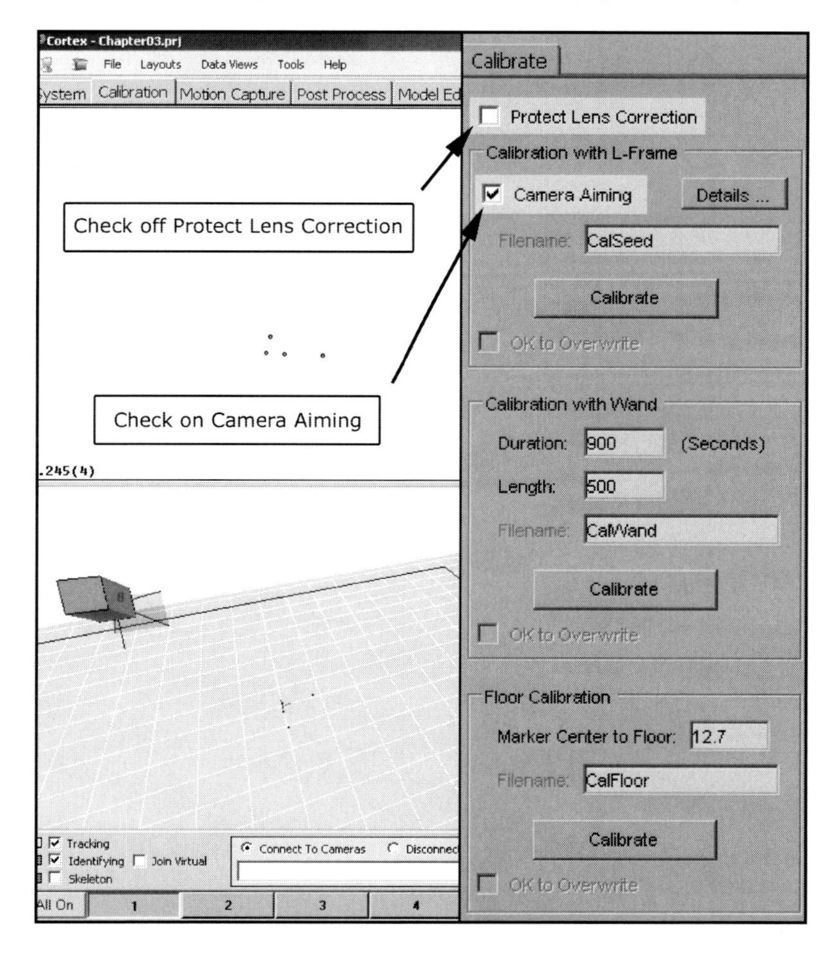

Figure 03_07.

Press the *Run* button at the bottom right of the screen to activate the cameras. Check every camera on the *2D Display* making sure that each camera is only seeing the 4 markers of the *L-Frame*. You can check cameras independently by pressing their button number at the bottom of the screen, you can check all of them at the same time by pressing the *All On* button (Figure 03_08).

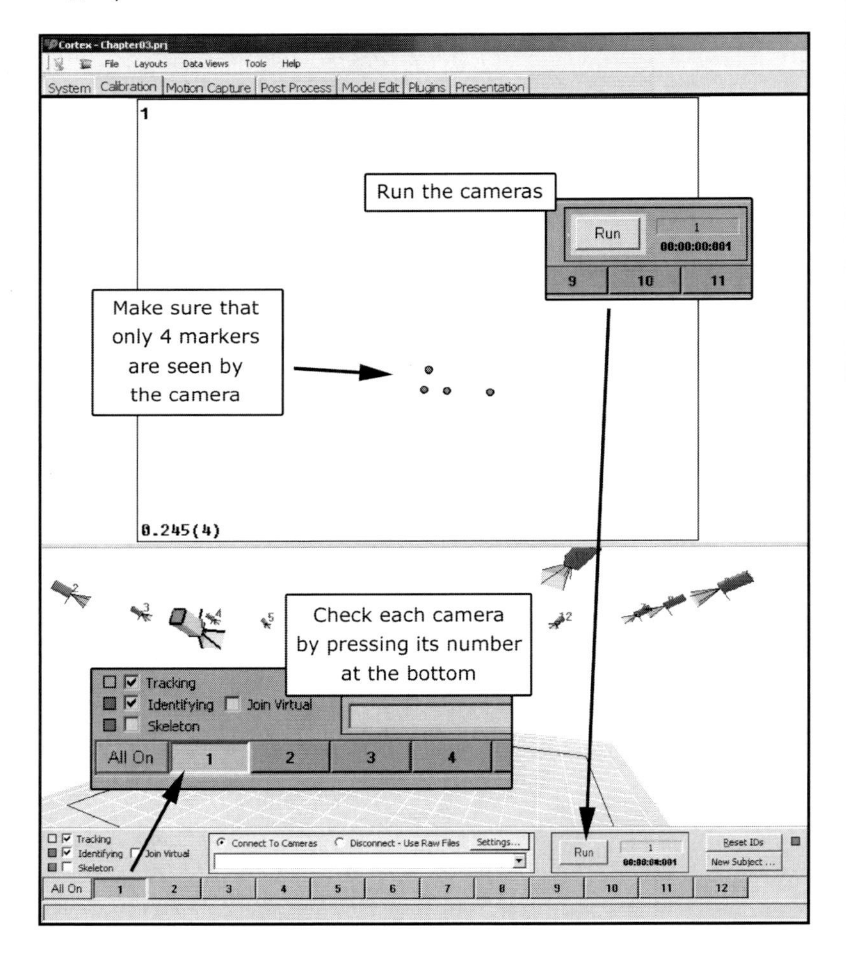

Figure 03_08.

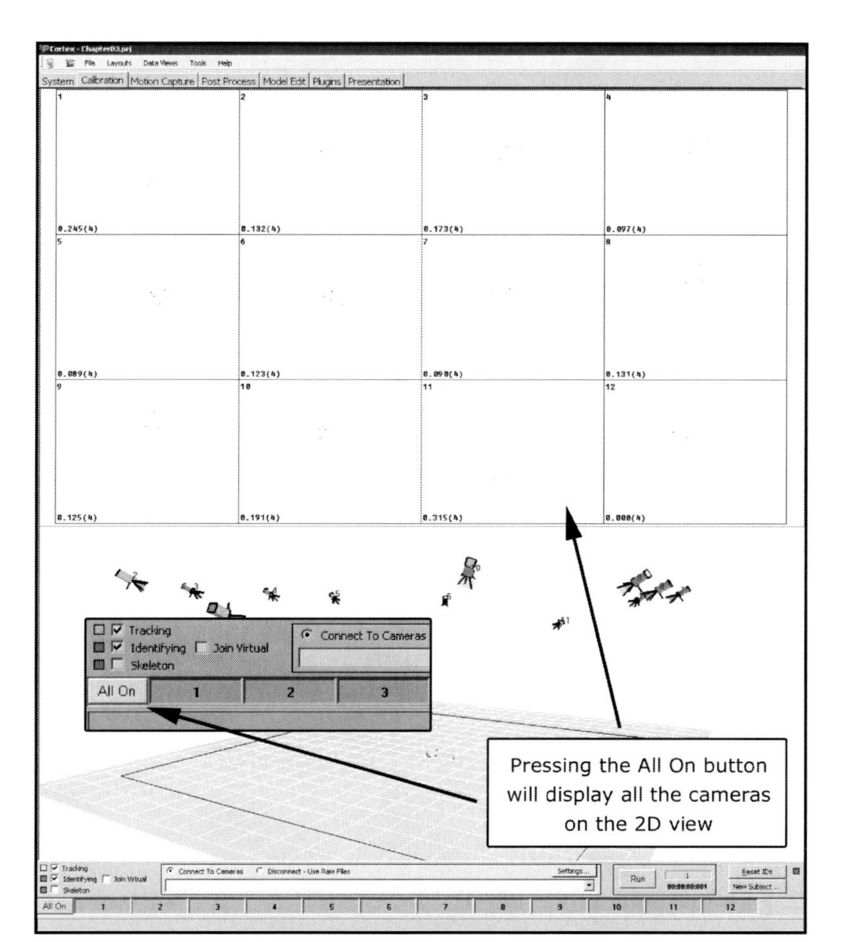
Pressing the All On button will display all the cameras on the 2D view
Tracking
Identifying
Join Virtual
Skeleton
Connect To Cameras
All On

Figure 03_09.

If there are more than 4 markers in any of the cameras you should look where the reflection is placed in the physical Volume and get rid of it. If it is impossible to dispose of the physical source of the reflection, then you can mask the camera by *middle mouse click + dragging* a box over the noise spec (Figure 03_10).

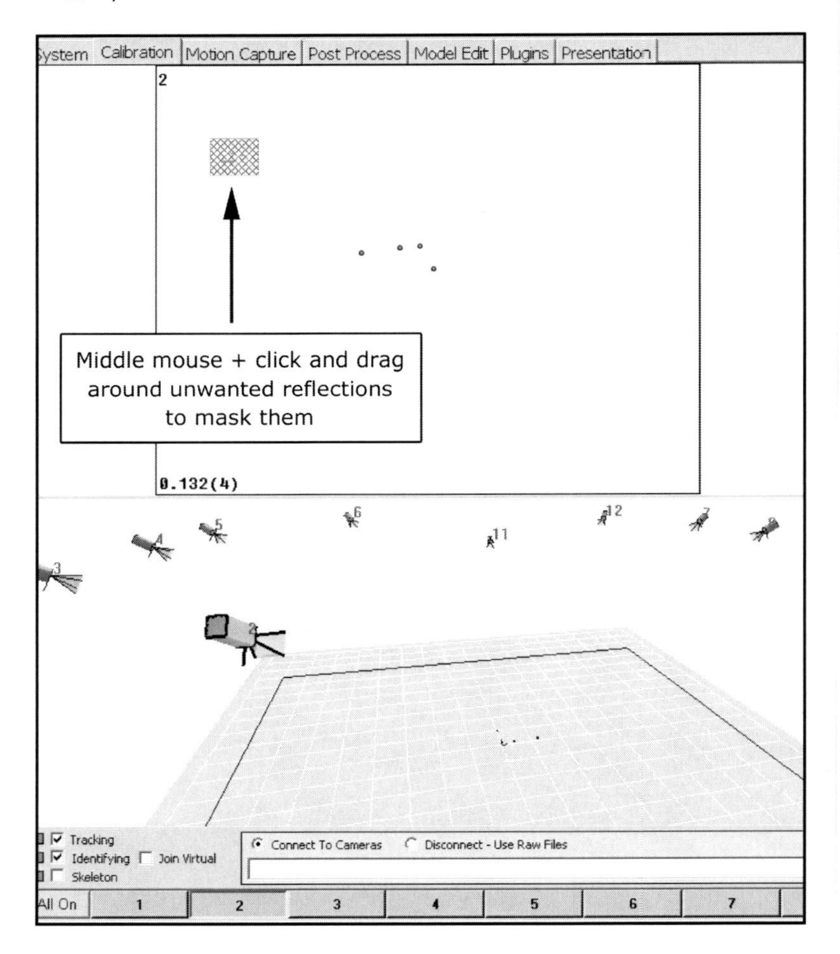

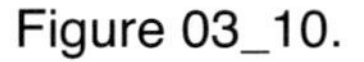

Figure 03_10.

Once you make sure that all cameras are only displaying the markers for the *L-Frame* and you have masked any noise reflections, press the *Collect and Calibrate* button in the *Calibration with L-Frame* area, you will see the cameras in the *3D Display* change positions. (Figure 03_11).

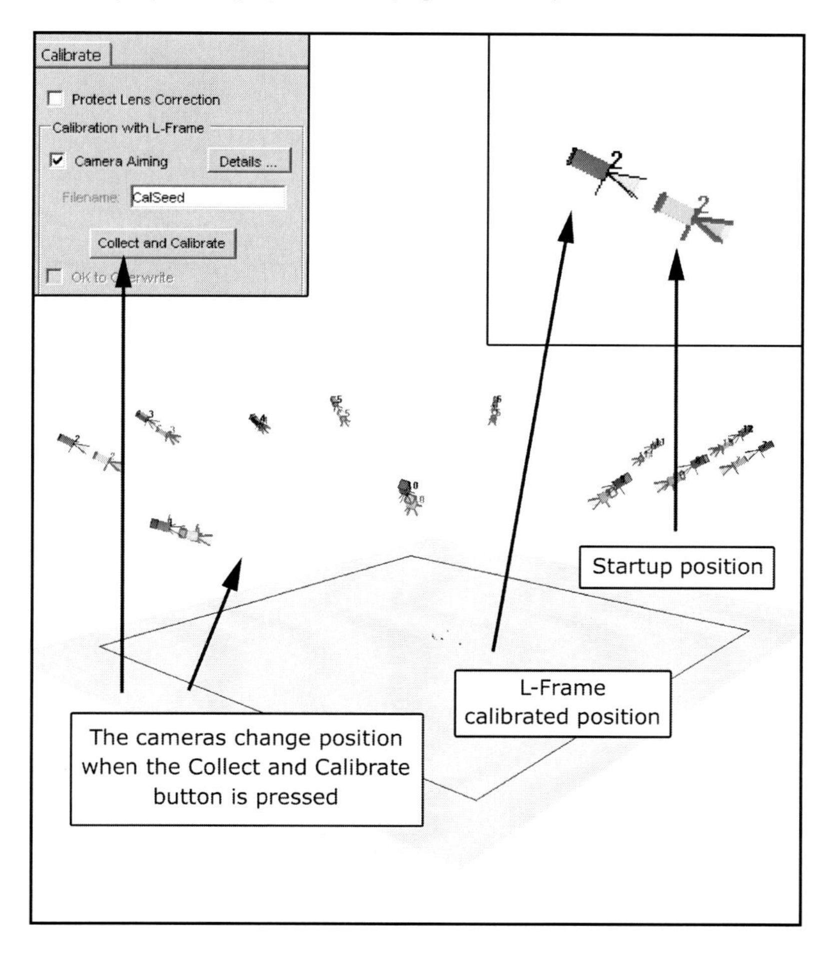

Figure 03_11.

The numbers of the cameras at the bottom section of the screen should turn yellow; this means that the cameras have been *seeded*. Double-check every camera to make sure it is only seeing the markers from the square. You have completed the first step of the system calibration process. Save the project and make sure to store the *L-Frame* in a secure place. If the *L-Frame* is bent or damaged a new one needs to be purchased or manufactured in order to be able to calibrate the system (Figure 03_12).

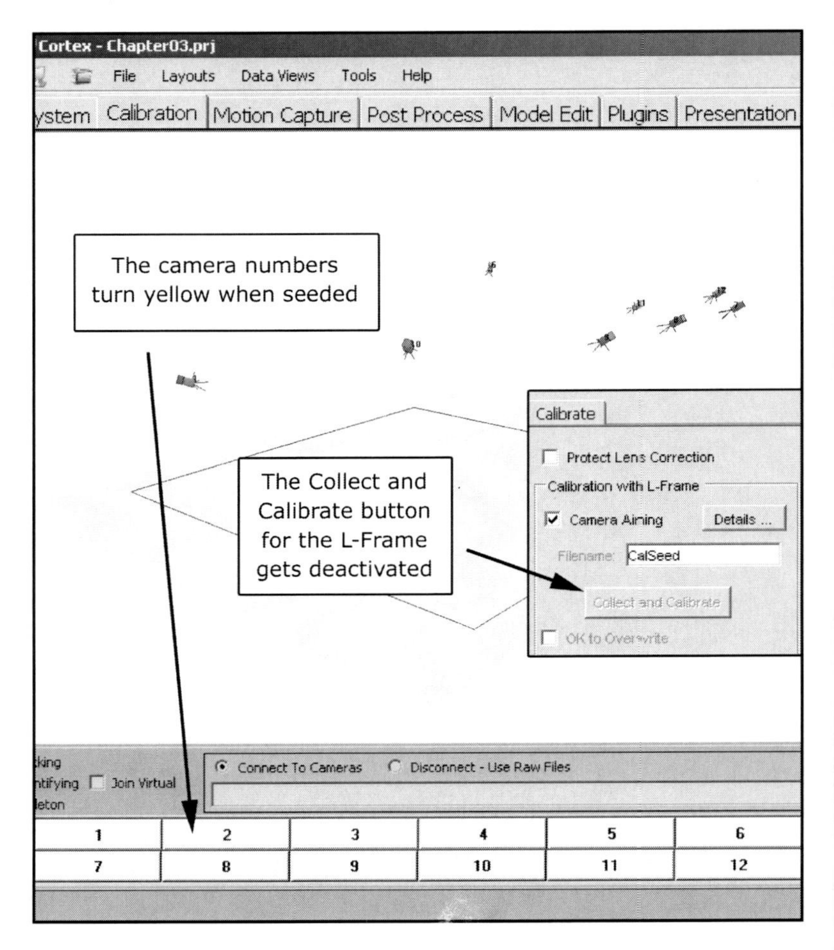

Figure 03_12.

Let us continue with the second step, the *Calibration with Wand.* Make sure that the cameras are paused. Under the *Calibration with Wand* section enter 900 in the *Duration* section. Enter the measurements of the *Wand* (from end marker to end marker) that will be used for the calibration in the *Length* section. Leave the default text in the *Filename* section (Figure 03_13).

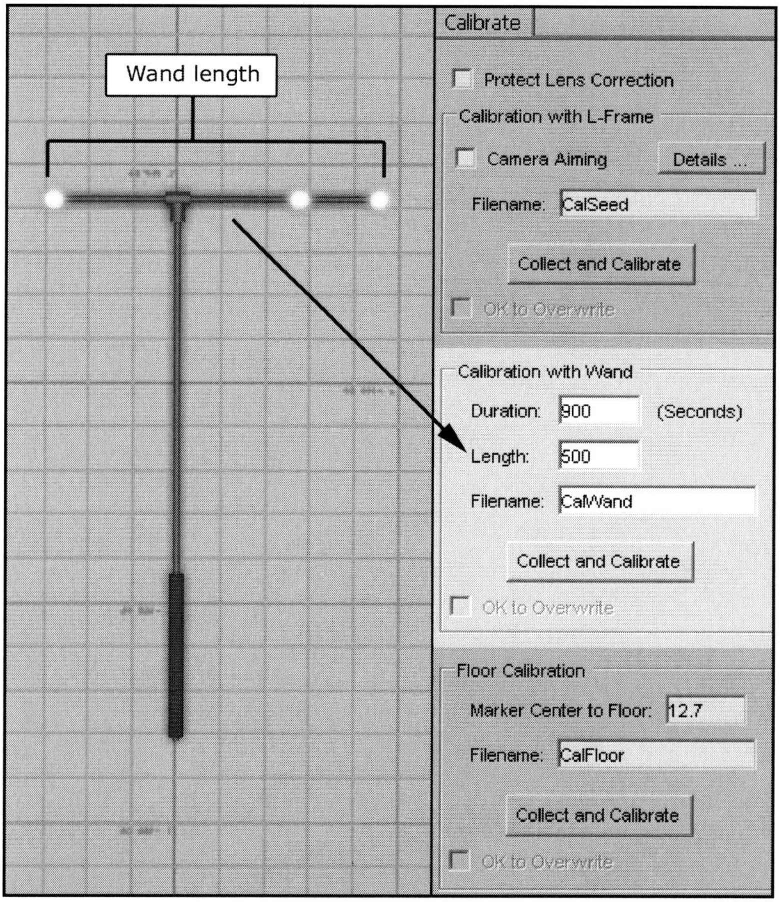

Figure 03_13.

Press the *Collect and Calibrate* button and walk around the *Volume* waiving the *Wand* up and down with its head parallel to the ground (Figure 03_14).

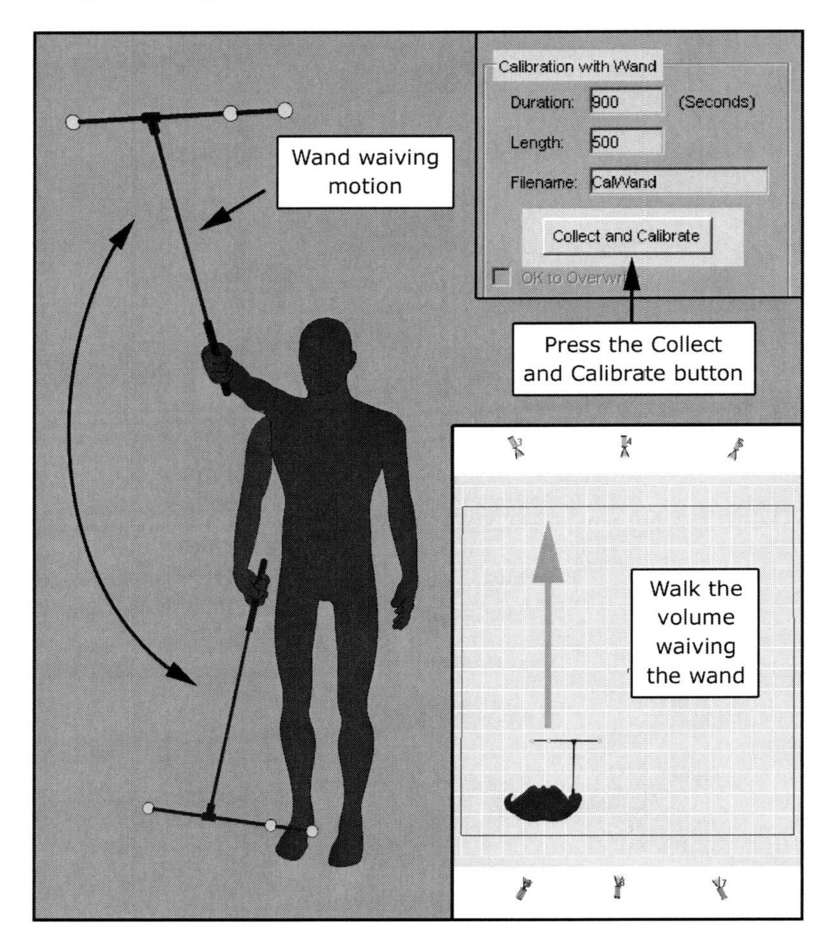

Figure 03_14.

Make sure to face only one side of the *Volume* until you cover it completely, then turn 90 clockwise degrees and walk the entirety of the *Volume*. Repeat the process until you are facing the side of the *Volume* that you started with (Figure 03_15).

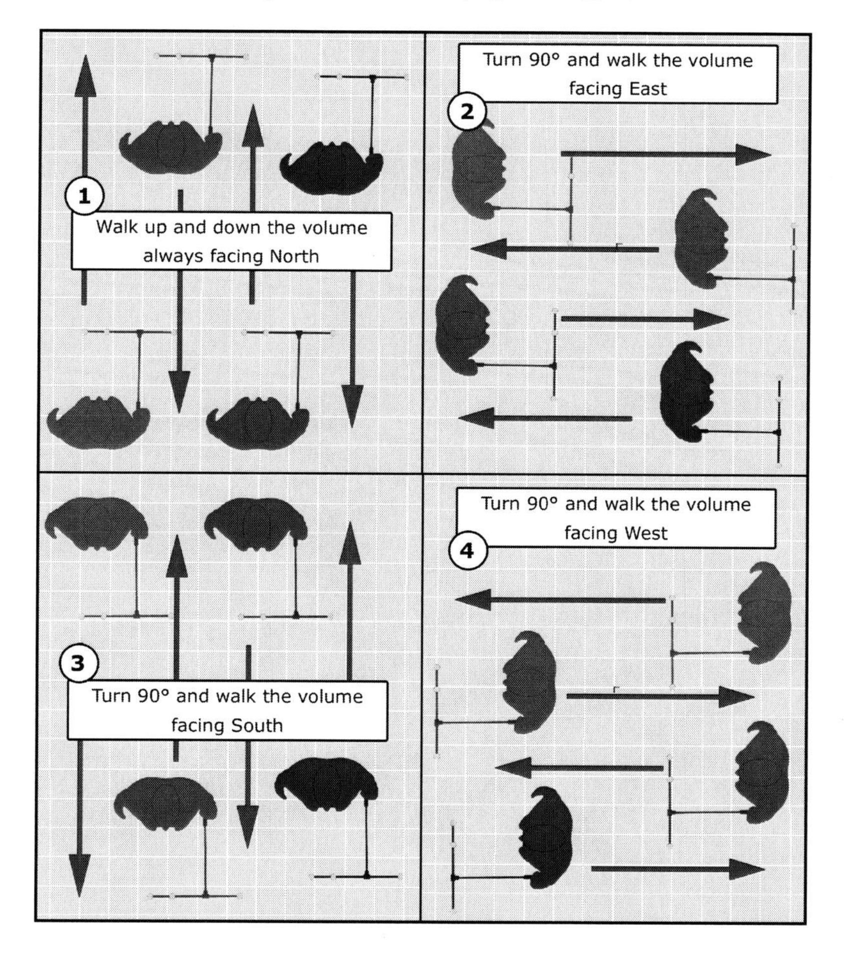

Figure 03_15.

Now walk the *Volume* waiving the *Wand* sideways with its head perpendicular to the ground (Figure 03_16).

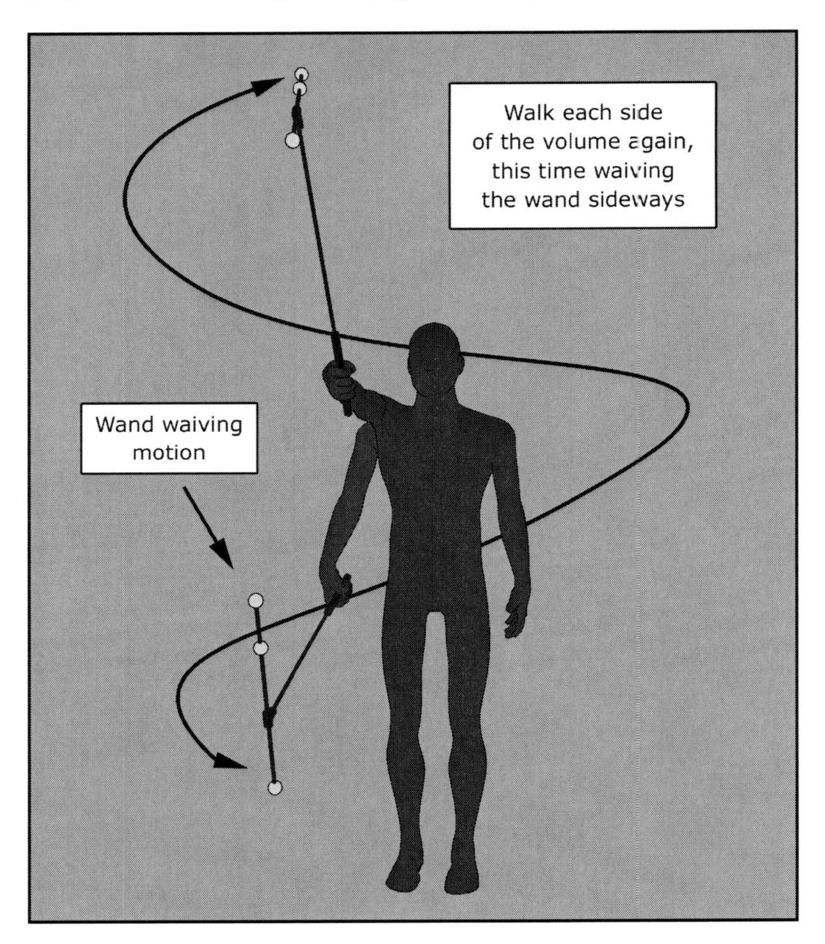

Figure 03_16.

Press the *Stop* button on the *Wand Calibration* section. The *RealTime* window appears warning you about the frames in which some cameras are seeing more than three markers for the *Wand*. Click *OK* to accept this warning (Figure 03_17).

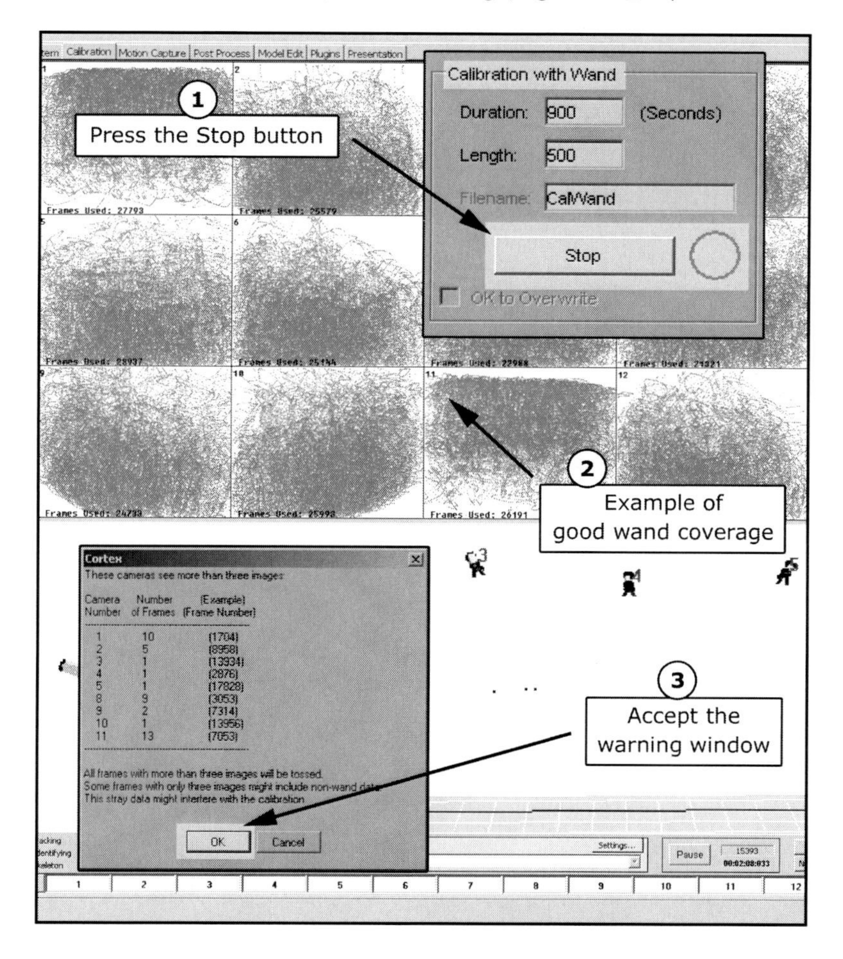

Figure 03_17.

The *Wand Processing Status* window gets initialized and starts processing the calibration data. You will see the numbers in the different sections of the window update as the operation progresses (Figure 03_18).

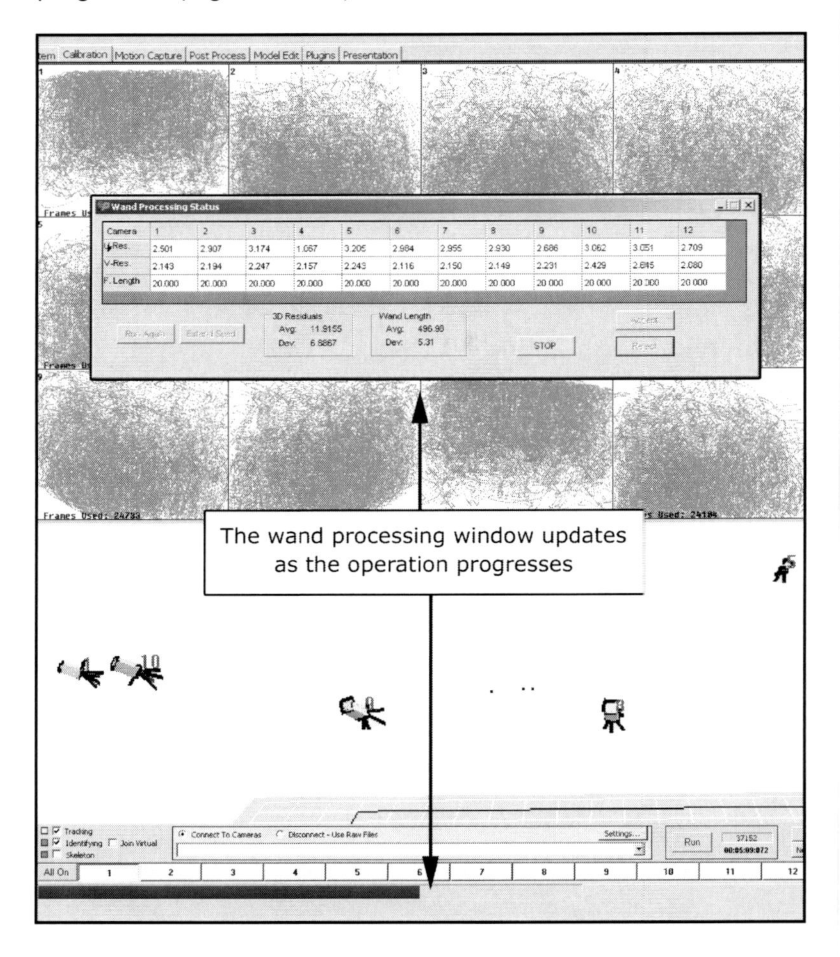

Figure 03_18.

Once the first run of the operation gets completed, press the *Extend Seed* button and then the *Run Again* button. You should see the numbers approximate to the physical specifications of your system. In the *Wand Length* section the "*Avg*" values should get closer to the *Length* of your Wand, the "*Dev*" values should lower as the processing progresses. In the *F. Length* section the numbers of watch camera should be getting close to the physical lenses installed in them (20 mm, 30 mm, 50 mm, etc). (Figure 03_19).

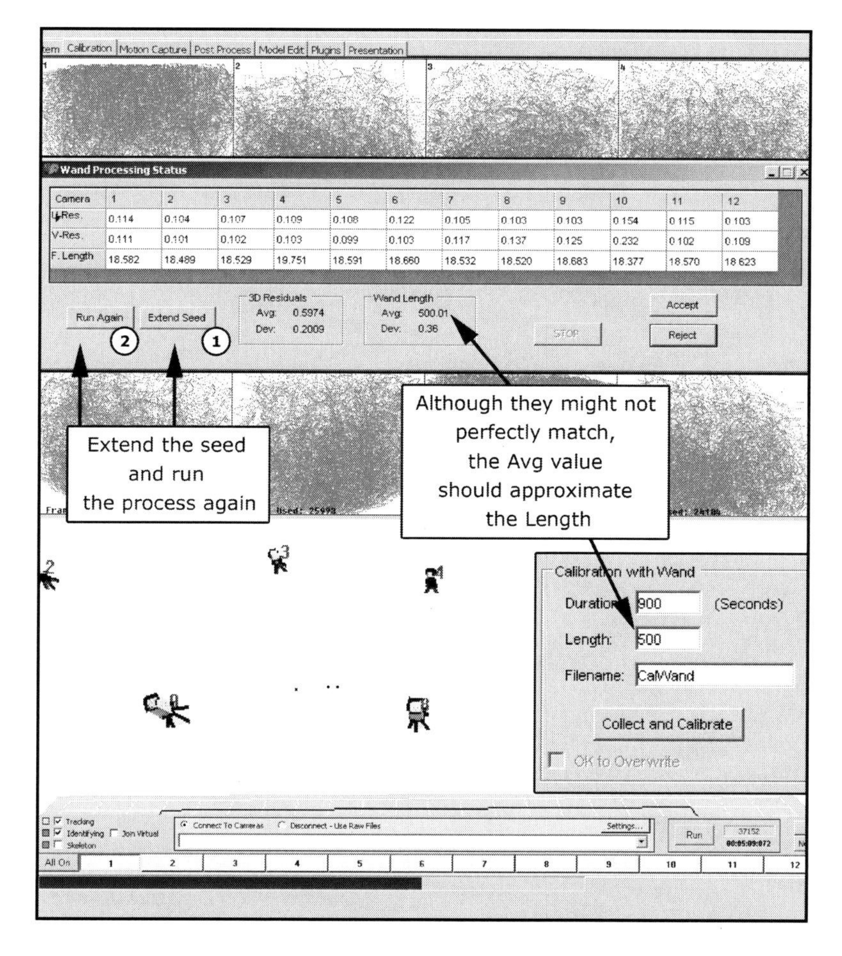

Camera	1	2	3	4	5	6	7	8	9	10	11	12
H-Res.	0.114	0.104	0.107	0.109	0.108	0.122	0.105	0.103	0.103	0.154	0.115	0.103
V-Res.	0.111	0.101	0.102	0.103	0.099	0.103	0.117	0.137	0.125	0.232	0.102	0.109
F. Length	18.582	18.489	18.529	19.751	18.591	18.660	18.532	18.520	18.683	18.377	18.570	18.623

3D Residuals: Avg: 0.5974, Dev: 0.2009

Wand Length: Avg: 500.01, Dev: 0.36

Figure 03_19.

Keep on refining the wand calibration by pressing the *Run Again* button as long as the numbers in the *Wand Processing Status Window* keep on changing. When the numbers in the window stop changing significantly and are fairly close to your physical values (*Wand Length* section and *F. Length* section) press the *Accept* button. You have completed the second step of the system calibration process. Save the project and store the *Wand* in a secure place. If the Wand is bent or damaged a new one needs to be purchased or manufactured in order to be able to calibrate the system (Figure 03_20).

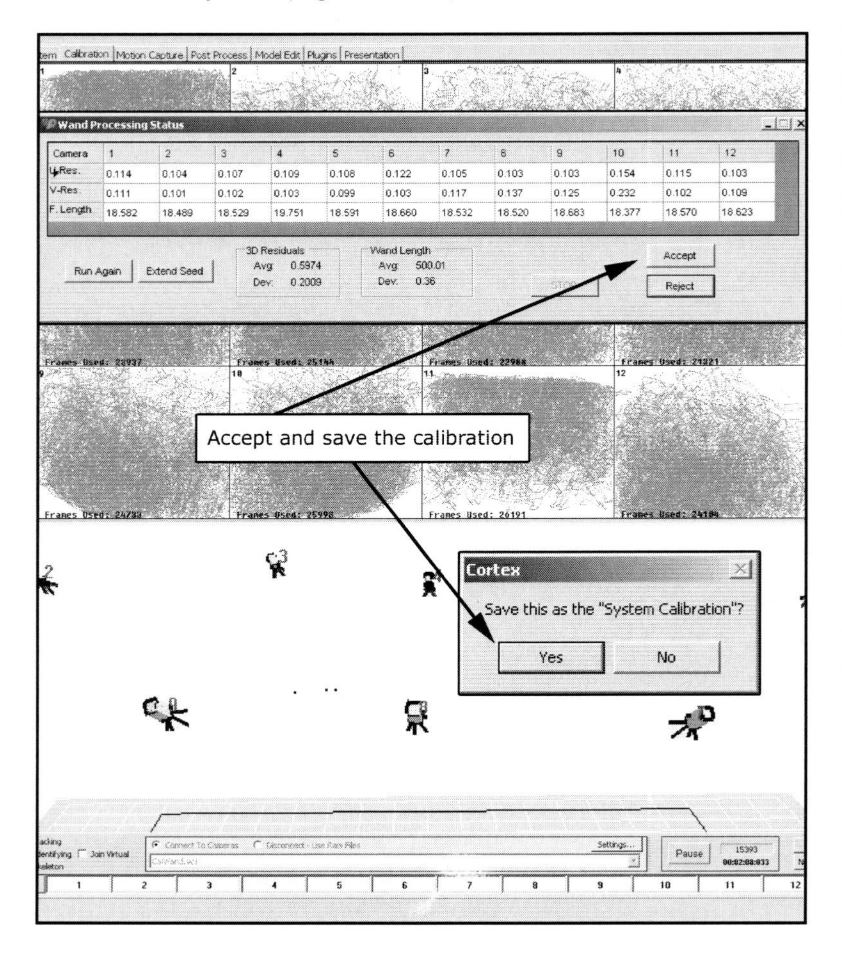

Figure 03_20.

If the numbers are far way from your physical values to an unacceptable degree, press the *Reject* button and start the calibration process again from the *L-Frame* step.

Most times only the L-Frame and the Wand calibrations are needed. The *Floor Calibration* is only needed if the floor of the motion capture room is uneven. To perform a *Floor Calibration* get a few identical markers and measure them from the base of the marker to the center of its circumference. Enter this measurement as the *Marker Center* value in the *Floor Calibration* section. Place the markers across the floor of the *Volume*, run the cameras and press the Calibrate button (Figure 03_21).

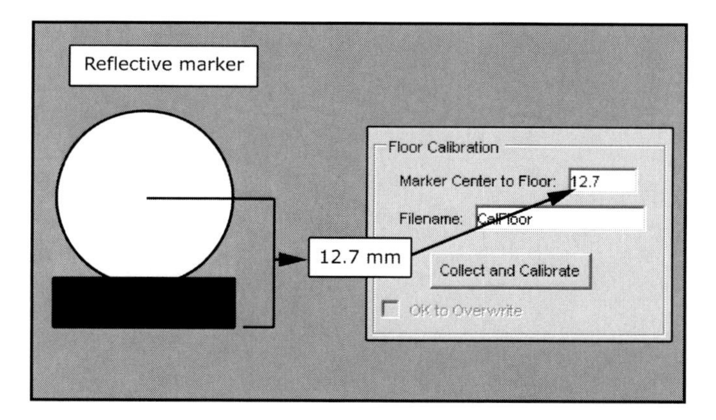

Figure 03_21.

The calibration of the system is done!

You will notice that the calibrate buttons in the calibration pane are grayed out and un-selectable. This is because there are now calibration files in the folder where your project is saved. When you need to calibrate, you need to activate the *OK to Overwrite* checkbox for those buttons to become active again.

Like it was noted at the beginning of the chapter, a good calibration is key to the accurate capture of markers. The system should be calibrated at the beginning of every capture session and hopefully during additional opportunities in the course of the capture day.

Chapter 04
Template Creation

Once the system is properly calibrated your cameras accurately track the reflective markers. You can get a performer in the *Volume* and watch the markers track his / her movements (Figure 04_01).

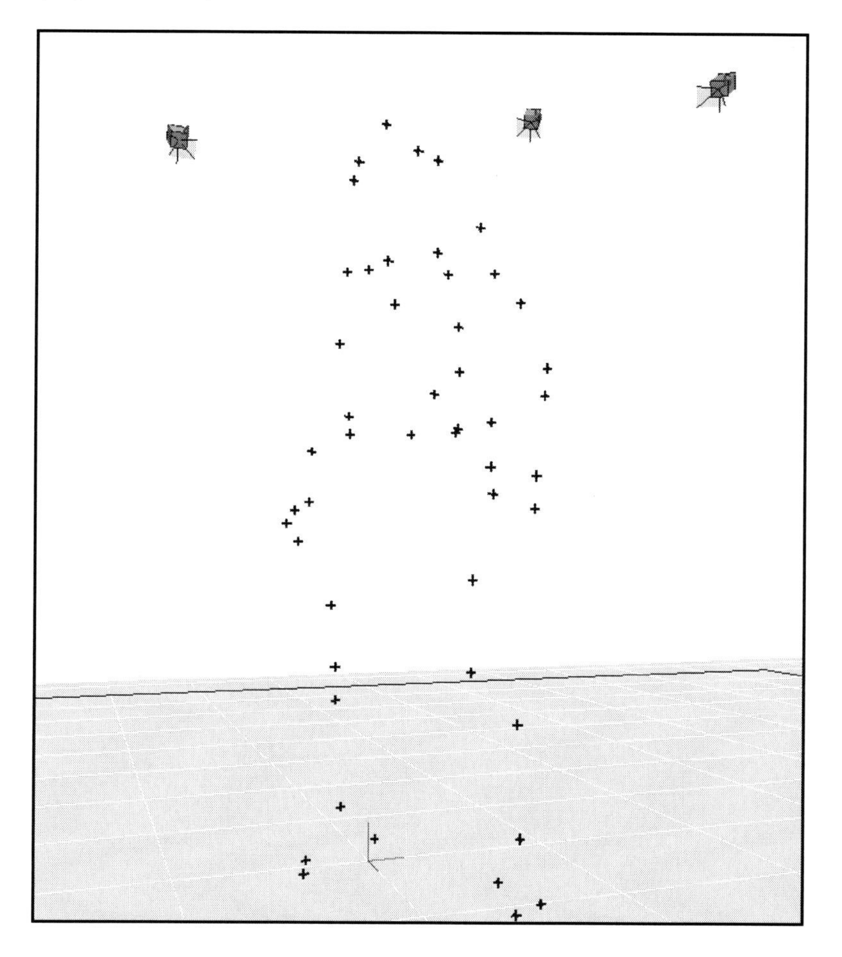

Figure 04_01.

Although your eye can understand the performer's movement, as far the system is concerned so far the markers are just reflections with nothing unique about them. Because of this the system just assigns a default name to the reflections (U_N with N being the number of markers seen in the scene) (Figure 04_02).

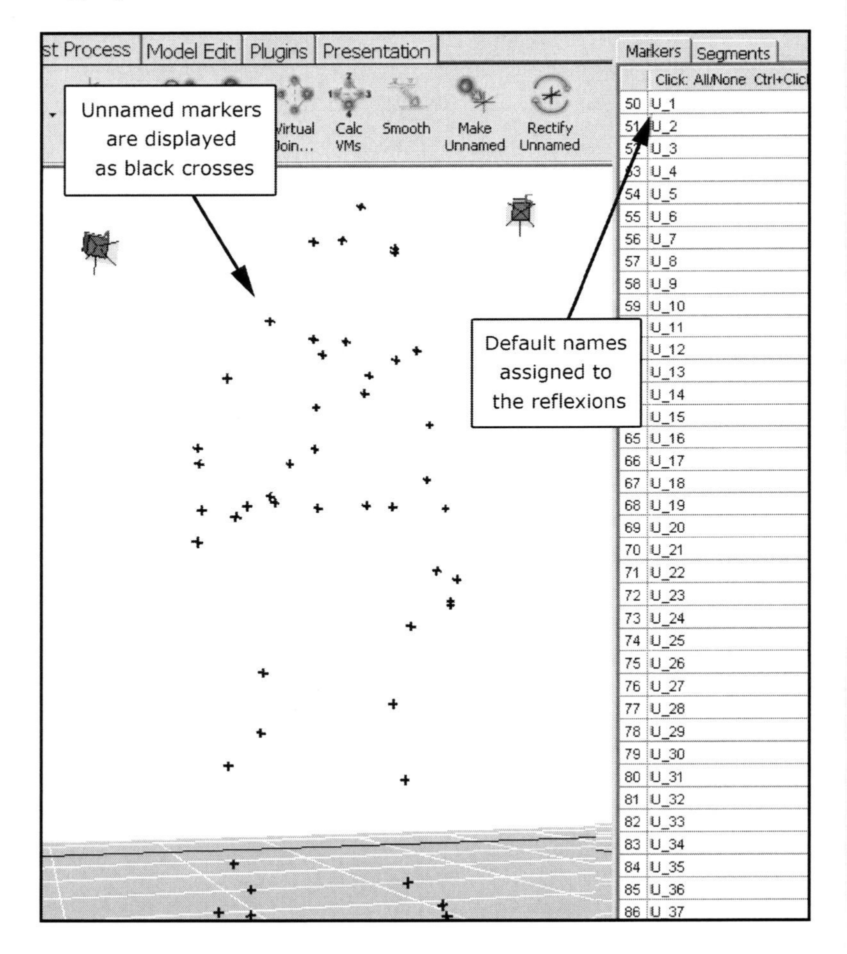

Figure 04_02.

In order to help the system to be consistent with the name-marker relationship so we can use those markers later to drive other CG elements we need to make them unique. To do this we need to give names to the markers and we need to measure them against each other (Figure 04_03).

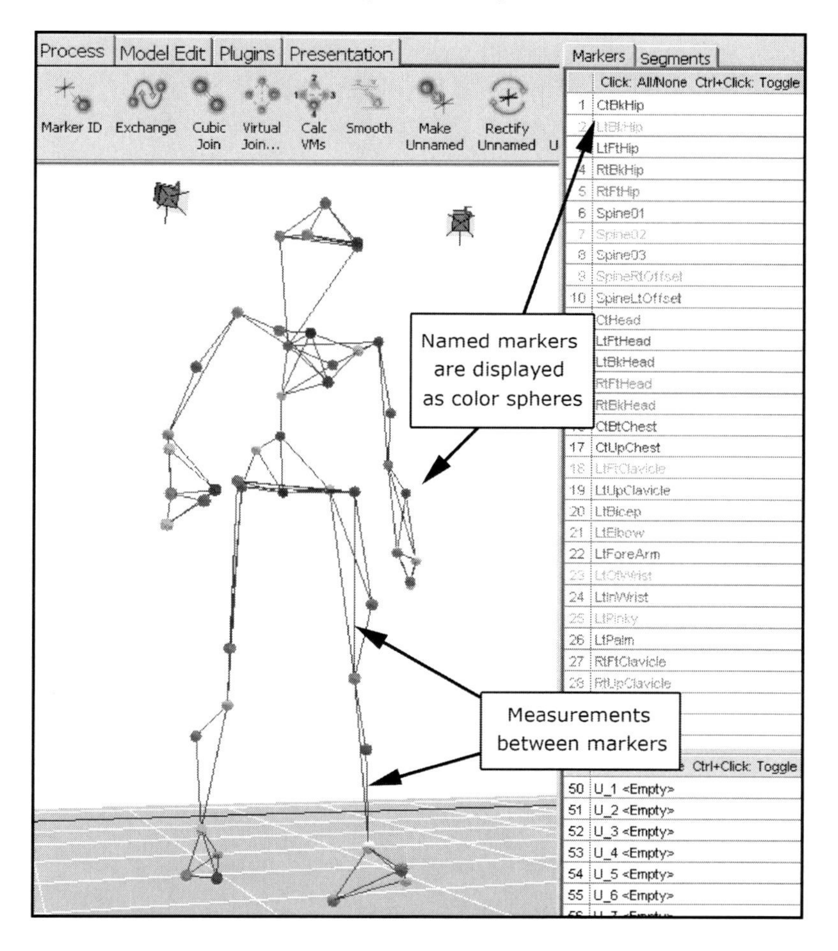

Figure 04_03.

RC Car Template Exercise

Lets approach the template creation process with a simple prop first. In this case an RC car.

Launch **Cortex** and Load the "Chapter04"[1] project.

If you are going to be following the tutorial from a computer that cannot connect to cameras, feel free to use the *Raw* files in the "Chapter04" folder (Figure 04_04).

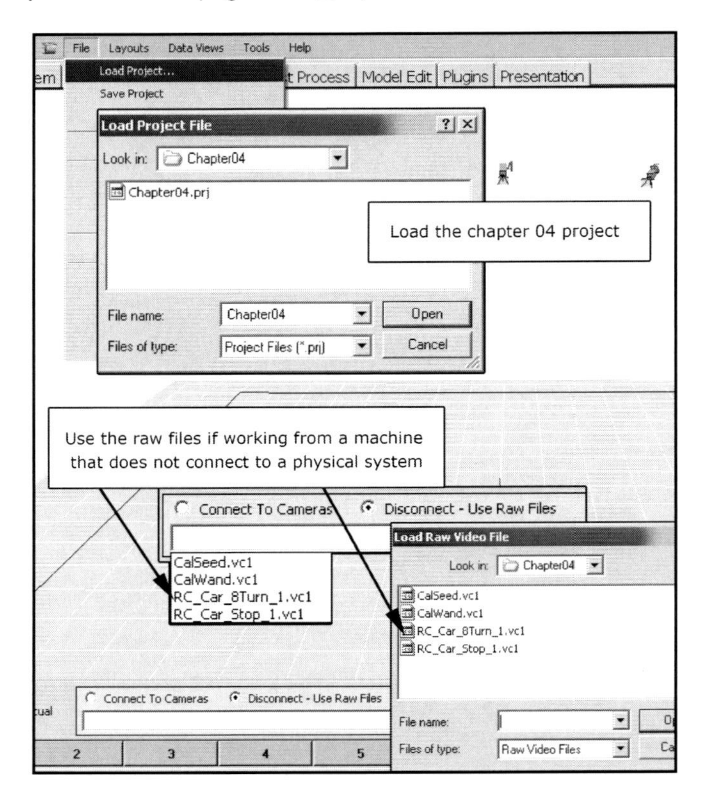

Figure 04_04.

Note: Raw Video files simulate the system being connected to the cameras.

[1] Download the files from www.MocapClub.com/TheMocapBook.htm

Calibrate the system and save the project. For more information on calibrating a system refer to Chapter 03.

Place reflective markers on an RC Car as shown in figure 04_05 (Figure 04_05).

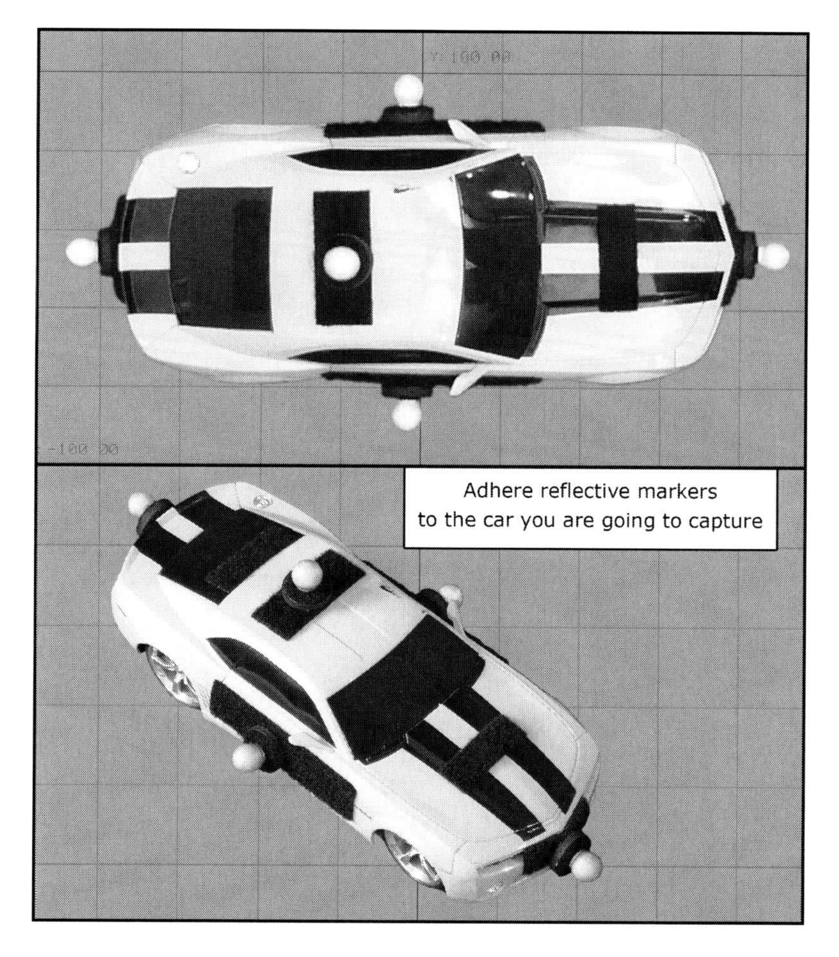

Figure 04-05.

We are going to record the car doing an 8-turn towards the edge of the *Volume*. Connect to cameras and accept the popup window.

Note: Chapter 03 has information on what to do is there is a discrepancy between the camera numbers in the dialogue and the physical number of cameras.

Note: If using Raw Video files load the "RC_Car_8Turn_1.vc1" instead (Figure 04_06).

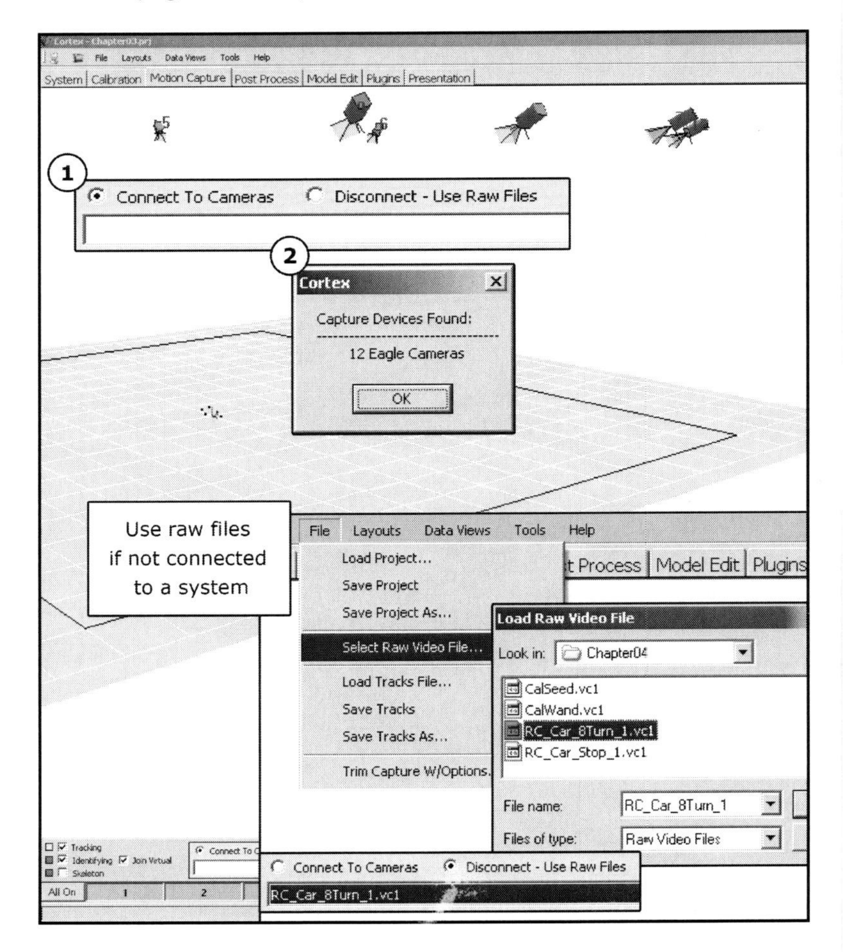

Figure 04_06.

In the *Motion Capture* pane, under the *Output files* section check on the *Tracked binary* box. Enter the name "RC_Car_8Turn_" in the *Name* box of the *Settings* section. Press the *Record* button and have the RC car do an 8-turn inside the *Volume* getting close to the edge. Stop the recording and press the *Load Last Capture* button (Figure 04_07).

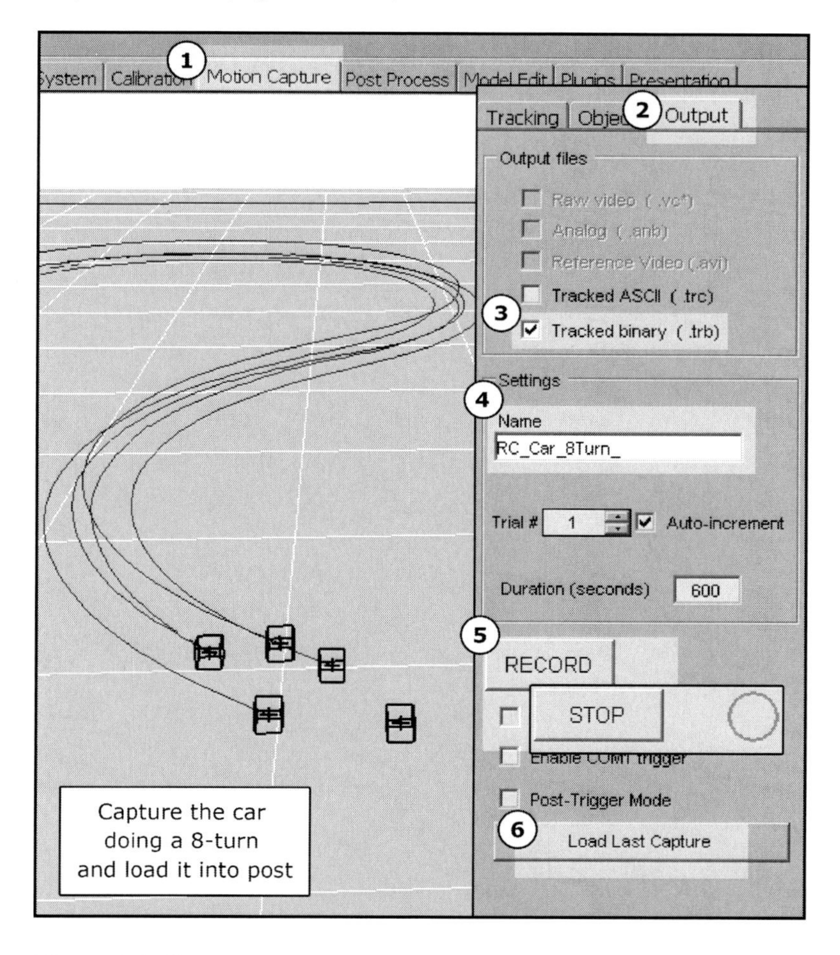

Figure 04_07.

The *Load Last Capture* button sends the user to the *Post Process* pane. Click on the marker for the left door in the 3D view and play the data. You can see that although the marker starts on the left door it jumps to the front as the car moves around the volume (Figure 04_08).

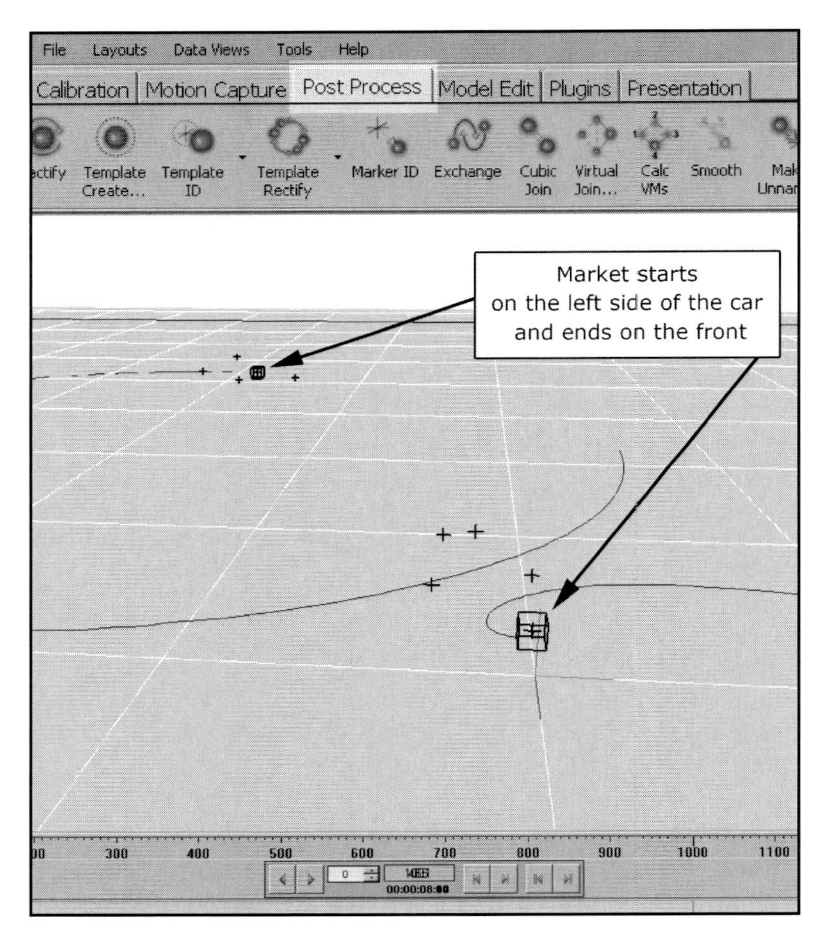

Figure 04_08.

This situation can be avoided by creating a proper template for the car. We will get back to the previous captured file after creating such template.

Place the car in the center of the volume. Change the name to "RC_Car_Stop_" in the *Name* box of the *Settings* section and record 2 seconds of the car standing still (Figure 04_09).

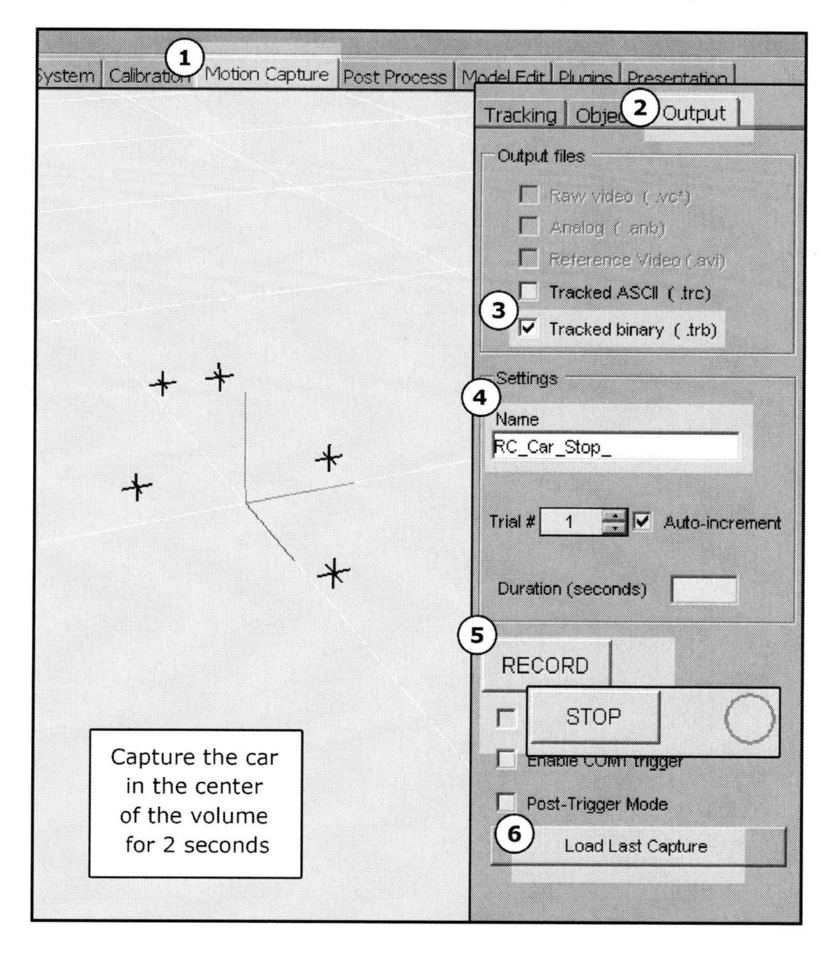

Figure 04_09.

Note: If using Raw Video, load the "RC_Stop_1.vc1" file.

Load Last Capture and access the *Model Edit* pane. Type "M_Front" in the space bellow the *Marker Names* section inside the *Markers* sub-pane and hit *Enter* to accept the name. Type "M_Back", "M_Top", "M_Right" and "M_Left" to complete the list. You have assigned the names to the markers that will compose the *Marker Set Template* (Figure 04_10).

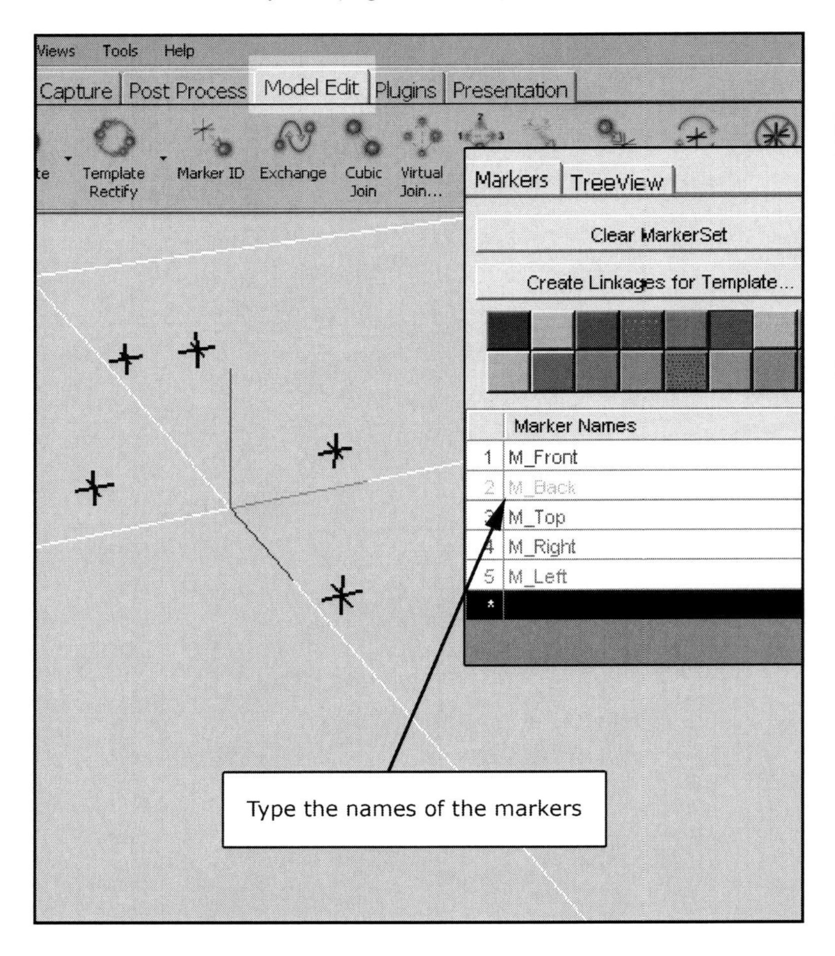

Figure 04_10.

To assign the list to the markers in the *3D Display* we need to identify the markers. Press the *Quick ID* button located in the *Post Process Tool Strip*. The *Identify Marker* window will appear with the first Marker from the list in the *Locate Marker* section (M_Front). The window is asking which one of the markers in the view is the "M_Front" As soon as you click on the marker you want to name "M_Front", the *Identify Marker* window loads the next item on the list. Assign the rest names of the list to the *Unnamed Markers* in the *3D Display* as the *Identify Marker* window asks for them (Figure 04_11).

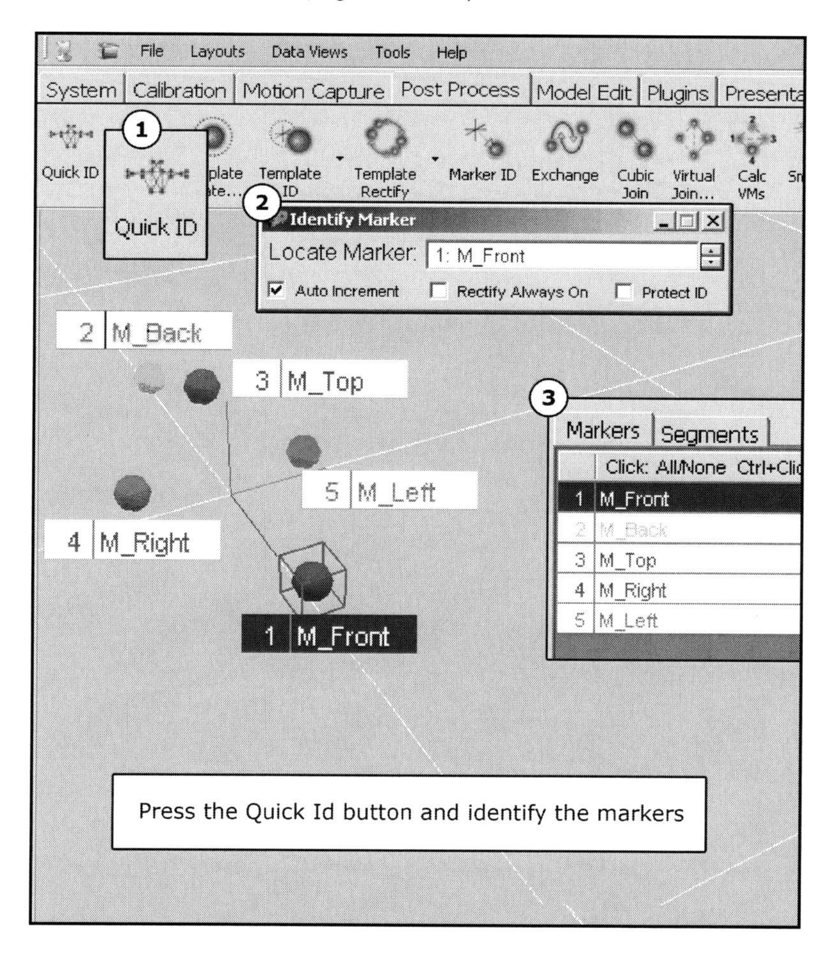

Figure 04_11.

Now we need to associate the markers to each other to make them unique. Since the measurements of one marker to the rest of the set are precise and exclusive, it makes the marker easier to identify by the system. We accomplish this by creating linkages between the markers. To do this press the *Create Linkages for Template* button in the *Markers* sub-pane and click and drag between the markers in the *3D Display*. Do this for every marker (Figure 04_12).

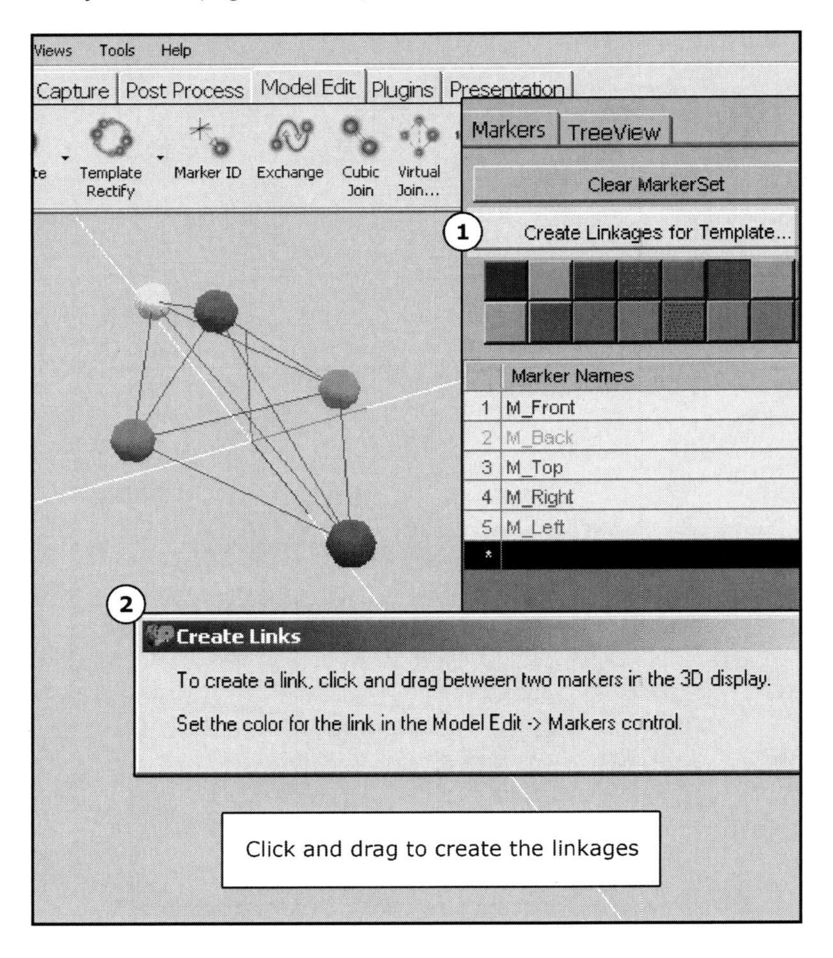

Figure 04_12.

The *TreeView* sub panel shows a list of the marker associations provided by the *Linkages* (Figure 04_13).

Markers TreeView

- Chapter04.prj
  - Markers (5)
    - M_Front
    - M_Back
    - M_Top
    - M_Right
    - M_Left
  - VMarkers (0)
  - Links (10)
    - M_Back To M_Front
    - M_Top To M_Front
    - M_Top To M_Back
    - M_Right To M_Front
    - M_Right To M_Back
    - M_Right To M_Top
    - M_Left To M_Front
    - M_Left To M_Back
    - M_Left To M_Top
    - M_Left To M_Right
  - SkB Segments (inactive)
  - Calcium Segments (inactive)

Figure 04_13.

Now that the markers have been named and *Linkages* have been created for them, we need to make sure that the system can use this information during the capturing process. To do this it is necessary to create a *Template* for the *Marker Set*. Press the *Template Create* button in the *Post Process Tool Strip*. The *Create Template* window appears. Check the *Template* option on and type "RC_Car_MSet" in the *MarkerSet* Name box, select the *Visible* option in the *Frames Range* section and check the *Include current frame as the Model Pose* option on. Press the *Create Template* button and accept the *Template has been created* message that follows (Figure 04_14).

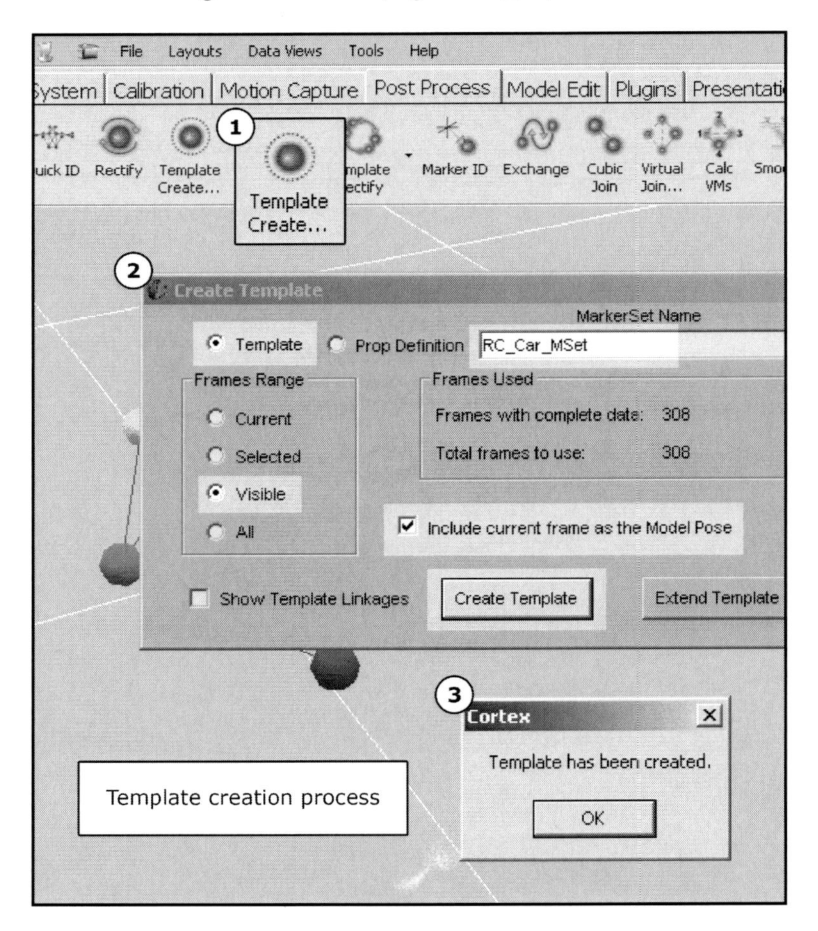

Figure 04_14.

The Marker Set Template for the RC Car is done. Now when you drive the car around the volume while *Running* the cameras you can see that all the markers and linkages are properly recognized in real time.
Load the "RC_CarCircle_1.vc1" file that you captured at the beginning of the tutorial, you can see that the markers and the linkages are also recognized In this file too (Figure 04_15).

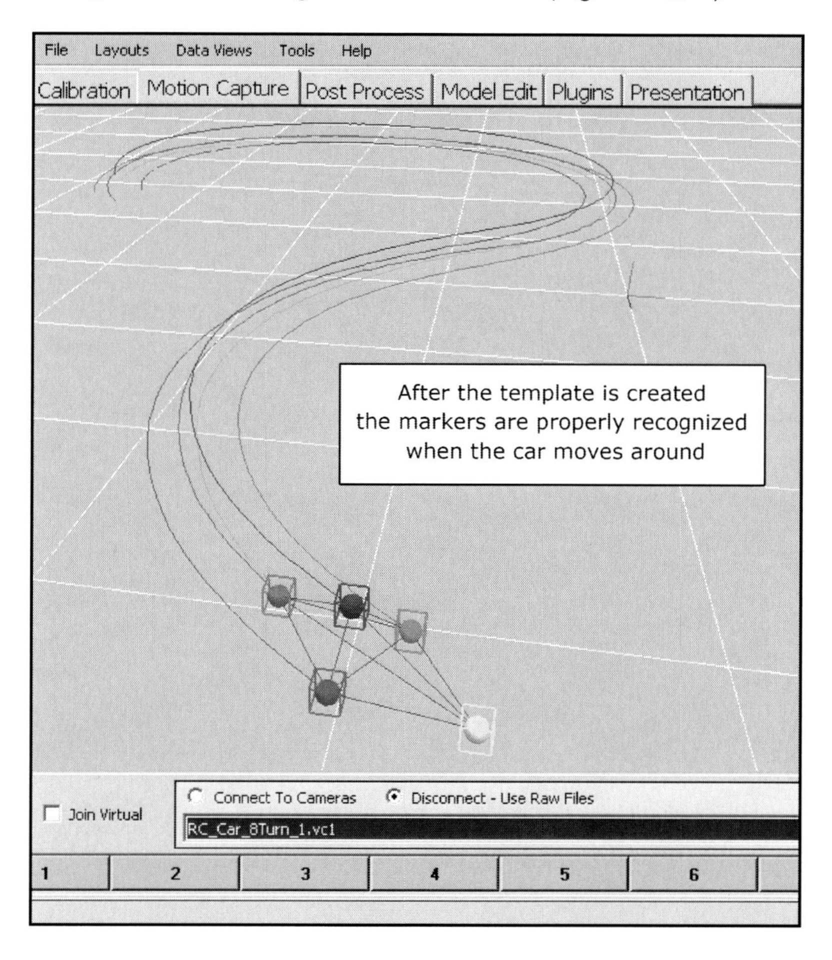

Figure 04_15.

Chapter 05
Human Template Exercise

For the most part, the measurements between the markers on an RC car are static. The distance from the front to the top of the car does not change as it drives around the *Volume*. On a human, things get a little more complicated as distances in some body-parts change when we move around. A great example of this situation is the distance from the chest to the top of the shoulder while the arm is down and while it is leveled with the ground (Figure 05_01).

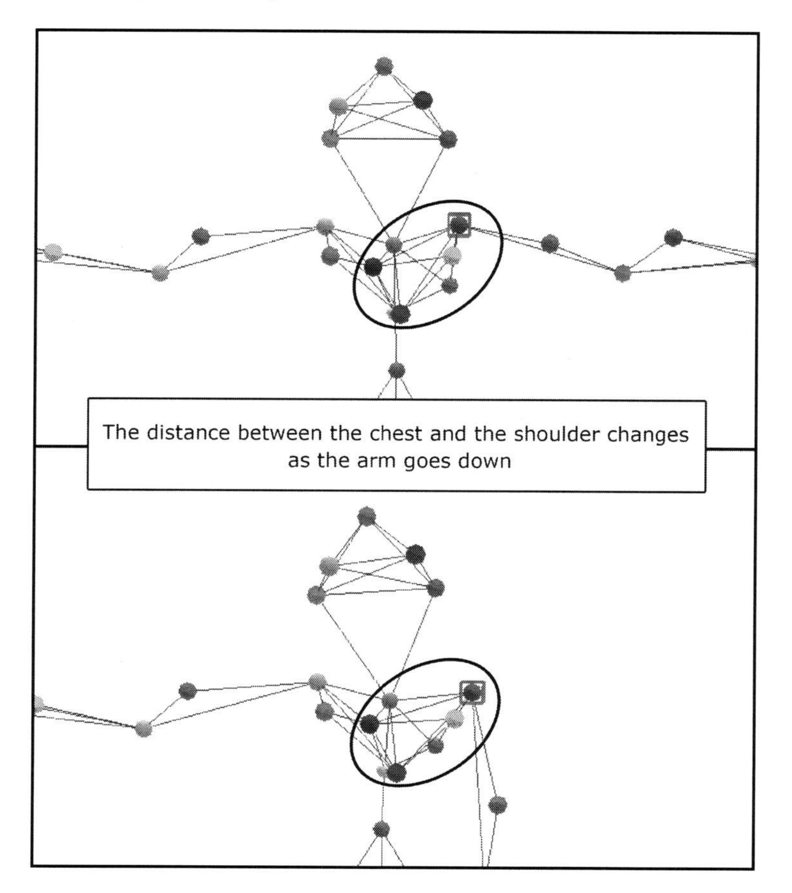

Figure 05_01.

This is one of the main reasons why the creation of a human template is a lot more complicated than a prop template. Instead of using set numbers, we want to use ranges for a template for a performer.

If you are going to be following the tutorial from a computer that cannot connect to cameras, feel free to use the *Raw* files in the "Chapter05" folder

Launch **Cortex** and Load the "Chapter05"[2] project.

Calibrate the system and save the project. For more information on calibrating a system refer to "Chapter 03".

Place reflective markers on the performer as shown in Figure 05_02 (Figure 05_02).

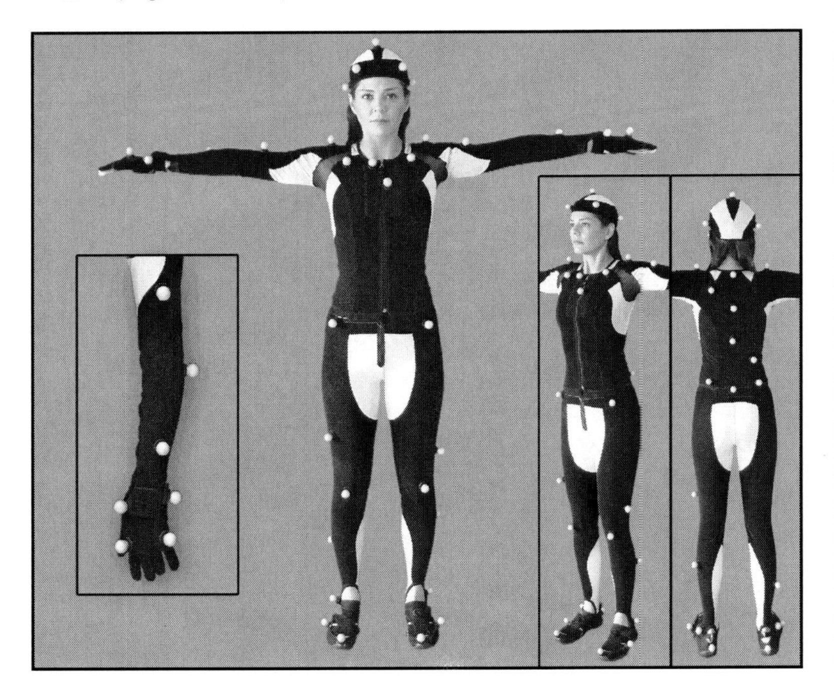

Figure 05_02.

---

[2] Download the files from www.MocapClub.com/TheMocapBook.htm

Record 2 seconds of the performer standing in *T-Pose* in the center of the *Volume*, facing the + *Z-axis*. Name the file "Actor_TPose_1". Having a *T-pose* will later help the process of transferring the motion from the optical markers to the digital character. Because of this you should start every motion that you want to capture In a *T-Pose* (Figure 05_03).

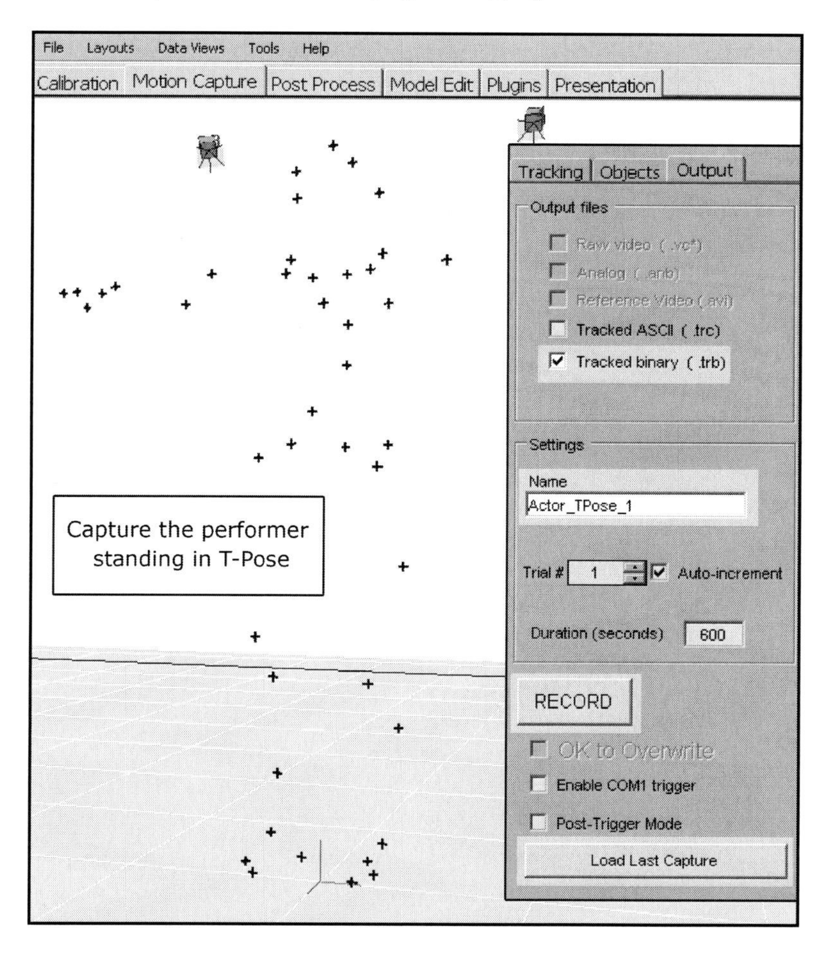

Figure 05_03.

Note: If using Raw Video files load the "TPose_1.vc1" instead.

Load Last Capture and access the *Model Edit* pane. Populate the Marker Names list in the *Markers* sub-pane by typing: "CtBkHip", "LtBkHip", "LtFtHip", "RtBkHip", "RtFtHip", "Spine01", "Spine02", "Spine03", "SpineRtOffset", "SpineLtOffset", "CtHead", "LtFtHead", "LtBkHead", "RtFtHead", "RtBkHead", "CtBtChest", "CtUpChest", "LtFtClavicle", "LtUpClavicle", "LtBicep", "LtElbow", "LtForeArm", "LtOtWrist", "LtInWrist", "LtPinky", "LtPalm", "RtFtClavicle", "RtUpClavicle", "RtBicep", "RtElbow", "RtForeArm", "RtOtWrist", "RtInWrist", "RtPinky", "RtPalm", "LtThigh", "LtKnee", "LtCalf", "LtAnkle", "LtHeel", "LtBall", "LtToe", "RtThigh", "RtKnee", "RtCalf", "RtAnkle", "RtHeel", "RtBall", "RtToe" (Figure 05_04).

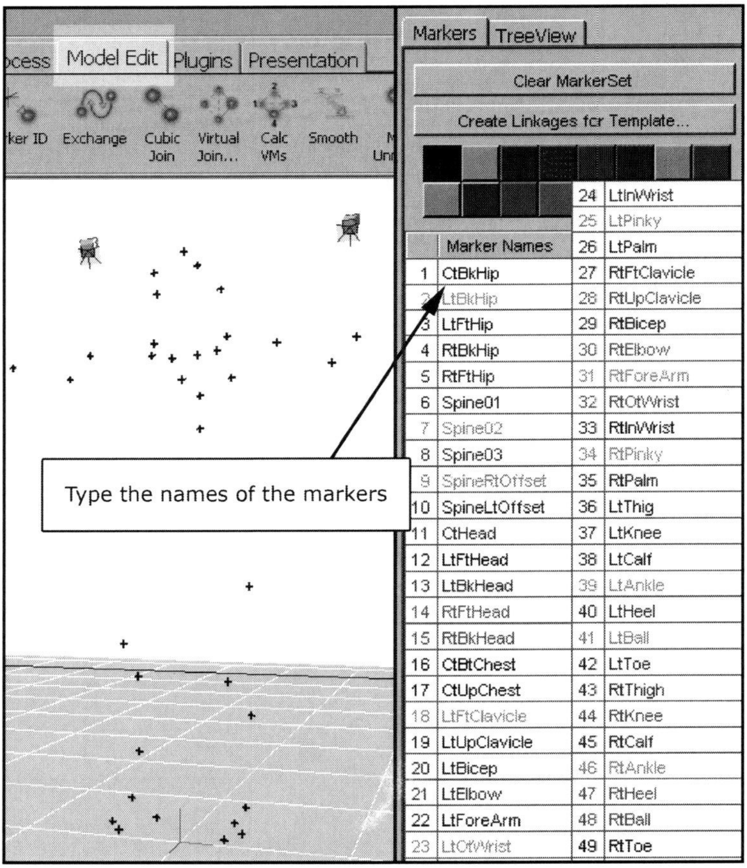

Figure 05_04.

Note: The marker names are arbitrary, we are using names that relate area where the marker is placed but you can change the names to fit specific requirements and pipelines.

Press the *Rectify Unnamed* and then the *Quick ID* button in the *Post Process Tool Strip*. Click on the markers In the *3D View* in the order that the *Identify Marker* window requires as shown on figure 05_05 (Figure 05_05).

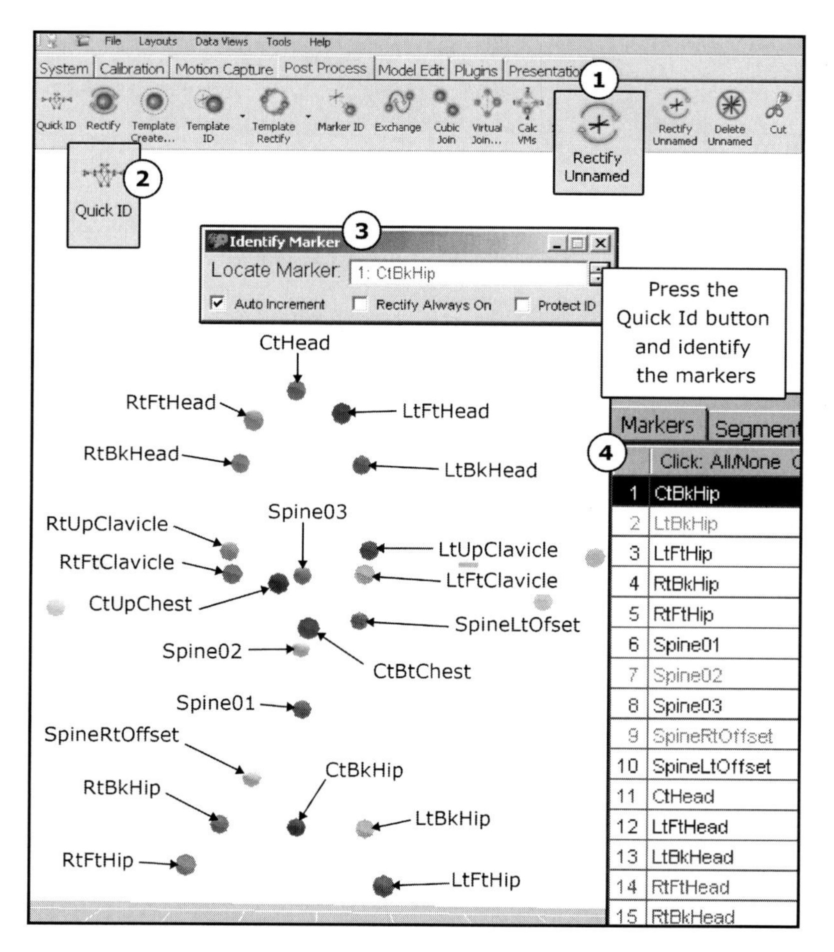

Figure 05_05.

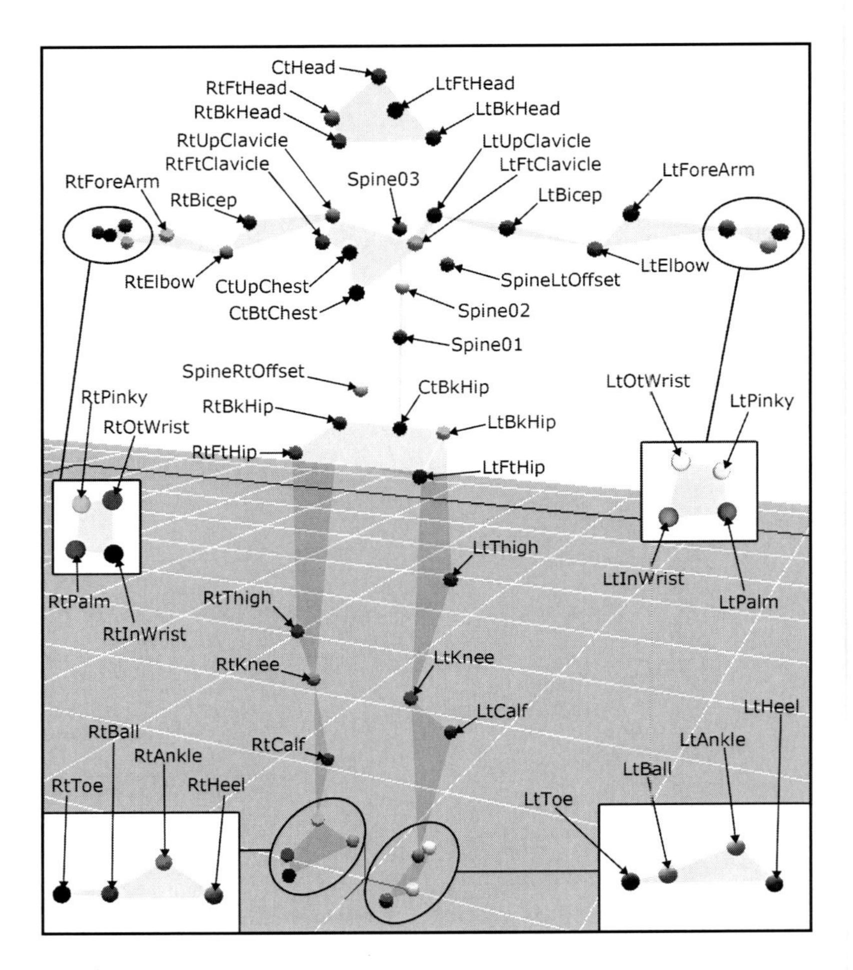
CtHead
RtFtHead
LtFtHead
RtBkHead
LtBkHead
RtUpClavicle
LtUpClavicle
RtFtClavicle
LtFtClavicle
RtForeArm
Spine03
LtForeArm
RtBicep
LtBicep
RtElbow
CtUpChest
SpineLtOffset
LtElbow
CtBtChest
Spine02
Spine01
SpineRtOffset
CtBkHip
LtOtWrist
RtPinky
RtBkHip
LtBkHip
LtPinky
RtOtWrist
RtFtHip
LtFtHip
LtThigh
LtInWrist
RtPalm
RtThigh
LtPalm
RtInWrist
RtKnee
LtKnee
LtCalf
LtHeel
RtBall
LtAnkle
RtAnkle
RtCalf
LtBall
RtToe
RtHeel
LtToe

Figure 05_06.

Press the *Create Linkages for Template* button in the *Markers* sub-pane and click and drag between the markers in the *3D Display*. Do this while trying to block the different body parts of the human body (Figure 05_07).

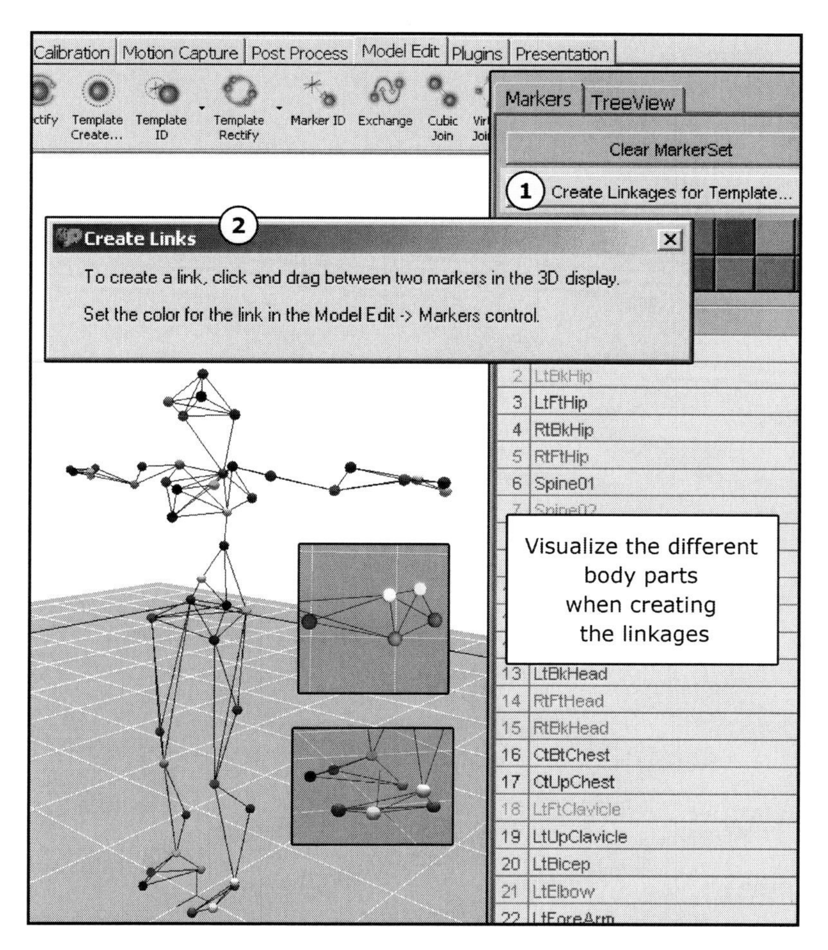

Fig 05_07.

Play the data to make sure that the markers are properly named throughout the timeline.

Press the *Template Create* button in the *Post Process Tool Strip*. Check the *Template* option on, type “Actor_MSet” in the *MarkerSet* Name box, select the *Visible* option in the *Frames Range* section and check the *Include current frame as the Model Pose* option on. Press the *Create Template* button and accept the *Template has been created* message that follows (Figure 05_08).

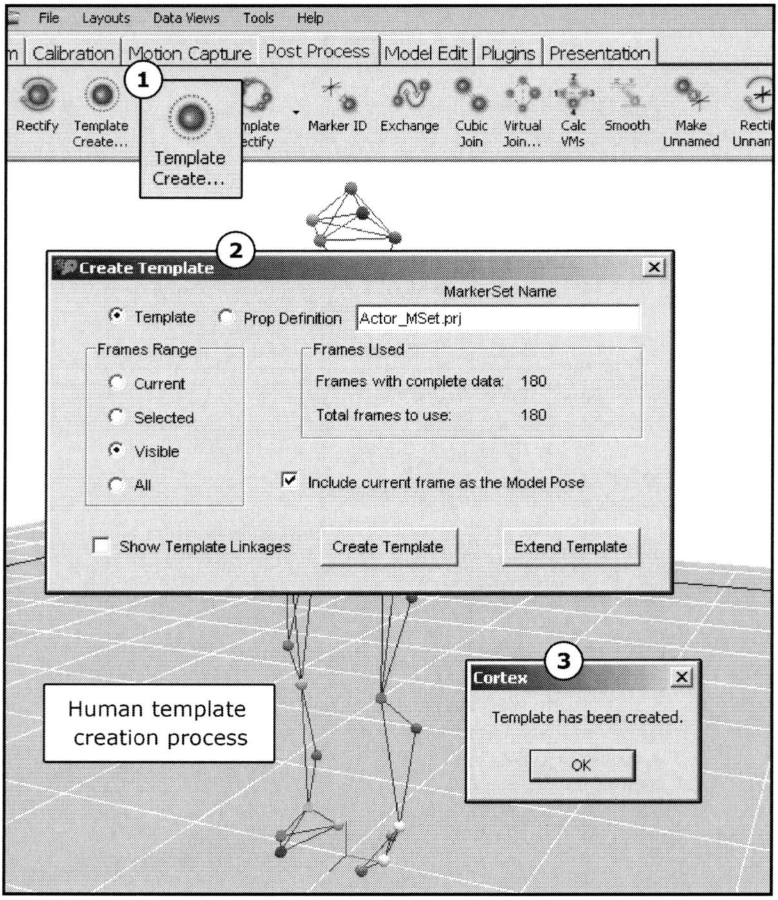

Figure 05_08.

The performer's *Template* has been created. There are a few issues with it though, when the actor stretches some parts of his body some of the markers are no longer recognized by the system (Figure 05_09).

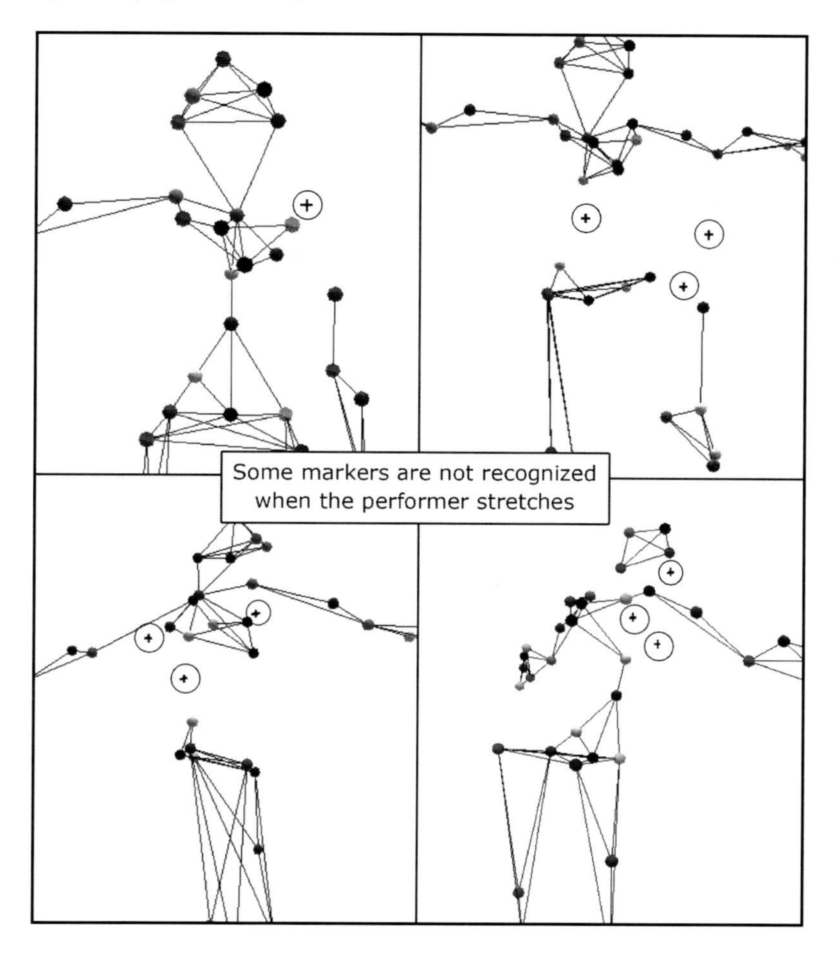

Figure 05_09.

In order to correct these problems we need to extend the *Template*.

Record the actor starting in *T-Pose* in the center of the volume facing the + *Z-axis*, then going through his / hers range of motion (ROM). Name the file “Actor_ROM_2” (Figure 05_10).

Figure 05_10.

Note: If using Raw Video files load the “Raw_Actor_ROM_2” instead.

Load Last Capture and *Rectify Unnamed* markers. Play the data and stop when the performer puts the arms down. The "LtUpClavicle" and the "RtUpClavicle" are not recognized by the system at this point in the range of motion. Select the "LtUpClavicle" in the *Markers* list in the *Post Process* pane. Press the *Marker ID* button. The Identify Marker window launches with the selected marker loaded in the *Locate Marker* section. Click the marker that is supposed to be the "LtUpClavicle" while the arms are down. The "LtClavicle" is now recognized when the left arm of the performer moves down. ID the "RtClavicle" when the arms are down by using the *Marker ID* button (Figure 05_11).

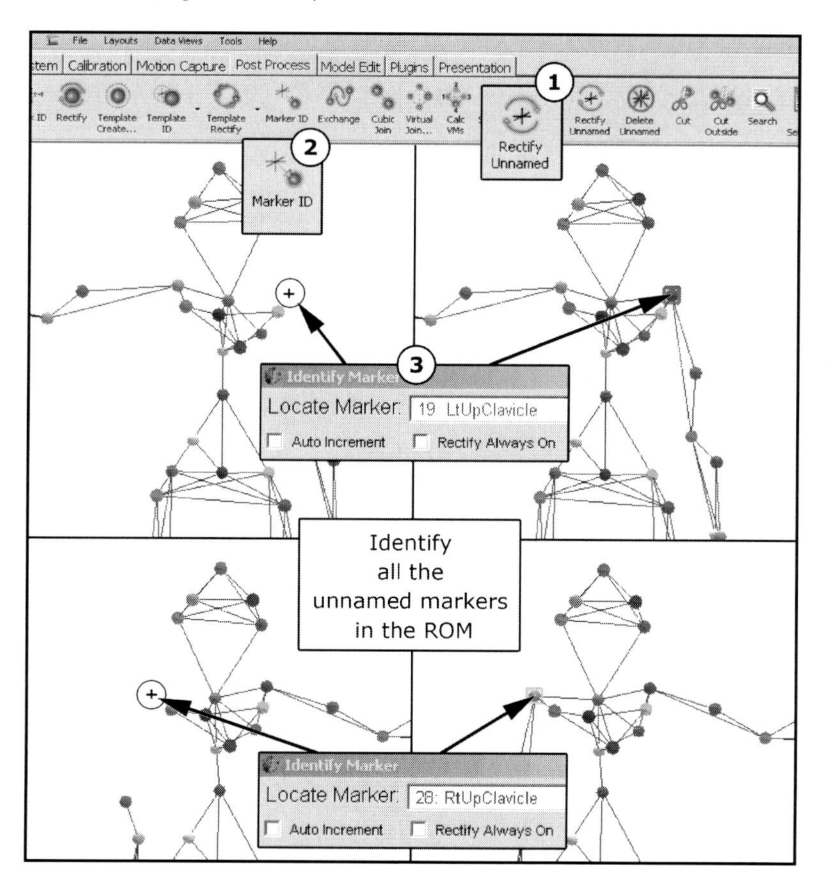

Figure 05_11.

Repeat the process for the rest of the markers that are not being recognized during sections of the motion.

Once all the markers have been properly identified, press the *Delete Unnamed* button in the *Post Process Tool Strip*. Click the *Template Create* button and use the same options as before but press the *Extend Template* instead of the *Create Template*. Accept the *Body template has been extended* message that follows (Figure 05_12).

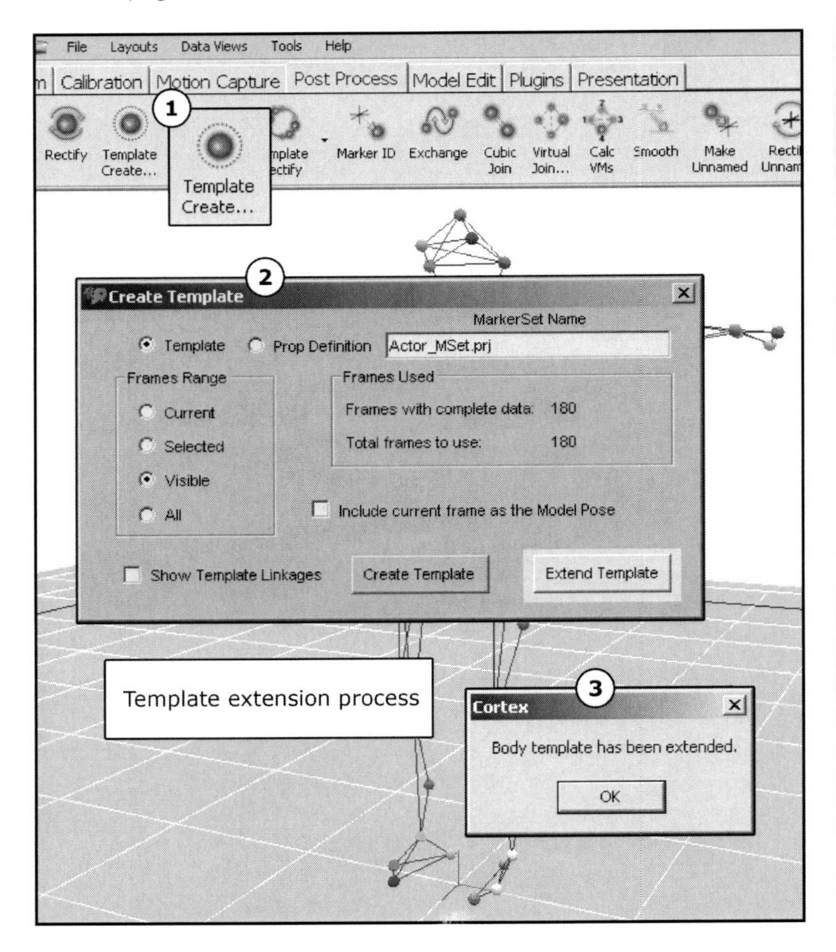

Figure 05_12.

You will now see that when the performer repeats the range of motion, the markers are properly recognized through the whole motion.

As you start capturing different movements you will find occasions where the *Marker Template* does not understand some markers. Manually ID the markers in the areas they are being lost and *Extend the Template*. The *Marker Set Template* will get stronger every time you do this.

Chapter 06
Tracking Basics

Tracking is the process of making sure that all the trajectories of the markers are accurate and that all markers are present during the take.

Note: A take is the motion that gets digitized from the time that the Record button is pressed to the time that the "Stop" button is hit (Figure 06_01).

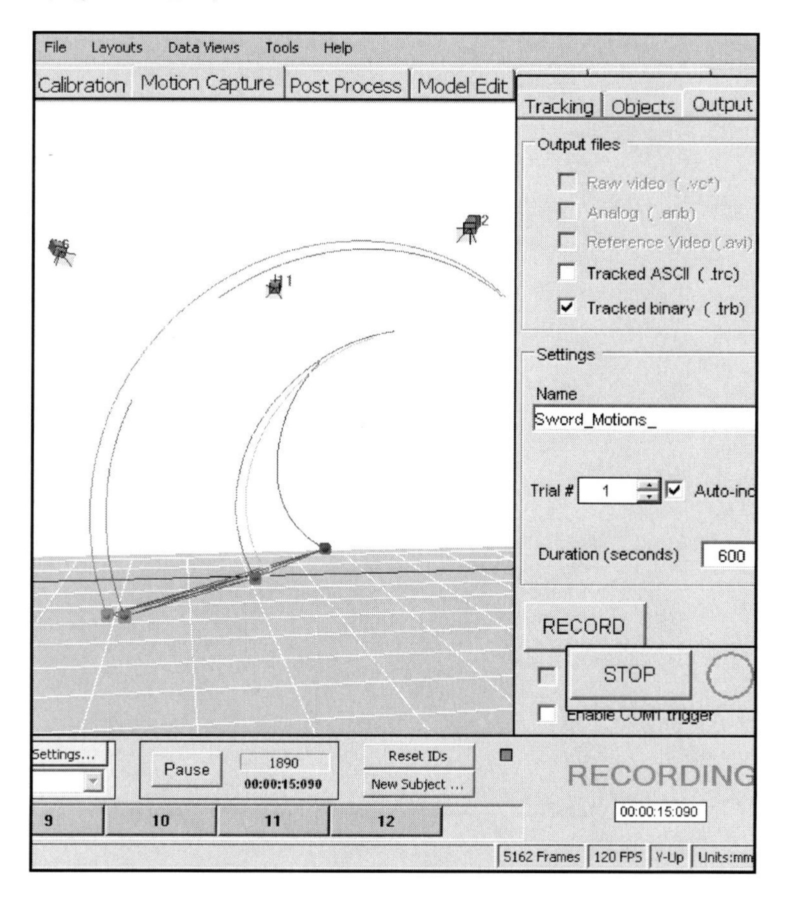

Figure 06_01.

This chapter is going to focus on illustrating the main concepts and operations involved in tracking optical data.

Load the “Chapter06”[3] project into **Cortex**.

Go to the *Post Process* tab and inspect the *Markers* section. Note how there are markers for every section of a sword (Figure 06_02).

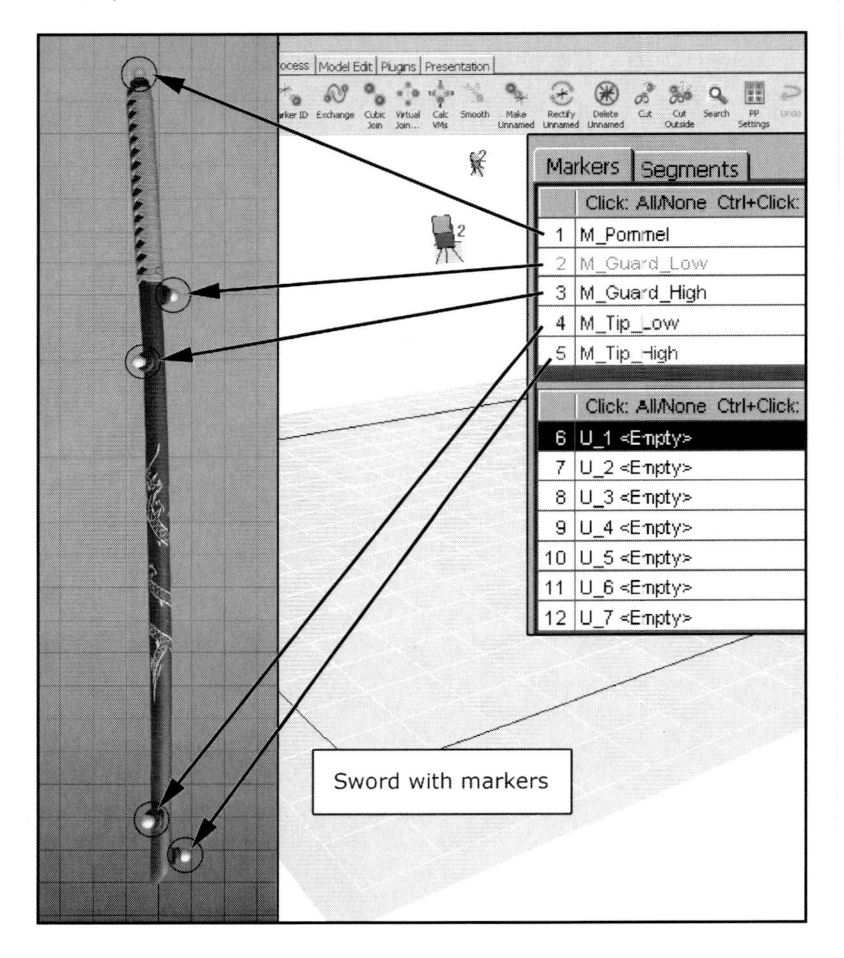

Figure 06_02.

[3] Download the files from www.MocapClub.com/TheMocapBook.htm

Load the “Sword_Motions_1.trb”.

Play the file to see the motions of the captured sword. Some markers become unnamed when the sword is swung about and some are completely occluded, there is even a spot at the end where two marker exchange trajectories (Figure 06_03).

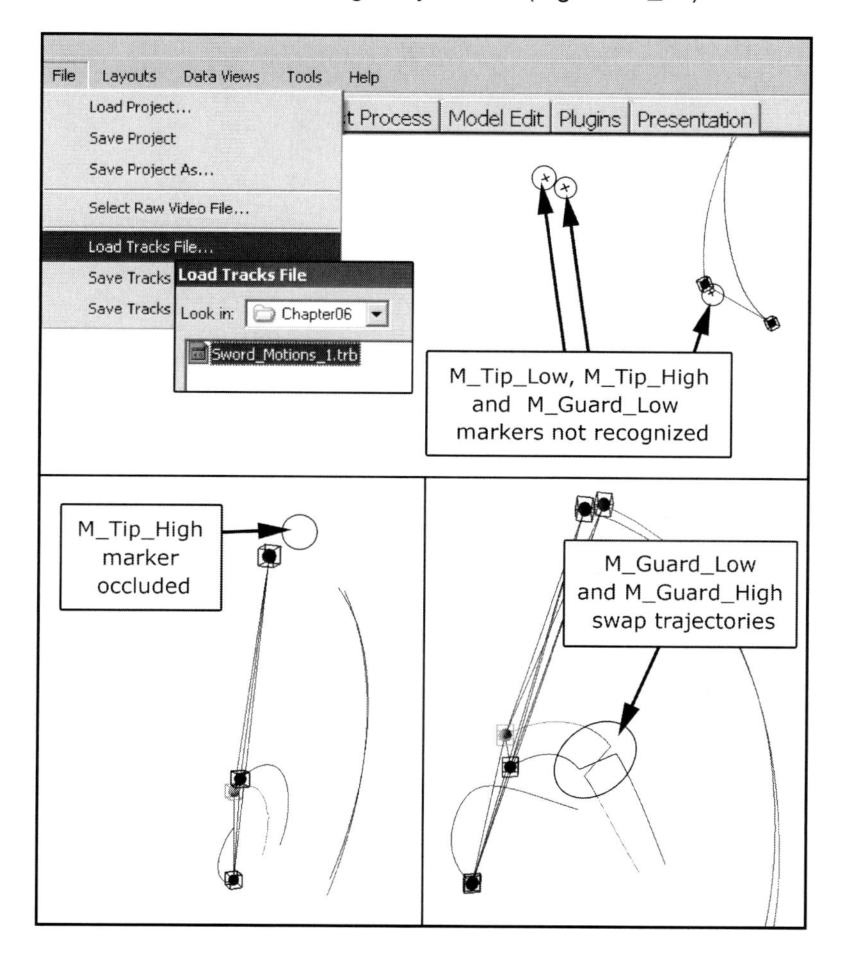

Figure 06_03.

To track this data we will start by identifying the unrecognized markers. Go to frame 484, select “M_Tip_High” from the *Markers* tab in the *Post Process* pane, press the *Marker ID* button and click on the unnamed marker in the *3D View*.
With the *Identify Marker* window still open, go to frame 492, select the “M_Guard_Low” marker from the tab list, and click on the unnamed marker in the *3D View* (Figure 06_04).

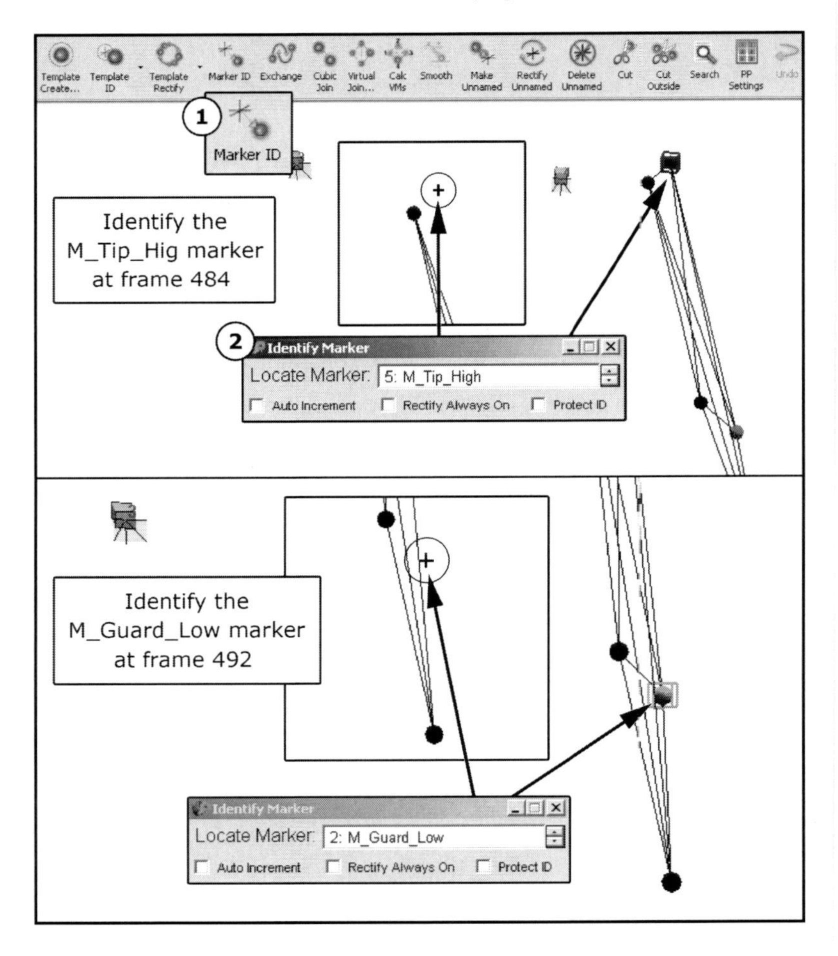

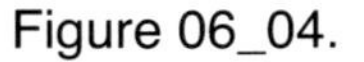
Figure 06_04.

Press the *Rectify Unnamed* button in the *Post Process Tool Strip*, this button consolidates unnamed markers.
With the *Identify Marker* window open, go to frame 570, select the "M_Tip_High" from the list and click on the unnamed marker in the *3D View*.
At frame 579, select the "M_Tip_Low" from the list and click on the unnamed marker in the *3D View* (Figure 06_05).

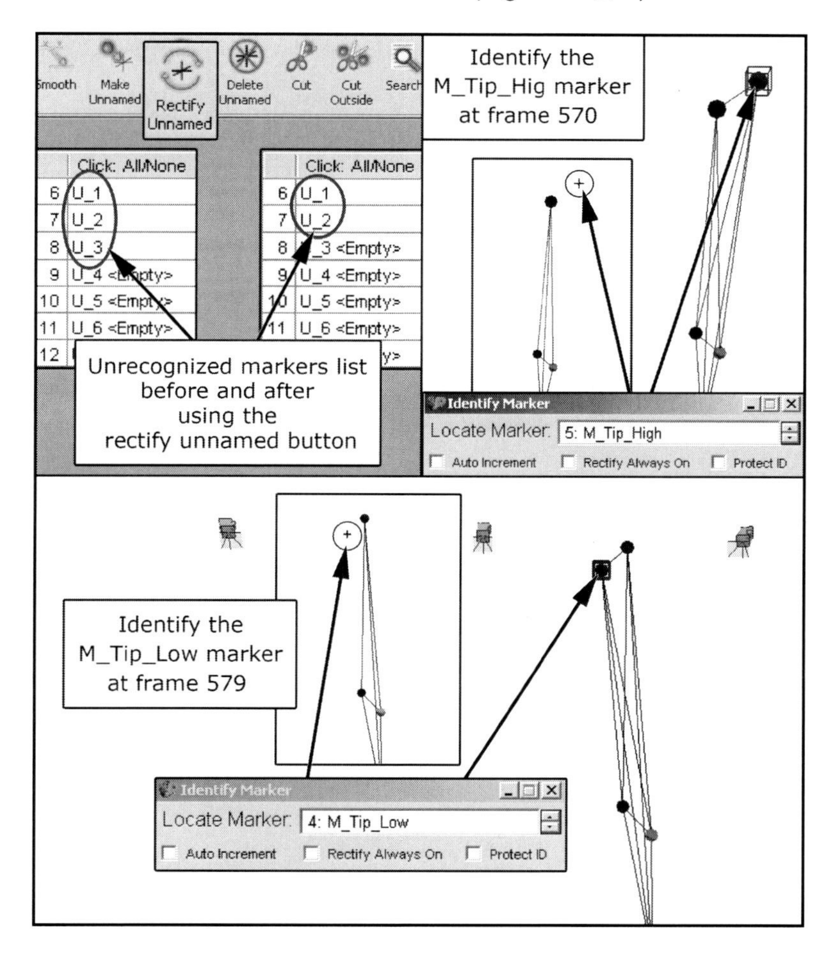

Figure 06_05.

Press the *Rectify Unnamed* button in the *Post Process Tool Strip* to consolidate unnamed markers. With the *Identify Marker* window open, go to frame 692, select the "M_Guard_Low" from the list and click on the unnamed marker in the *3D View*. Go to frame 977, select the "M_Guard_Low" from the list and click on the unnamed marker in the *3D View*.
Press the *Rectify Unnamed* button in the *Post Process Tool Strip* to consolidate unnamed markers. You can see that there are no more unnamed markers to identify in the markers list. (Figure 06_06).

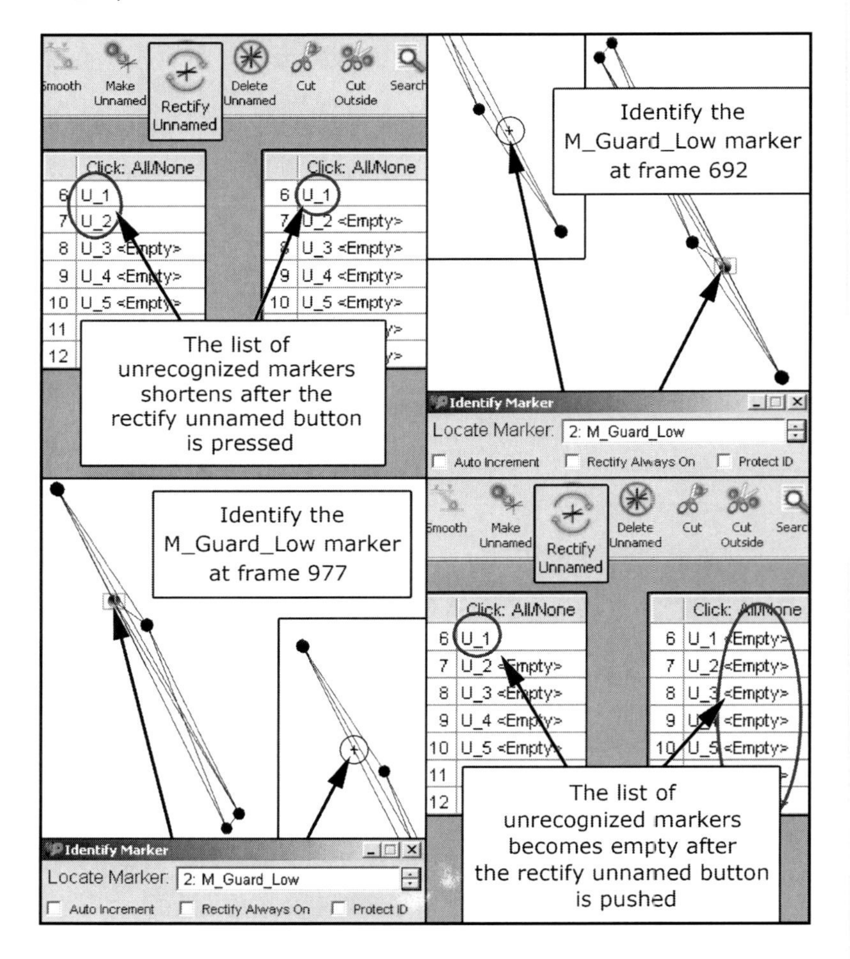

Figure 06_06.

Now that all the unnamed markers have been identified we need to take care of the gaps in the data because of occluded markers. Make sure that the *Identify Marker* window is closed and that your have a two pane view with *Markers XYZ Graphs* on the top and *3D View* on the bottom (Figure 06_07).

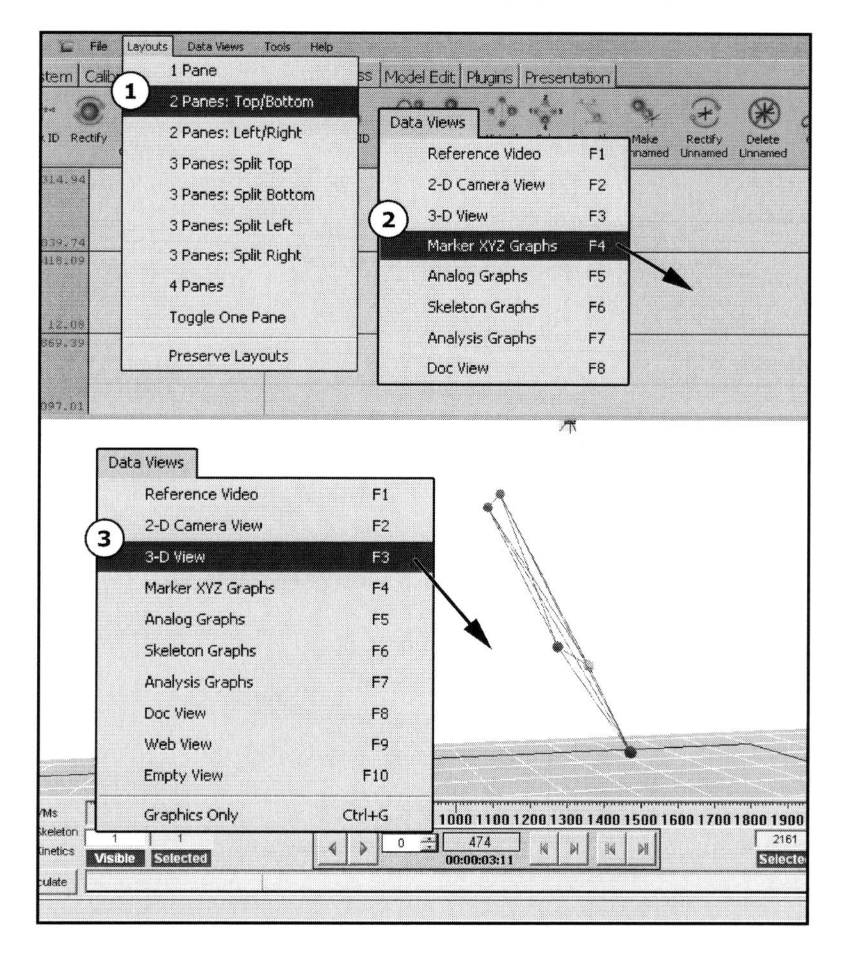

Figure 06_07.

Select the "M_Tip_Low" marker and go to frame 105. With the cursor in the *Markers XYZ Graphs* area press the "*I*" key repeatedly until only the gap from frame 104 to frame 113 is seen. Press the "*L*" key. The gap is filled by a blue straight line drawn between the frame 104 and frame 113. Press the "*O*" key repeatedly until the entirety of the XYZ graph is seen. Note how only the seen gap got filled (Figure 06_08).

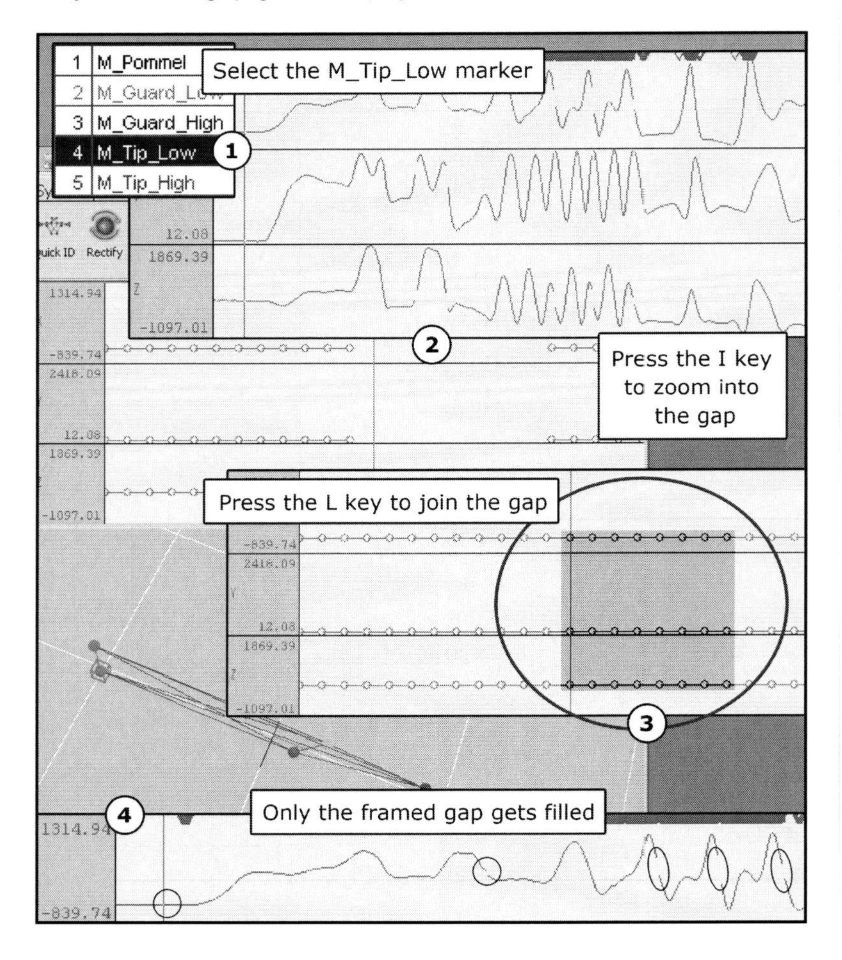

Figure 06_08.

Select the "M_Guard_High" from the list, go to frame 240, zoom into the gap with the "*I*" key and press the "*L*" key to fill it. Repeat the process for frame 445 (Figure 06_09).

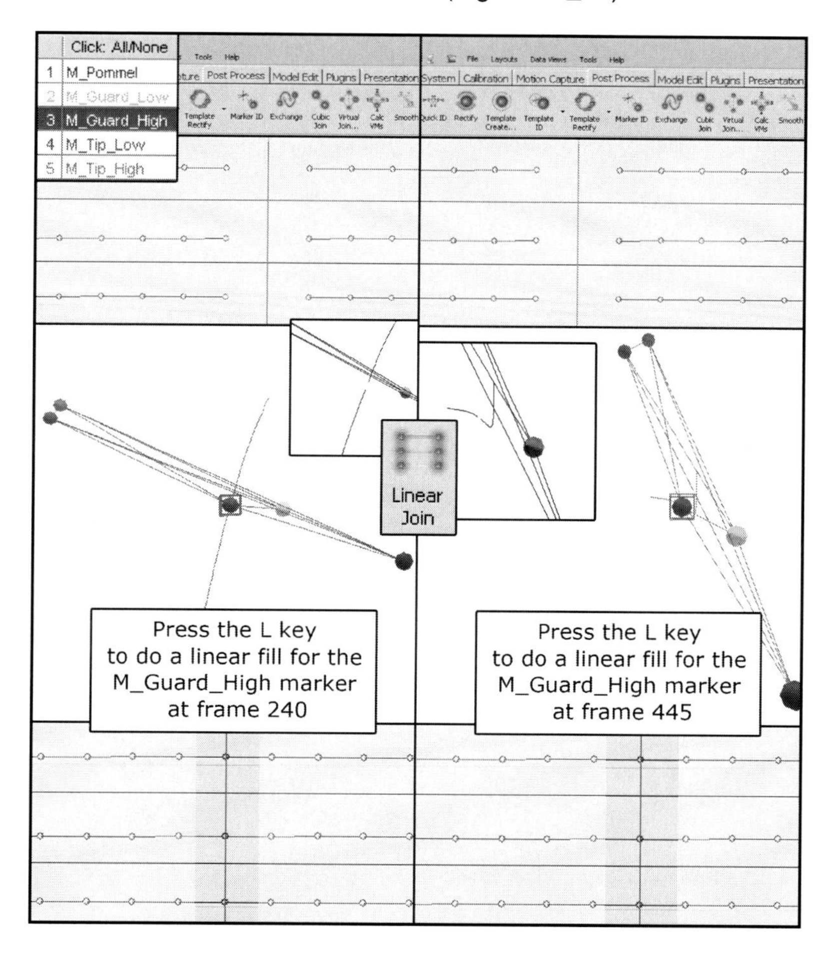

Figure 06_09.

Select the "M_Tip_Low" from the list, go to frame 693, zoom into the gap with the *"I"* key and press the *"L"* key to fill it. Step forward to around frame 701 to analyze the new trajectory. We can see that the new trajectory is not accurate, instead of continuing the arc the fill operation drew a straight line between the frames. This is exactly what the "Join Linear" operation does; it draws a straight line between the gaps in the data.
Undo the operation by pressing *Ctrl + Z* and press the *"C"* key to do a cubic fill for the gap. The *Cubic Join* operation tries to fill gap by continuing the curvature of the animation curve (Figure 06_10).

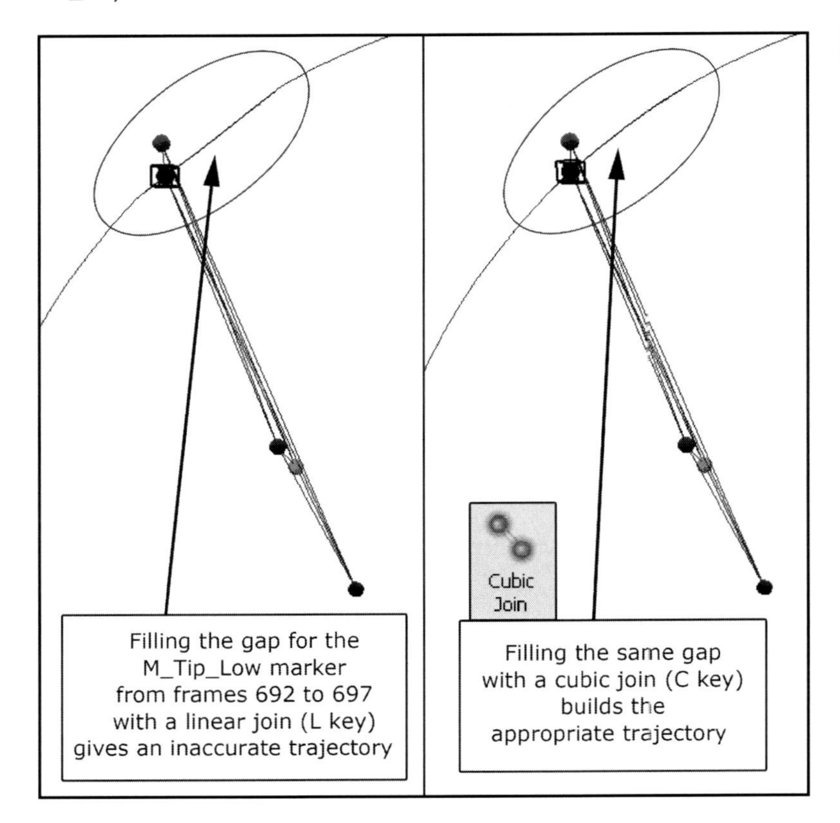

Figure 06_10.

With the “M_Tip_Low” still selected go to frame 799, zoom into the gap and execute a *Cubic Join* operation by pressing the *“C”* key. Also do a *Cubic Join* for the “M_Pommel” at frame 934 (Figure 06_11).

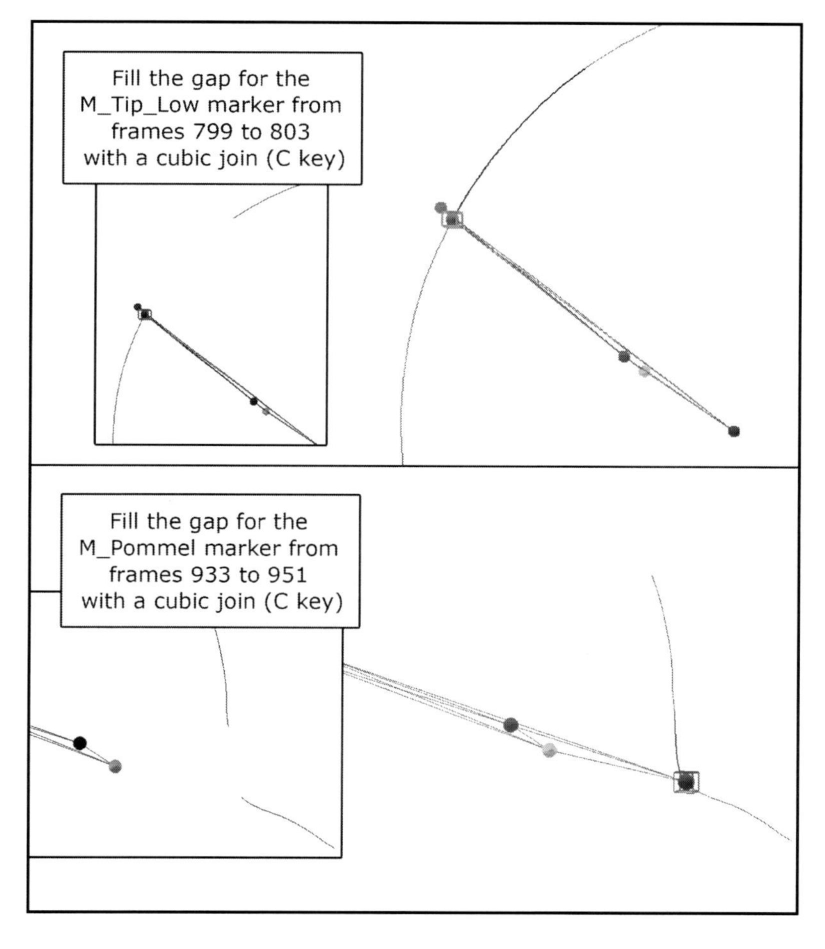

Figure 06_11.

Select the “M_Tip_High” marker and go to frame 1166 and frame the gap. Do a *Cubic Join*. The “M_Tip_High” does not follow a similar trajectory to the other markers from the sword after the fill operation.
We need a more complex operation to extract the accurate trajectory of the “M_Tip_High” marker at this gap, Undo the operation by pressing *Ctrl + Z*. We need to fill the gap using a Virtual Join fill. With the “M_Tip_High” marker still selected, press the *“V”* key to access the *Virtual Marker Join* window. Click on the *Origin Marker* section and select the “M_Pomel” marker in the *3D View*. Then Click on the *Long Axis* section and select the “M_Guard_High” marker from the *3D View*. Finally, Click on the *Plane Marker* section and select the “M_Guard_Low” marker from the *3D View*. Hit the *“V”* key to execute the *Join Virtual* operation. Note how the “M_Tip_High” now follows the proper trajectory along side the rest of the markers of the sword (Figure 06_12).

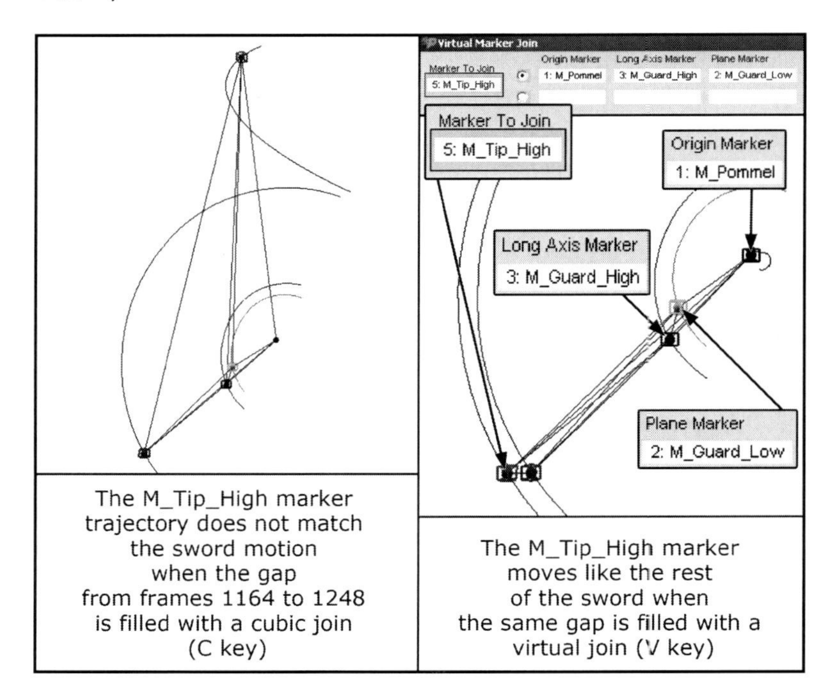

The M_Tip_High marker trajectory does not match the sword motion when the gap from frames 1164 to 1248 is filled with a cubic join (C key)

The M_Tip_High marker moves like the rest of the sword when the same gap is filled with a virtual join (V key)

Figure 06_12.

The *Join Virtual* fill operation is one of the most reliable tracking operations to deal with occluded markers because it uses the trajectories of 3 selected markers to reconstruct the gap of the lost marker. This is why we are going to use this operation to fill the rest of the gaps in this exercise. With the "M_Tip_High" still selected make sure that the entire timeline is framed in the XYZ graphs by pressing and holding the *"O"* key. Press the *"V"* key and note how the rest of the gaps in the "M_Tip_High" curves are filled (Figure 06_13).

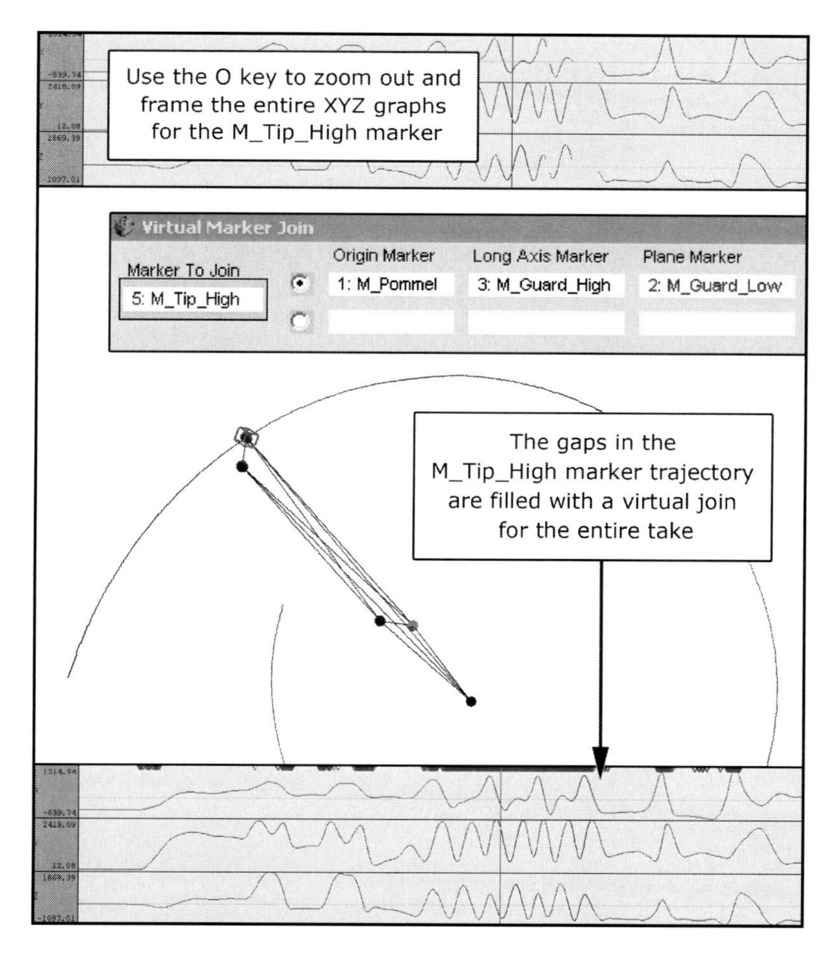

Figure 06_13.

Select the "M_Pommel" marker and. Hit the *"V"* key to call the *Virtual Marker Join* dialog and populate the window as follows: *Origin Marker* = "M_Tip_Low", *Long Axis Marker* = "M_Guard_Low", *Plane Marker* = "M_Guard_High". Press the *"V"* key to execute the operation.
Also fill the gaps in the "M_Guard_Low" marker by populating the *Virtual Marker Join* window as follows: *Origin Marker* = "M_Tip_High", *Long Axis Marker* = "M_Guard_High", *Plane Marker* = "M_Tip_Low". Press the *"V"* key to execute the operation. (Figure 06_14).

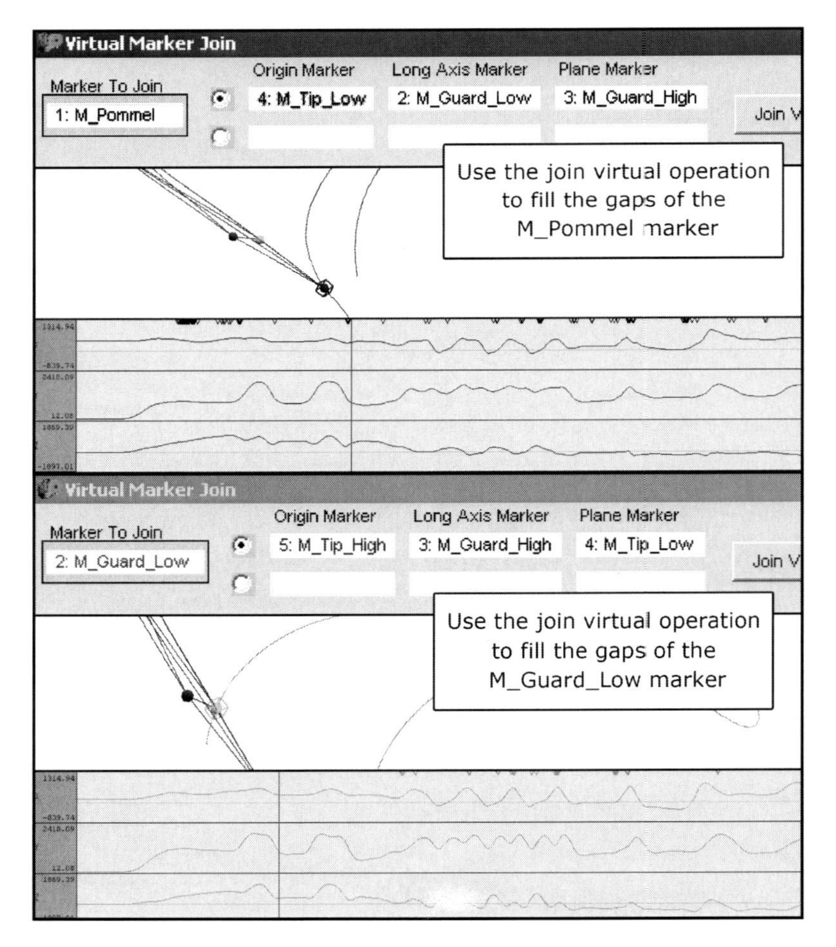

Figure 06_14.

Select the "M_Guard_High" marker and populate the *Virtual Marker Join* window as follows: *Origin Marker* = "M_Pommel", *Long Axis Marker* = "M_Tip_High", *Plane Marker* = "M_Guard_Low". Press the *"V"* key to execute the operation. (fig image of virtual definition window plus the highlighted filled gaps).
Fill the gaps in the "M_Tip_Low" marker by populating the Virtual Marker Join window as follows: *Origin Marker* = "M_Pommel", *Long Axis Marker* = "M_Guard_High", *Plane Marker* = "M_Guard_Low". Press the *"V"* key to execute the operation (Figure 06_15).

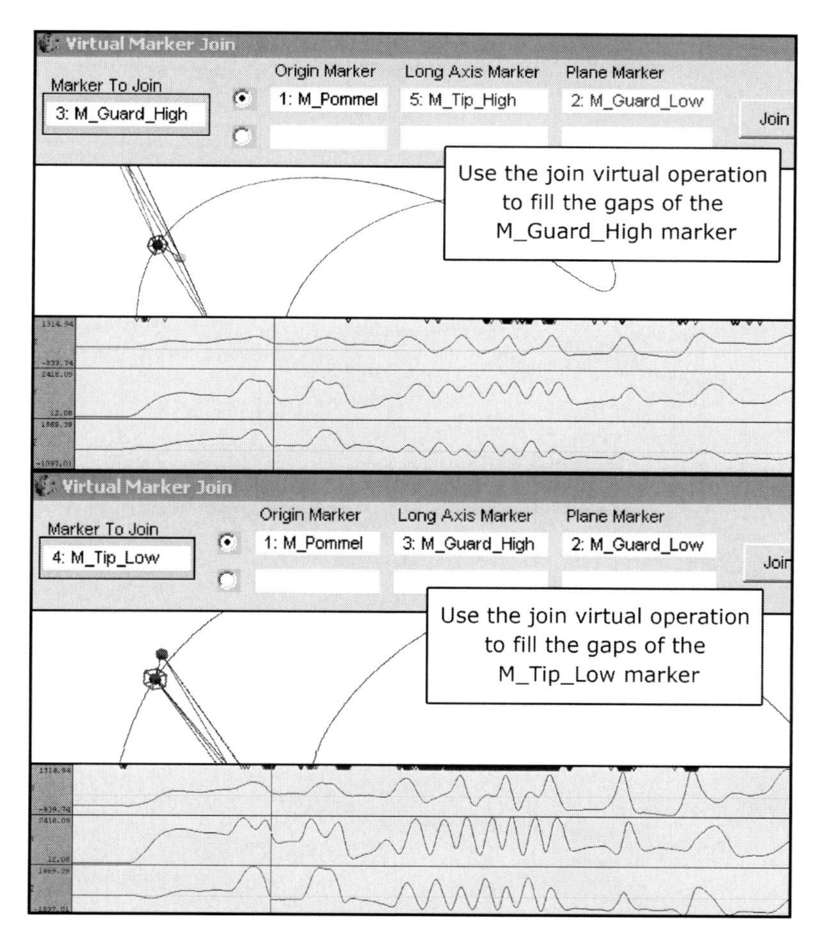

Figure 06_15.

Now that all gaps have been filled we need to address the last problem of the sword data, the marker swaps. Go to frame 1620 and select the "M_Guard_Low" and the "M_Guard_High" markers. Step forward and backwards a few frames; you will be able to see that both markers exchange trajectories at frame 1621. Make sure that you are back at frame 1620 and press the *"2"* key to select the frames from this point until the end of the motion. Press the *"G"* key to exchange the markers trajectories. Note how both markers now follow their appropriate trajectories (Figure 06_16).

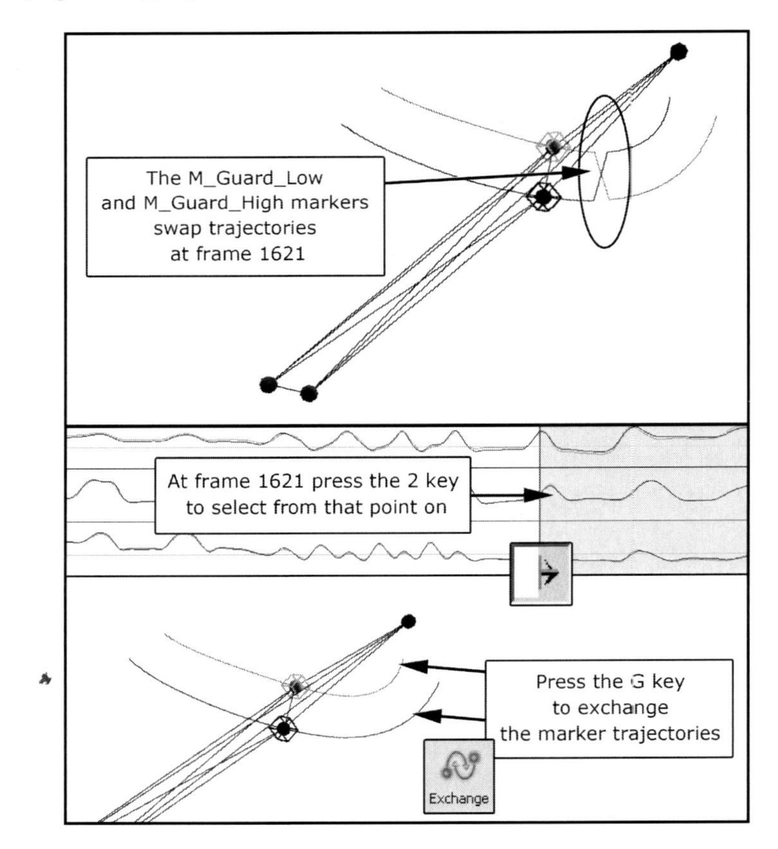

Figure 06_16.

You are done with tracking of the sword exercise, this data is now clean and ready to control the motion of a CG prop in the 3D software package of your choice.

Chapter 07
Tracking A Human Motion

In this chapter we will apply the concepts we learned in the sword exercise, this time in order to track a human motion. Specifically, we will be tracking the motion of someone doing a back flip. Tracking human motions is more complicated than tracking props not only because of the amount of markers involved, but also for the intricacy of what the human body can do (Figure 07_01).

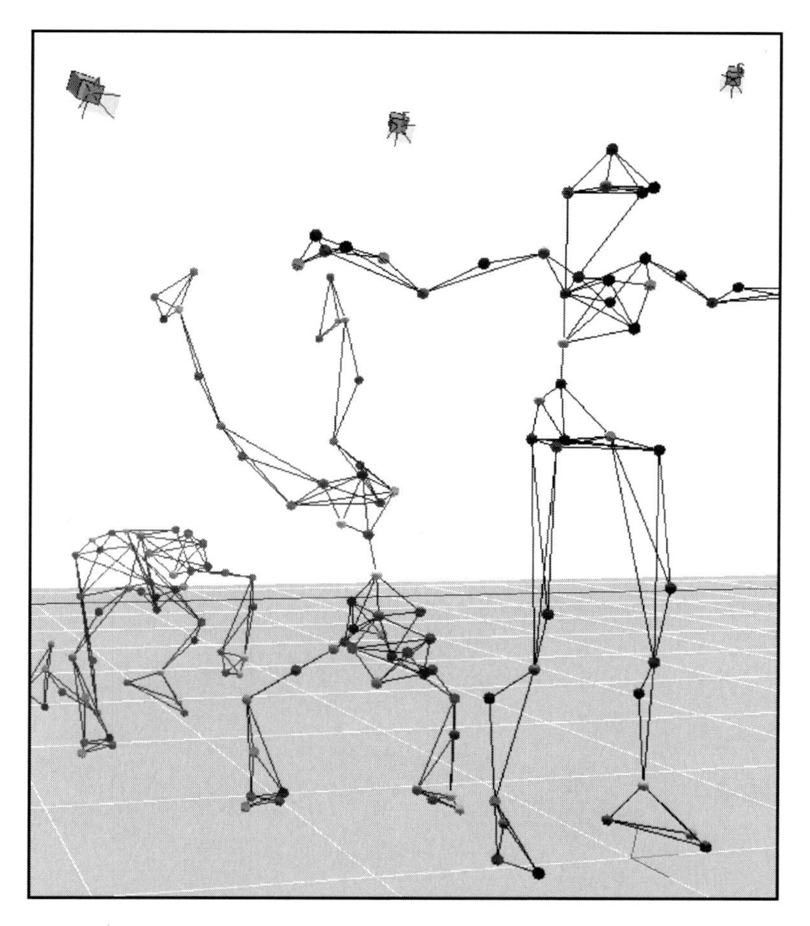

Figure 07_01.

Load the "Chapter07"[4] project into **Cortex**. Also Load the "Human_Motion_1.trb".

Play the file to see the performer do a back flip. When the performer brings his arms forward some of the markers start to get occluded. At the point where he is arched backwards the whole back it's lost from view. When the flip is landed most markers from the chest are unseen. By the time the performer gets up most of the upper body is unrecognized and some markers are identified in the wrong places (Figure 07_02).

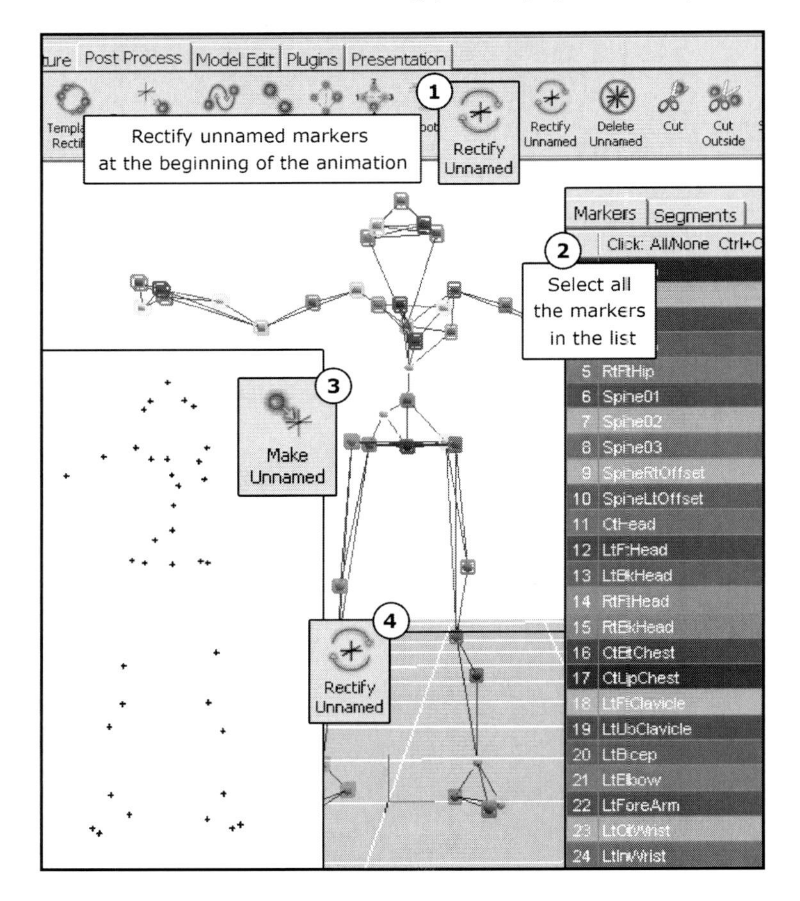

Figure 07_02.

---

[4] Download the files from www.MocapClub.com/TheMocapBook.htm

Sometimes instead of trying to correct every single minor miss identification and swap of the markers, it is easier to start from scratch and do a fresh ID of the template.
Go to the beginning of the animation and press the *Rectify Unnamed* button. Select all the markers from the *Template* by clicking at the top of the list and press the *"Y"* key to turn them into unnamed markers. Press the *Rectify Unnamed* button again. You will see all the markers in the *3D View* turn into black crosses (Figure 07_03).

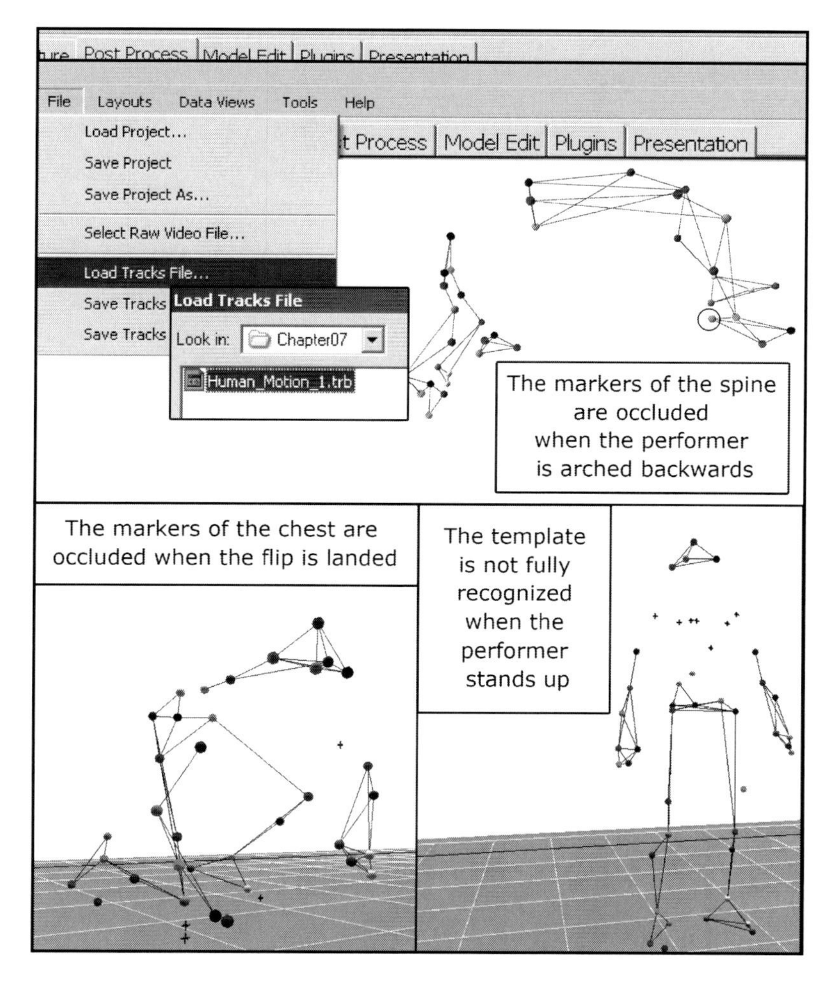

Figure 07_03.

Press the *Quick ID* button and click on the markers in the *3D View* in the order that the *Identify Marker* window prompts you to. Press the *Rectify Unnamed* button after you are done identifying the markers (Figure 07_04).

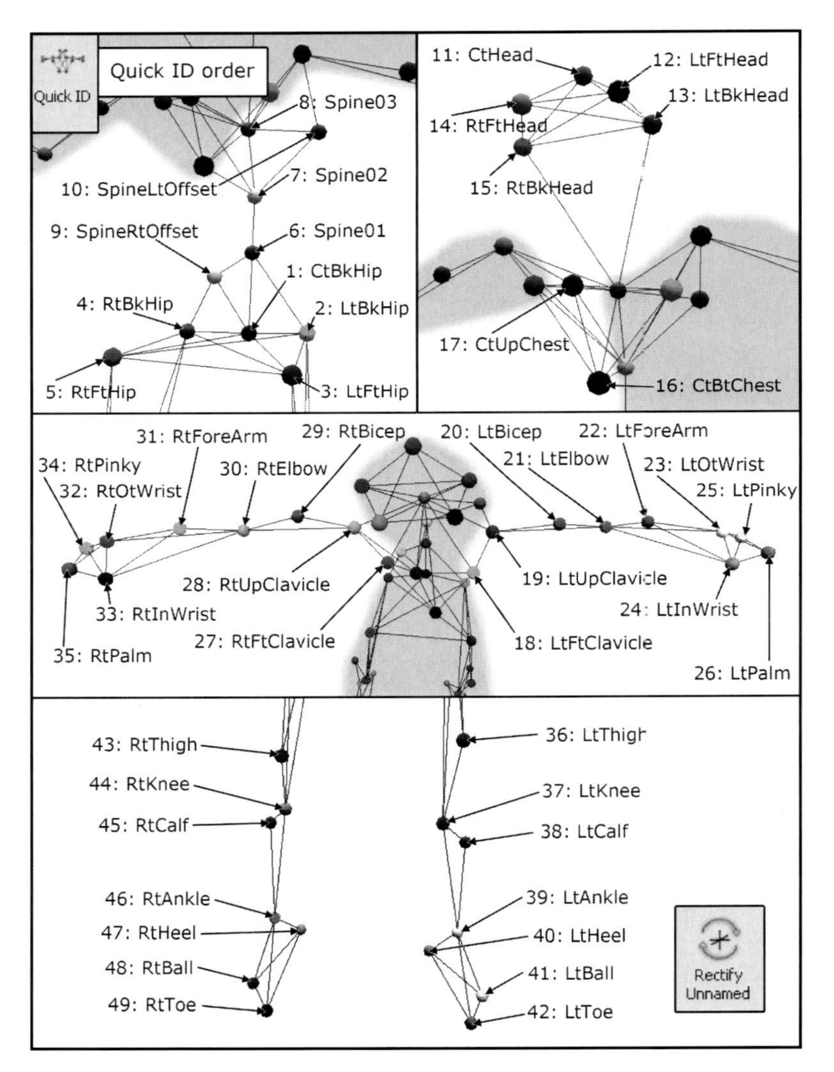

Figure 07_04.

Go to the end of the timeline. Here you can see that most of the markers are unrecognized by the template, but more importantly there are a few that are miss-identified. This is the case of the "Spine02" and "RtAnkle" markers. Go to frame 1509 and press the *"2"* key to select the frames from here until the end of the timeline. Select the "SpineRtOffset" from the list, push the *Marker ID* button and click on the "Spine02" marker on the *3D View*. This forces to apply the "Spine02" trajectory from frame 1509 on, to the "SpineRtOffset" marker (Figure 07_05).

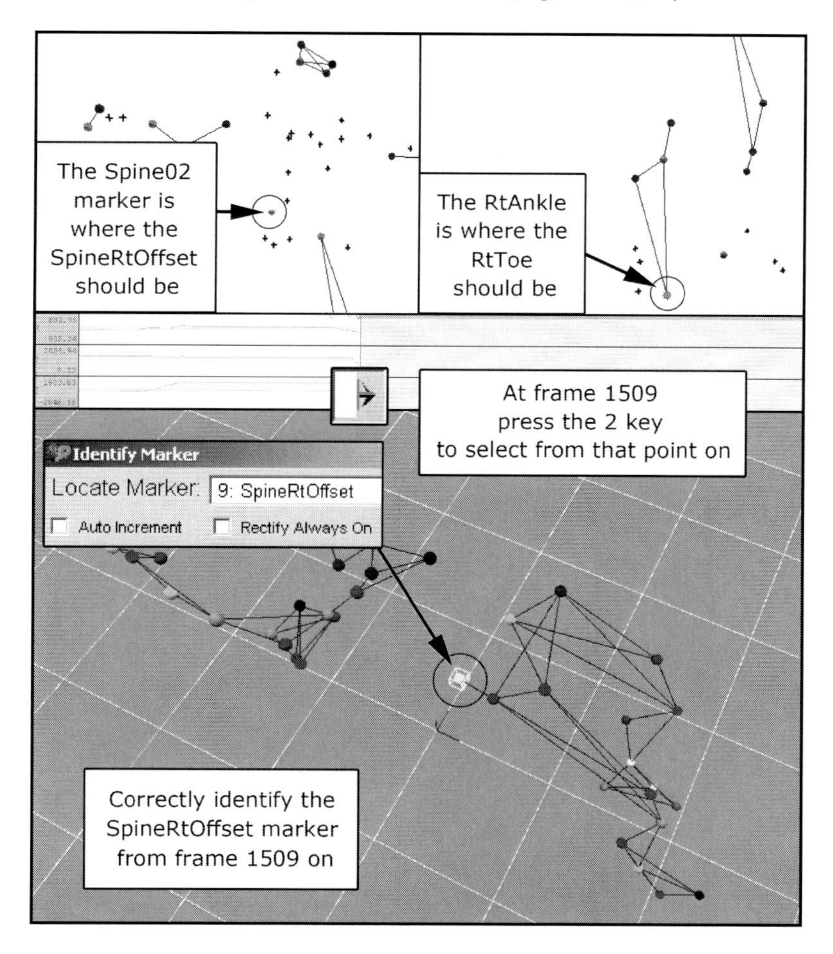

Figure 07_05.

Go to frame 1595 and press the *"2"* key. With the *Identify Marker* window still open, select the "RtToe" marker from the list and click on the "RtAnkle" marker on the *3D View*. The "Rtoe" takes the trajectory that was previously mislabeled "RtAnkle" from frame 1595 on (Figure 07_06).

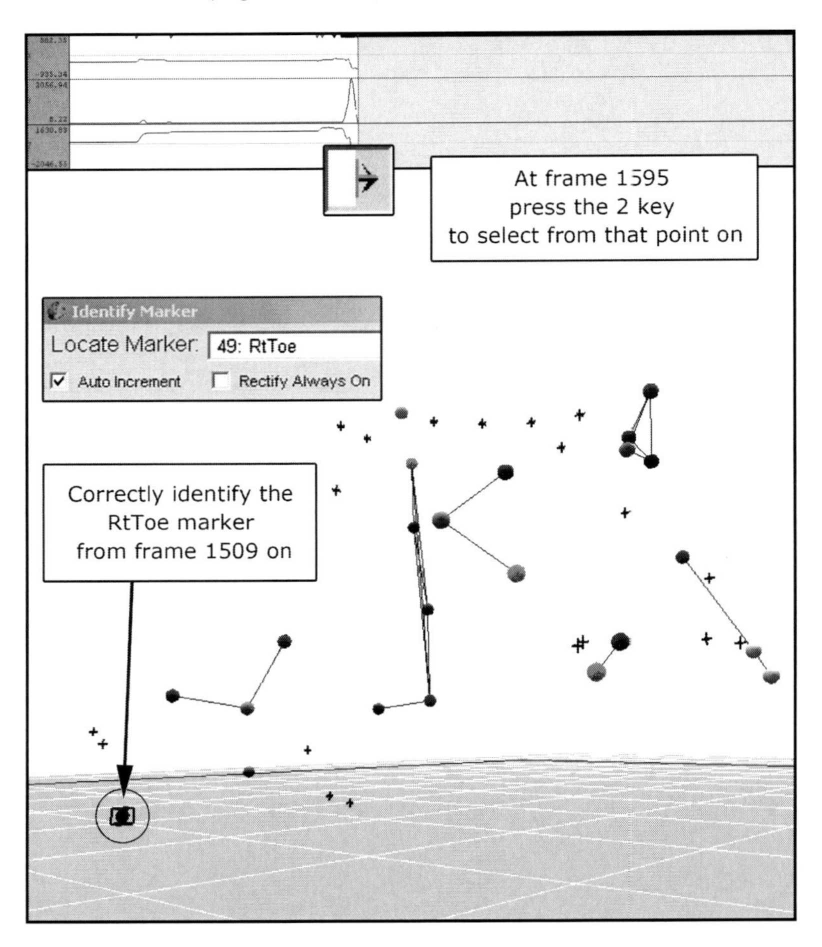

Figure 07_06.

We have corrected all the miss-identified markers; lets now take care of the unnamed ones. Press the *"4"* key to select the entire timeline. Go to the end of the timeline and *Rectify Unnamed* markers. Launch the *Identify Marker* window and click on the markers that are unidentified according to the file template. You can navigate the marker list by using the *"E"* and *"D"* keys (Figure 07_07).

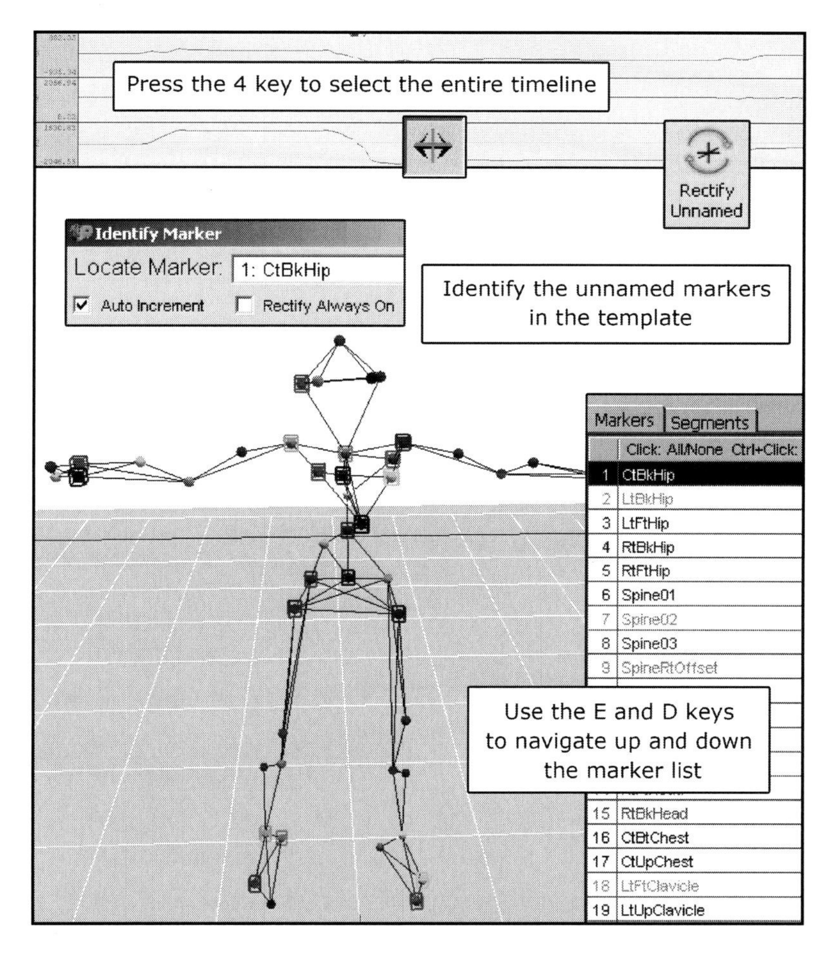

Figure 07_07.

Lets now take care of the unnamed markers in the middle of the back flip motion. We will start at the land of the flip and work our way towards the takeoff. Press the *Rectify Unnamed* button to consolidate unnamed markers. With the *Identify Marker* open, go to frame 1719, select the "Spine03" marker from the list and click on the unnamed marker in the *3D View*.
Press the *Rectify Unnamed* markers button again, go to frame 1648, select the "CtUpChest" marker from the list and click on the unnamed marker in the *3D View* (Figure 07_08).

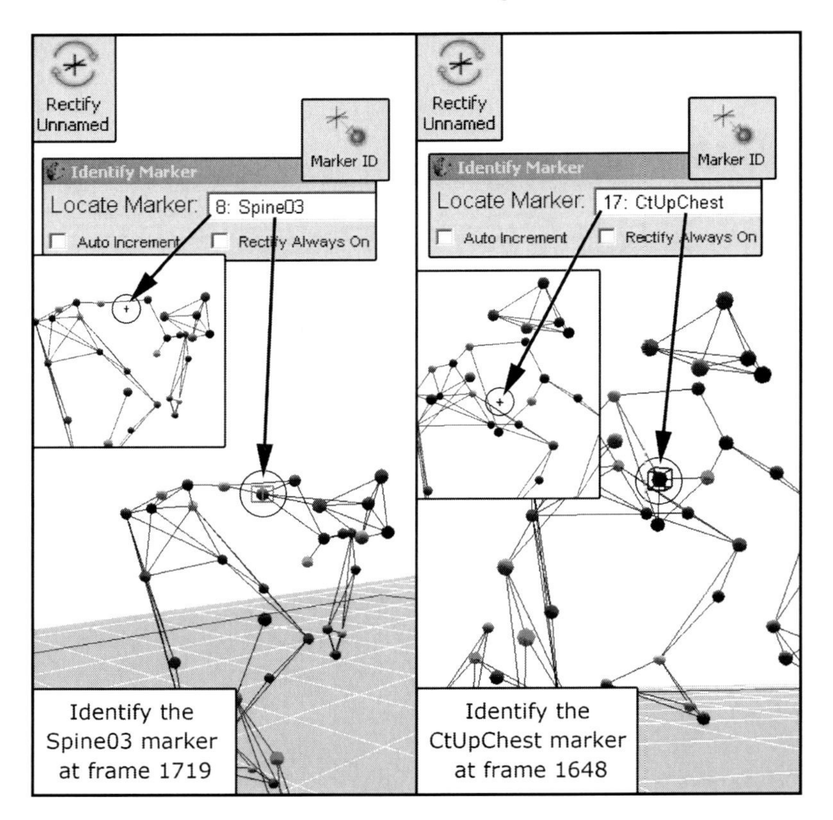

Figure 07_08.

Press the *Rectify Unnamed* markers button, go to frame 1623, select the "LtUpClavicle" marker from the list and click on the unnamed marker that corresponds to it in the *3D View*. Select the "LtFtClavicle" marker from the list and click on the unnamed marker that corresponds to it in the *3D View* Press the *Rectify Unnamed* markers button.
Go to frame 1611, select the "Spine03" marker from the list and click on the unnamed marker in the *3D View* (Figure 07_09).

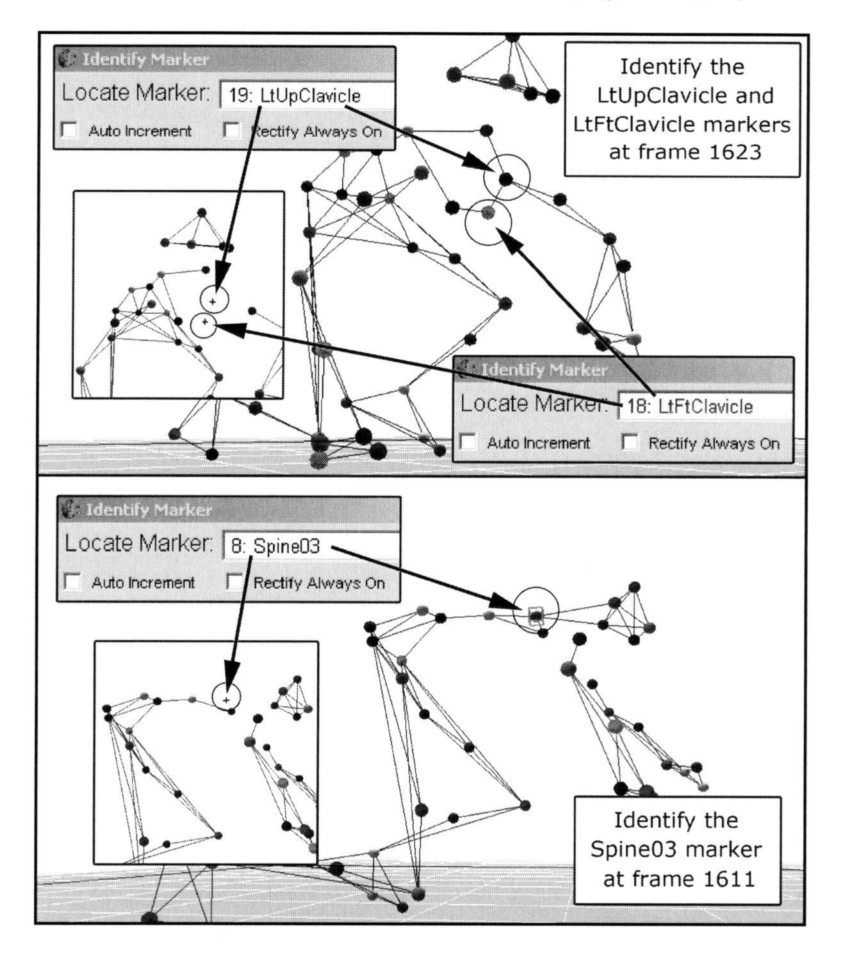

Figure 07_09.

Press the *Rectify Unnamed* markers button. At frame 1561 identify the unnamed marker as "LtUpClavicle" and *Rectify Unnamed* markers again. At frame 1551 identify the unnamed marker as "CtUpChest" and *Rectify Unnamed* markers (Figure 07_10).

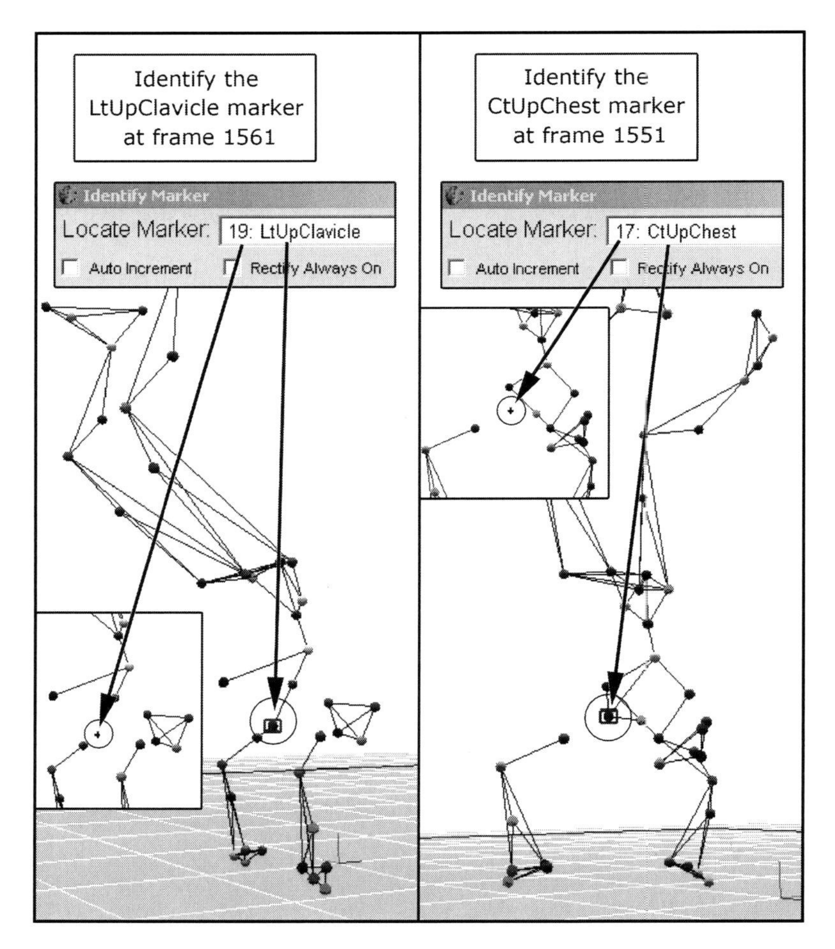

Figure 07_10.

For the most part the unnamed markers have been identified. We now have enough information to fill the remaining gaps using the *Join Virtual*, *Join Cubic* and *Join Linear* operations. Make sure that the *Identify Marker* window is closed. Select the "CtBkHip" at the top of the list, go to the beginning of the timeline and hit the *"N"* key to find the first gap in the data, this will take us to frame 1506. Hit the *"V"* key to launch the *Virtual Marker Join* window. Populate the *Virtual Marker Join* window as follows: *Origin Marker* = "LtFtHip", *Long Axis Marker* = "RtFtHip", *Plane Marker* = "LtBkHip". Press the *"V"* key to execute the operation. Most of the gaps for the "CtBkHip" marker have been filled. However there is still a one-frame gap at 1518 and a two-frame gap from 1530 to 1531. Press the *"C"* key to fill these two gaps with a *Cubic Join* operation (Figure 07_11).

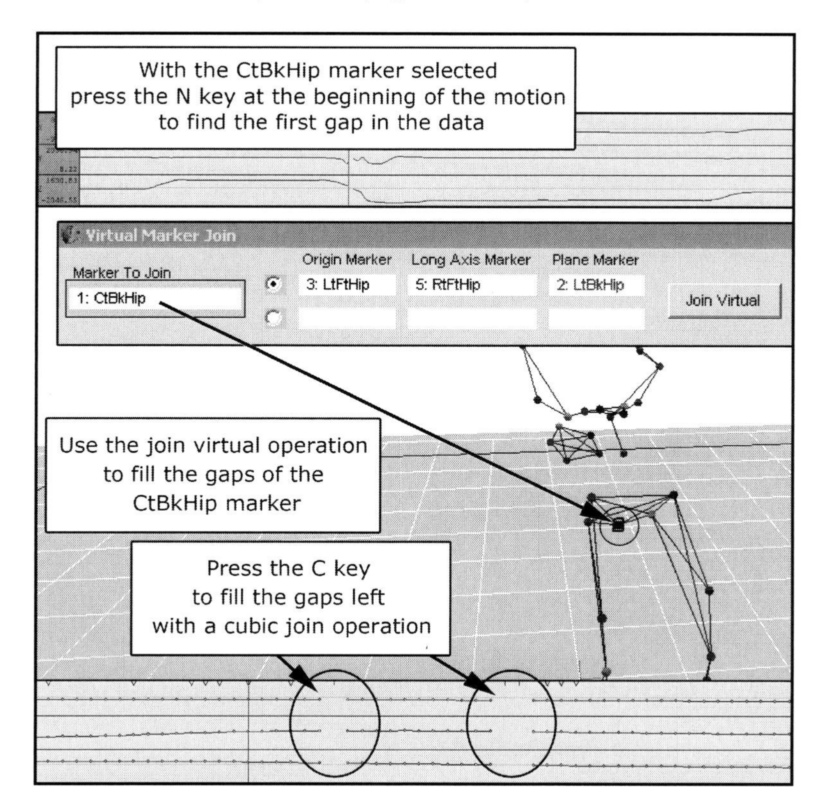

Figure 07_11.

Continue using the *"N"* key to find new gaps in the data and the *Virtual Join* operation to fill those gaps. Figure 07_12 shows how to populate the *Virtual Marker Join* window for the "LtBkHip" and "LtFtHip".
Fill the leftover gaps for the "LtFtHip" marker trajectory with a *Cubic Join* (C key). Press the *"N"* key to go to the next gap (Figure 07_12).

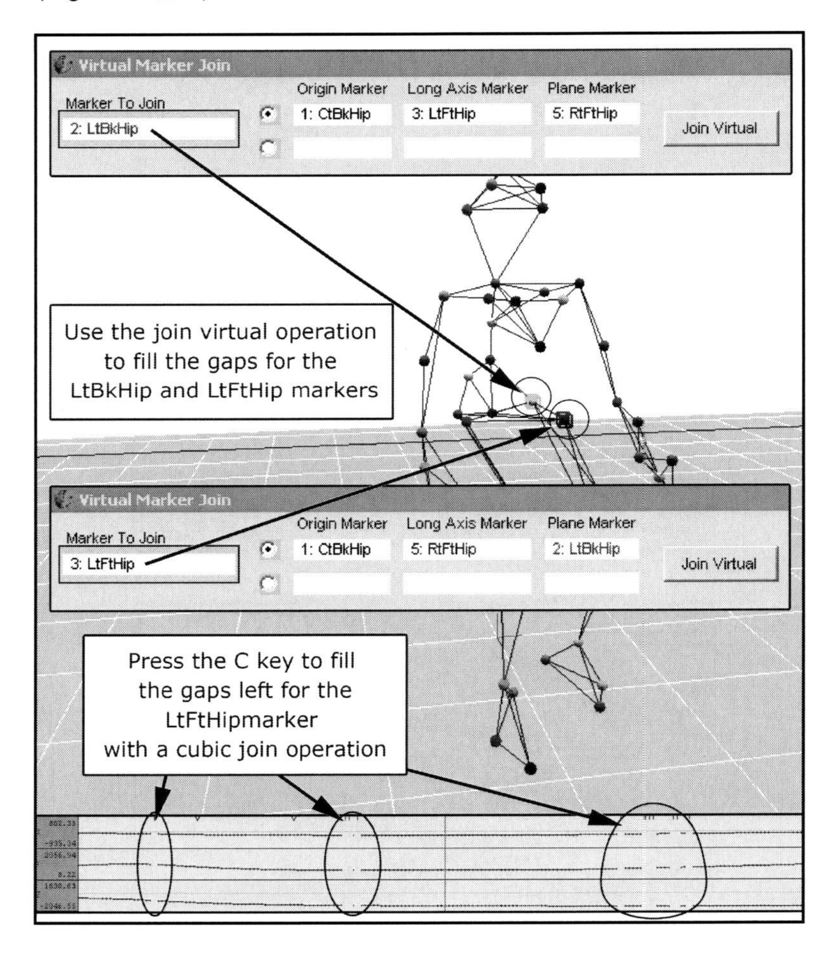

Figure 07_12.

Continue filling the gaps in the data using the *Virtual Join* operation. Populate the *Virtual Marker Join* window for the “RtBkHip”, “RtFtHip”, “Spine01” and “Spine02” as figure 07_13 indicates (Figure 07_13).

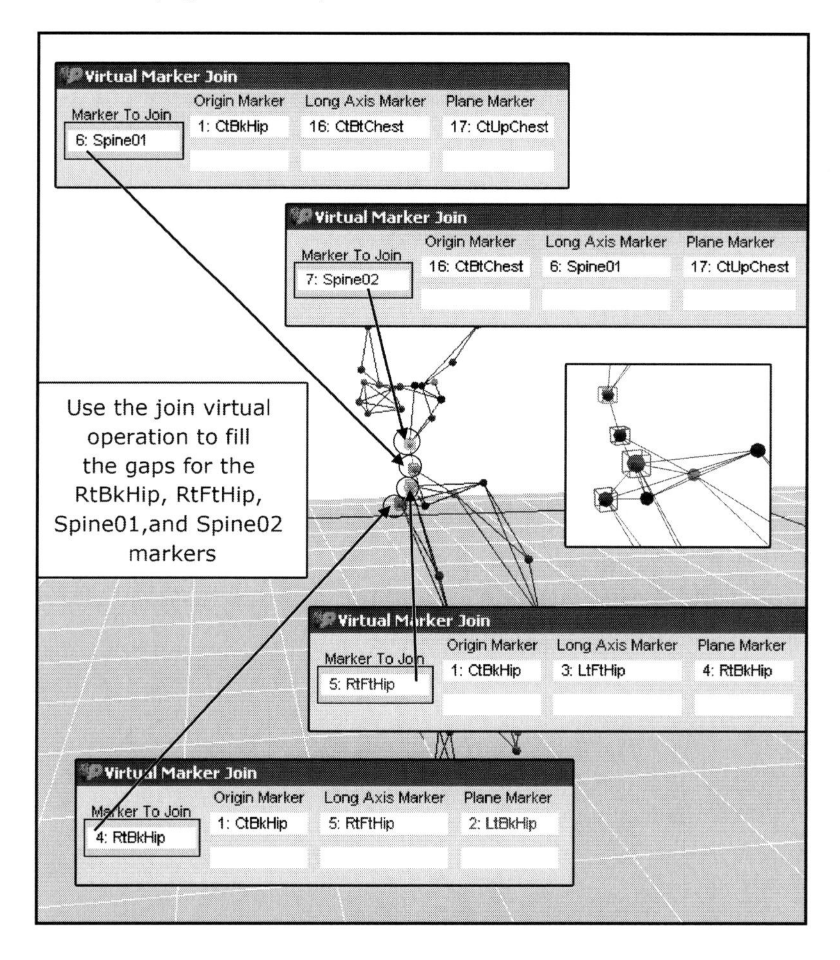

Figure 07_13.

We do not have enough information to do a fill the gaps in the "Spine03" marker's data yet, so lets move on to the next few markers and come back to "Spine03" later. Select the "SpineRtOffset" marker and do a *Join Virtual* operation using figure 07_14 as a guide of how to fill the different sections of the *Virtual Marker Join* window (Figure 07_14).

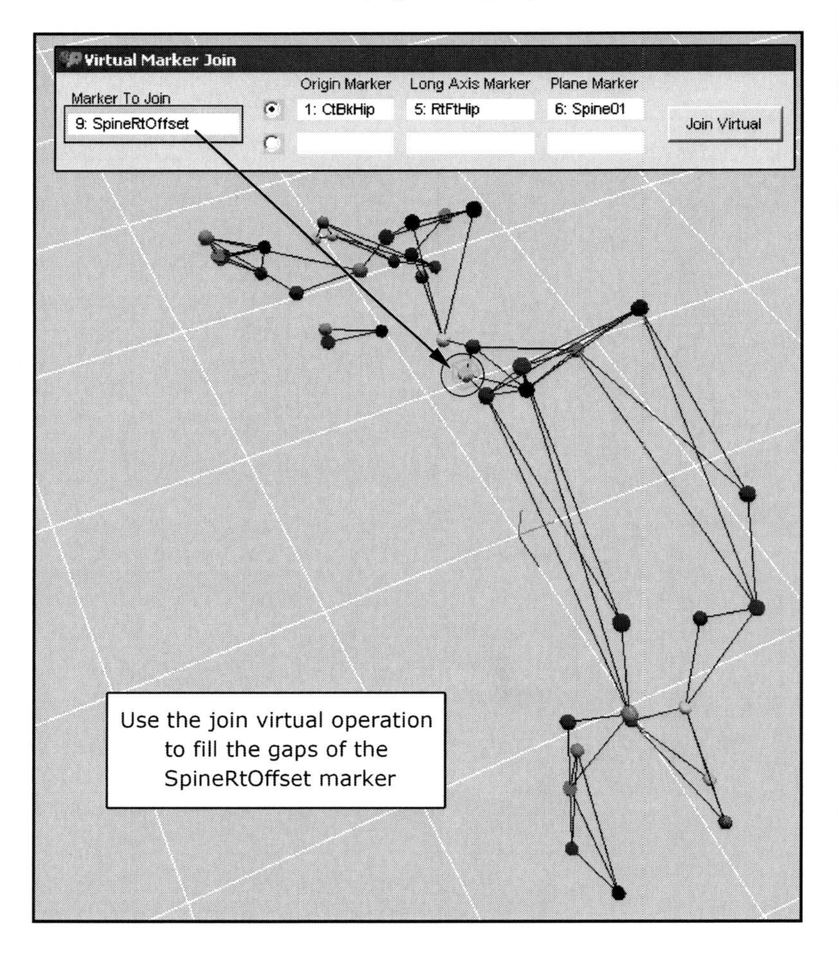

Figure 07_14.

Select the "SpineLtOffset" marker. When examining it during the back flip section of the motion, we can see that the marker exchanges trajectories with the "LtBicep" at frames 1514, 1515 and 1519. Executing a *Join Virtual* without correcting this issue will result in a very problematic trajectory for the marker. To fix this problem go to frame 1514 and press the *"3"* key to exclusively select the current frame. Select the "LtBicep" marker from the list, press the *Marker ID* button and click on the "SpineLtOffset" in the *3D View.* The marker gets identified as the "LtBicep" only for frame 1514. Go to frame 1515 and identify the "SpineLtOffset" marker as the "LtBicep" and repeat the same process for frame 1519. Close the *Identify Marker* window (Figure 07_15).

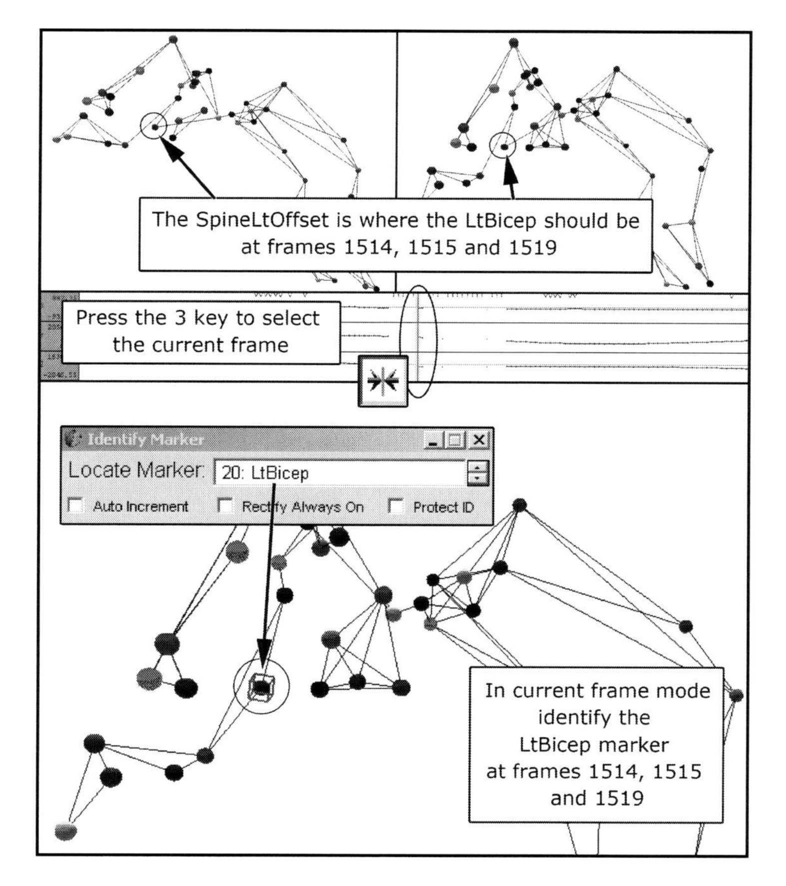

Figure 07_15.

We can know fill the gaps in the "SpineLtOffset" marker data using the *Virtual Join* operation. Press the *"4"* key to select the entire timeline and populate the *Virtual Marker Join* window using figure 07_16 as a guide
Now that we have tracked the "SpineLtOffset" marker, we have enough information to fill the gaps of the "Spine03" marker. Select the "Spine03" marker and use figure 07_16 as a guide to fill the slots of the *Virtual Marker Join* window (Figure 07_16).

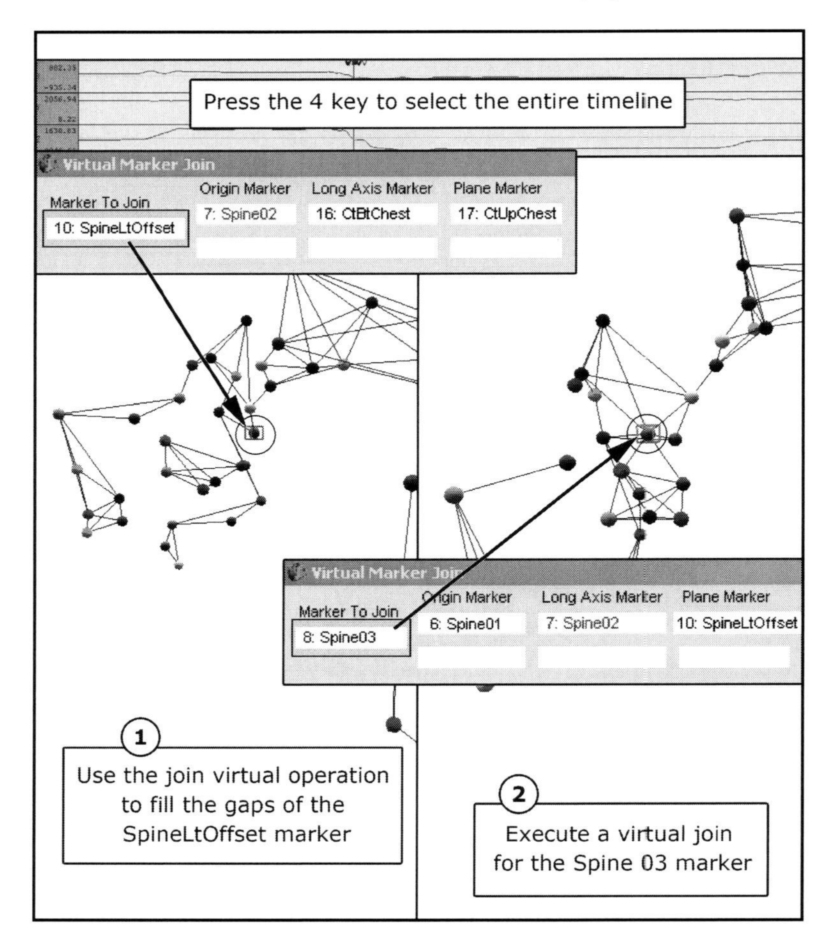

Figure 07_16.

Press the *"N"* key to go to the next gap. The "CtHead" marker is automatically selected, since this marker has only on frame missing so you can press the *"L"* key to do a *Linear Join* for the gap (Figure 07_17).

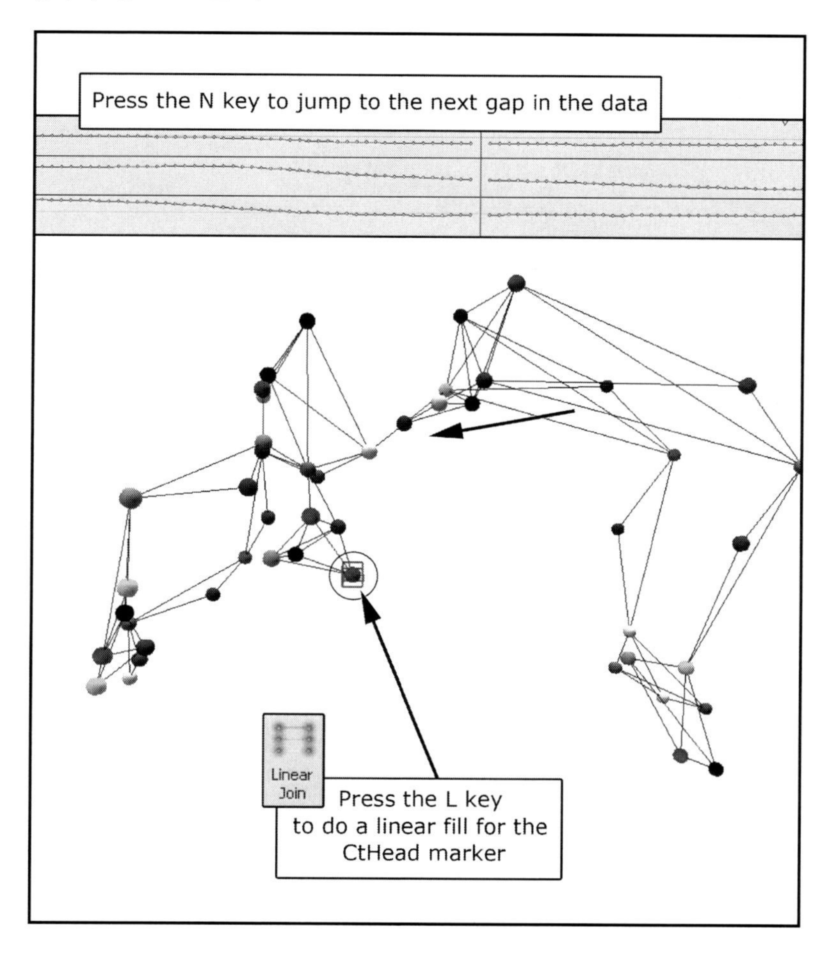

Figure 07_17.

Continue filling the gaps in the data using the *Virtual Join* operation. Populate the *Virtual Marker Join* window for the "LtFtHead", "LtBkHead", "RtFtHead", "RtBkHead", "CtBtChest" CtUpChest", LtFtClavicle", LtUpClavicle" and "LtBicep" markers as figure 07_18 indicates (Figure 07_18).

Marker To Join	Origin Marker	Long Axis Marker	Plane Marker	
12: LtFtHead	11: CtHead	14: RtFtHead	15: RtBkHead	Join Virtual
13: LtBkHead	11: CtHead	14: RtFtHead	15: RtBkHead	Join Virtual
14: RtFtHead	11: CtHead	15: RtBkHead	12: LtFtHead	Join Virtual
15: RtBkHead	11: CtHead	12: LtFtHead	13: LtBkHead	Join Virtual
16: CtBtChest	7: Spine02	10: SpineLtOffset	8: Spine03	Join Virtual
17: CtUpChest	7: Spine02	10: SpineLtOffset	8: Spine03	Join Virtual
18: LtFtClavicle	8: Spine03	7: Spine02	16: CtBtChest	Join Virtual
19: LtUpClavicle	7: Spine02	16: CtBtChest	8: Spine03	Join Virtual
20: LtBicep	21: LtElbow	19: LtUpClavicle	22: LtForeArm	Join Virtual

Figure 07_18.

The "LtBicep" still has a few gaps in its trajectory after the *Virtual Join* operation. Press the *"L"* key to fill them using the *Join Linear* operation. Press the *"N"* key to go to the next gap (Figure 07_19).

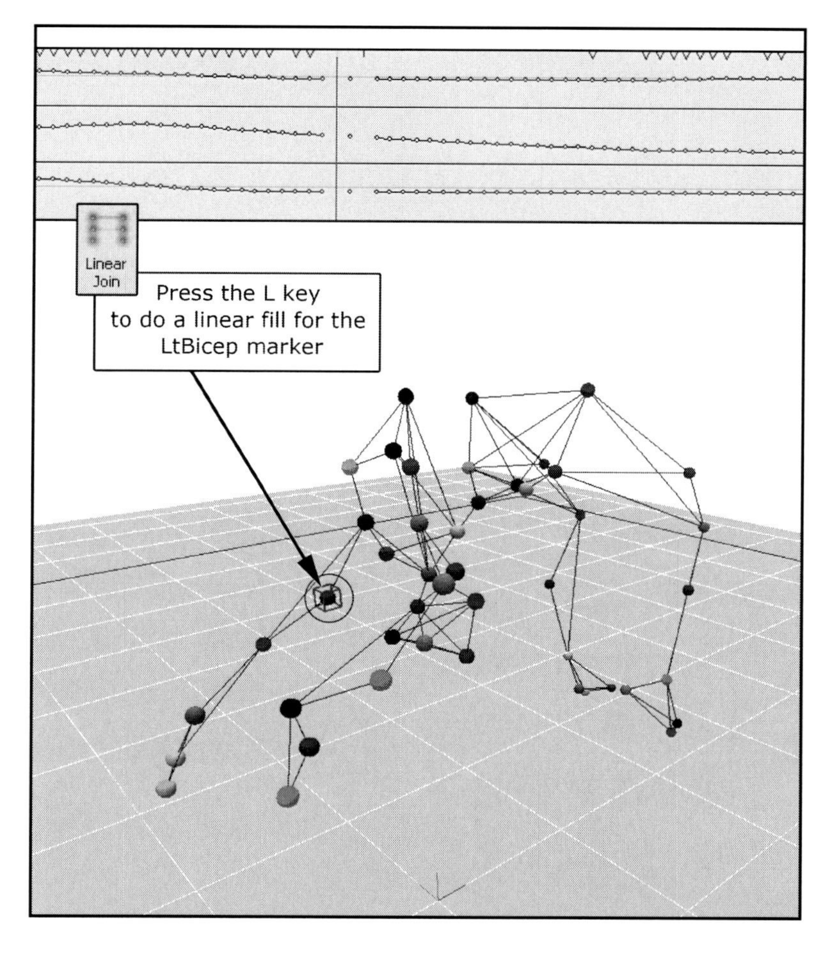

Figure 07_19.

Go on filling the rest of gaps using *Virtual Join*. Fill the different sections of the *Virtual Marker Join* window for the "LtForeArm", "LtOtWrist", "LtInWrist", "LtPalm", "RtFtClavicle" and "RtUpClavicle" like figure 07_20 indicates (Figure 07_20).

Marker To Join	Origin Marker	Long Axis Marker	Plane Marker
22: LtForeArm	23: LtOtWrist	21: LtElbow	20: LtBicep
23: LtOtWrist	22: LtForeArm	25: LtPinky	21: LtElbow
24: LtInWrist	23: LtOtWrist	21: LtElbow	25: LtPinky
26: LtPalm	23: LtOtWrist	22: LtForeArm	25: LtPinky
27: RtFtClavicle	8: Spine03	7: Spine02	17: CtUpChest
28: RtUpClavicle	7: Spine02	29: RtBicep	30: RtElbow

Figure 07_20.

The trajectory of the "RtUpClavicle" marker still has a one-frame gap. Use the *"L"* key shortcut to perform a *Linear Join* for this marker, then press the *"N"* key to go to the next gap. The "RtBicep" marker is automatically selected, and it has only on frame missing so you can press the *"L"* key to do a *Linear Join* for the gap (Figure 07_21).

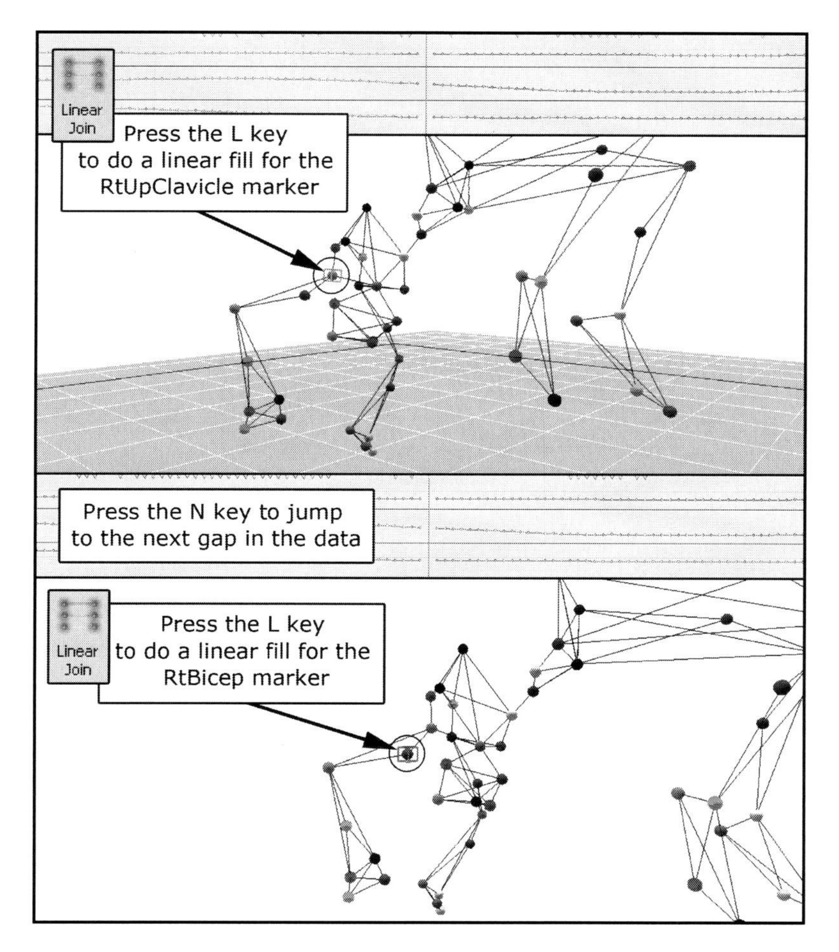

Figure 07_21.

Continue filling the gaps in the data using the *Virtual Join* operation. Populate the *Virtual Marker Join* window for the “RtOtWrist”, “RtInWrist”, “RtPinky”, “RtPalm”, “LtAnkle”, “RtTThigh”, “RtBall” and “RtToe” markers as figure 07_22 indicates (Figure 07_22).

Marker To Join	Origin Marker	Long Axis Marker	Plane Marker	
32: RtOtWrist	31: RtForeArm	30: RtElbow	33: RtInWrist	Join Virtual
33: RtInWrist	31: RtForeArm	30: RtElbow	32: RtOtWrist	Join Virtual
34: RtPinky	31: RtForeArm	30: RtElbow	35: RtPalm	Join Virtual
35: RtPalm	32: RtOtWrist	31: RtForeArm	33: RtInWrist	Join Virtual
39: LtAnkle	37: LtKnee	40: LtHeel	38: LtCalf	Join Virtual
43: RtThigh	4: RtBkHip	44: RtKnee	5: RtFtHip	Join Virtual
48: RtBall	46: RtAnkle	45: RtCalf	47: RtHeel	Join Virtual
49: RtToe	46: RtAnkle	45: RtCalf	48: RtBall	Join Virtual

Figure 07_22.

Now when you press the *"N"* key a warning dialog that reads *Nothing found* appears. This means that all the gaps have been filled. Press the *Delete Unnamed* button to clear the unidentified markers list (Figure 07_23).

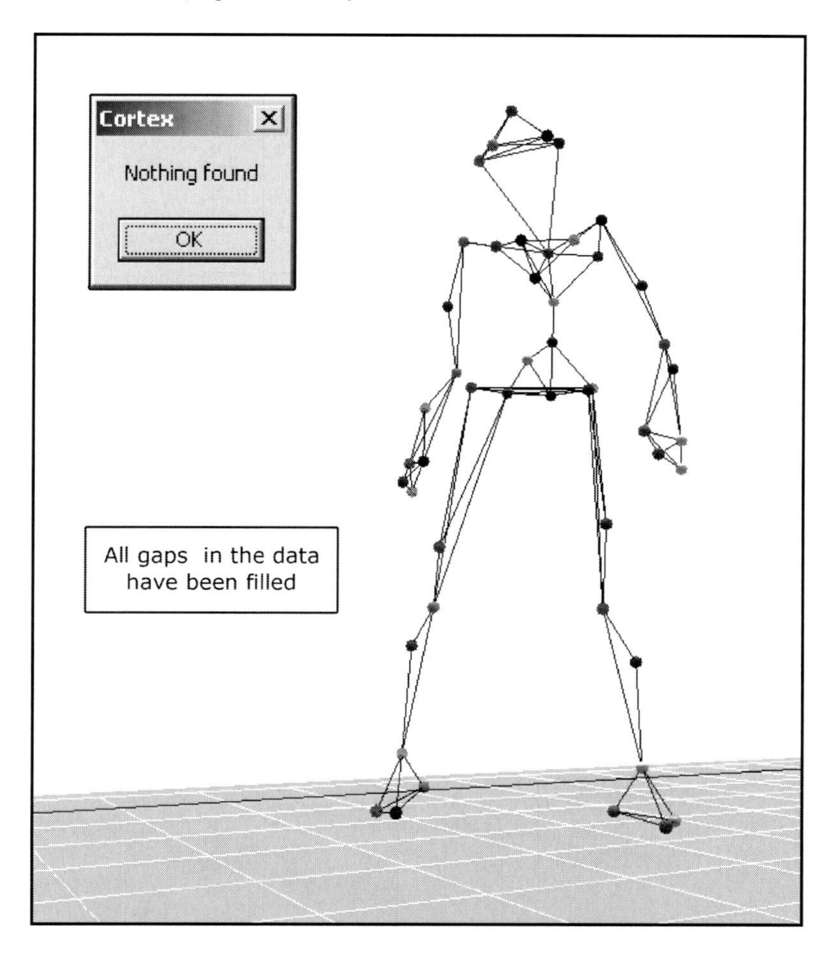

Figure 07_23.

Note: If the Nothing found warning window does not appear go to the first marker in the list and start searching and filling gaps using linear, cubic or virtual operations.

Now we are going to smooth the data a little bit in some areas where we have jittery markers. Go to the *Tools* menu and under click on *Settings*. Select the *Smoothing* tab under *Post Process Tools* and make sure that *Butterworth* is selected with a *Frequency of 3*. In the *Post Process Dashboard* type 323 in the *High Visible Frames* (last visible frame). Select the "LtOtWrist" and "LtForeArm" markers and press the *"M"* key to execute the smooth operation (Figure 07_24).

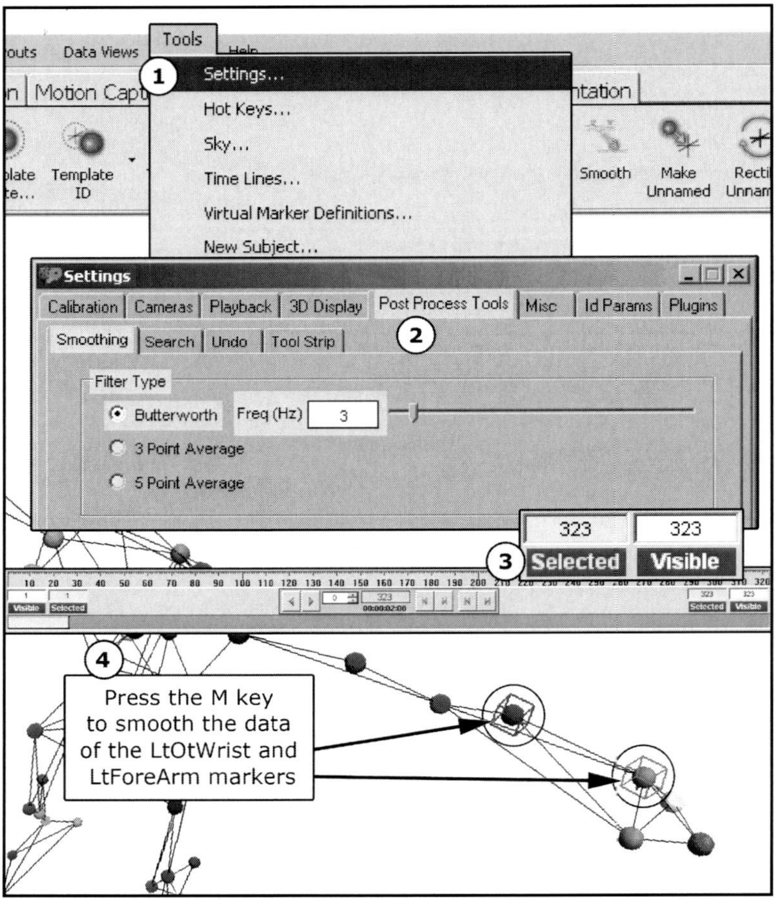

Figure 07_24.

Note: the smooth operation only affects selected frames.

The smooth operation drastically reduced the noise of the "LtOtWrist" and the "LtForeArm" while in *T-Pose*. Under the *Settings* window change the *Filter Type* from *Butterworth* to *5 Point Average*. Type 1846 in *High Visible Frames* and 1495 in *Low Visible Frames*. Select the "Spine01", "Spine02", "Spine03", "SpineLtOffset", "CtBtChest", "CtUpChest", "LtFtClavicle", "LtUpClavicle", "RtFtClavicle" and "RtUpClavicle" markers. Press *"M"* to smooth the trajectories (Figure 07_25).

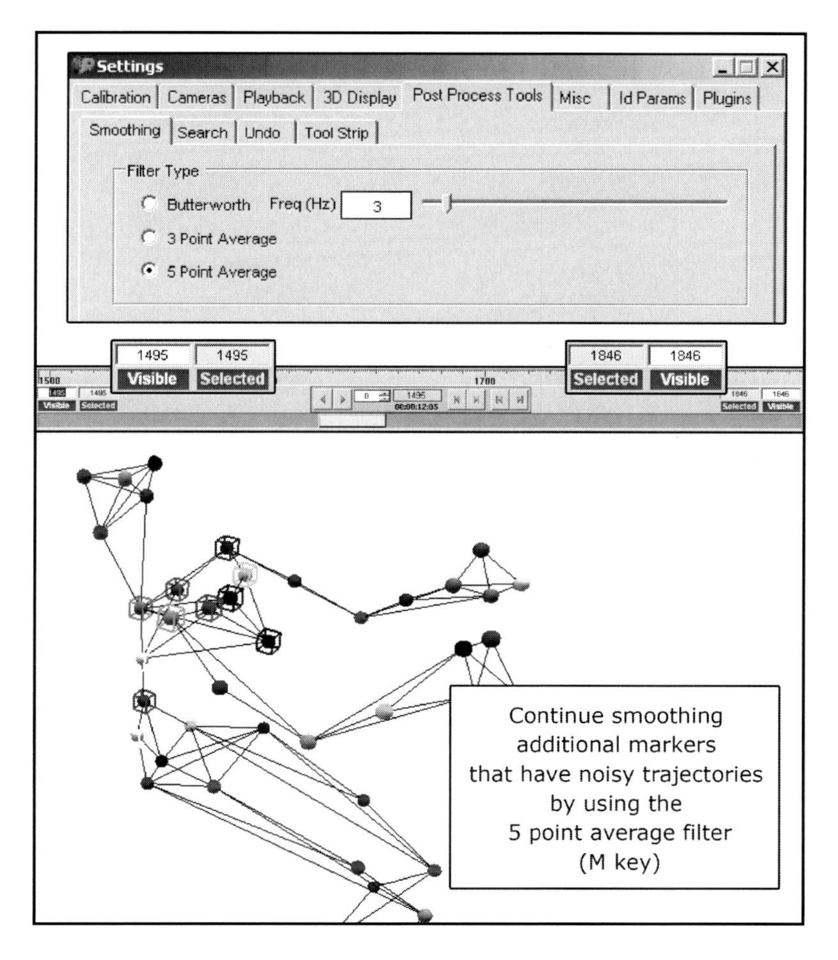

Figure 07_25.

We are done tracking the motion. The only thing left to do is to export the data as a *TRC* file so it can be taken into **Motion Builder** and applied into a digital character. Go to the *File* menu and click on *Trim Capture W/Options*, this will launch the *Trim Capture Options* window. Check on *Tracked ASCII* in the *Output Files* section, *Discard Unnamed Markers* in the *Tracked Markers* section and *Save All Frames* in the *Frames* section. Press the *Export Trimmed Capture* button and name the file "Man_BackFlip.trc" (Figure 07_26).

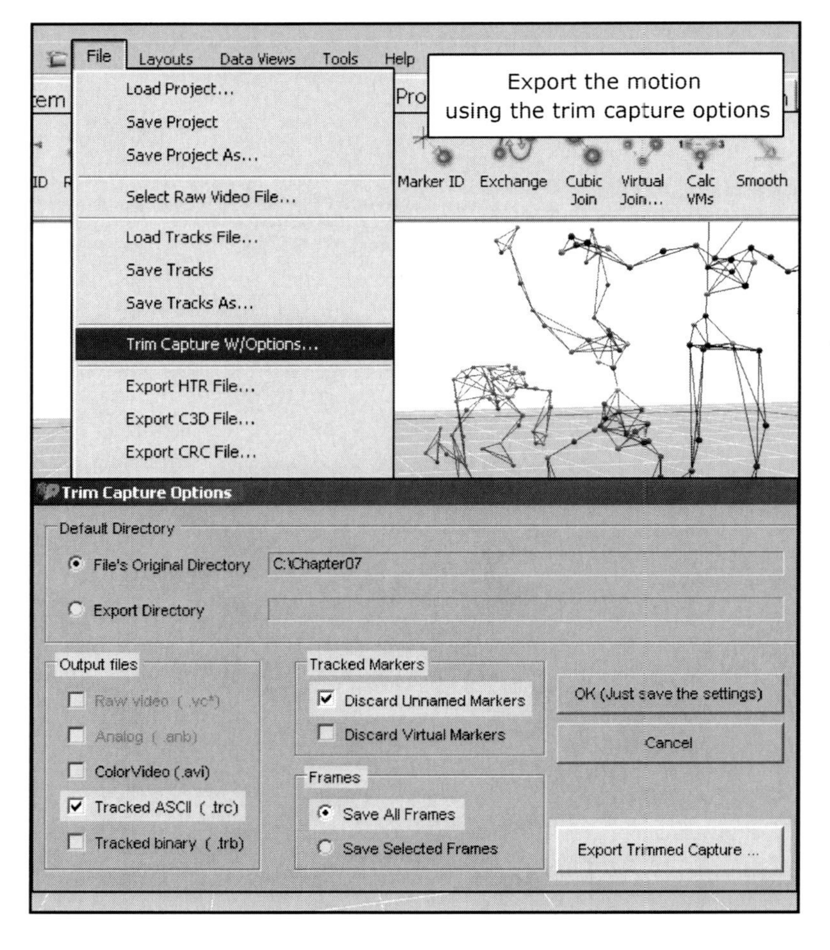

Figure 07_26.

Note: If you want to export only a portion of the motion, type the range that you want to use in the Low Visible Frames and High Visible Frames boxes and activate the Save Selected Frames option in the Trim Capture Options window.

Part 03
Solving In Motion Builder

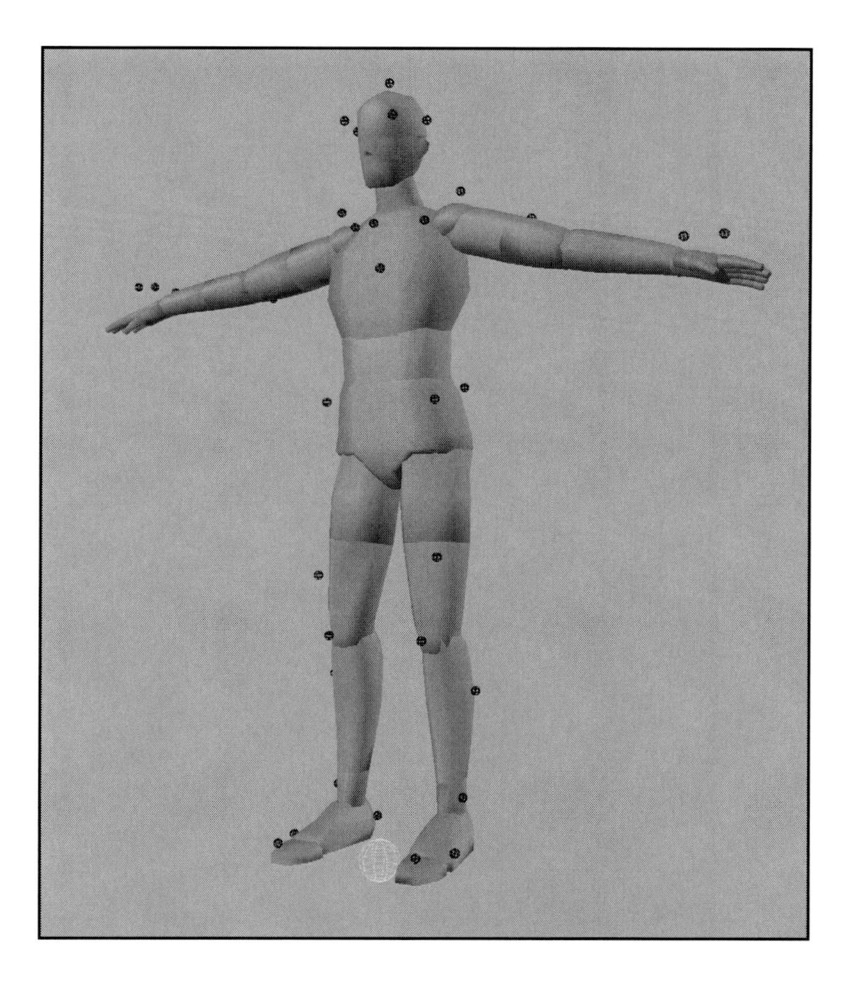

Chapter 08
Solving Optical Data

Solving is the process of converting the translational data from the optical markers into rotational data for a skeletal hierarchy. In this chapter we are going to explore the process of *Solving* optical data in **Motion Builder** through the use of the *Actor* tool. (Figure 08_01).

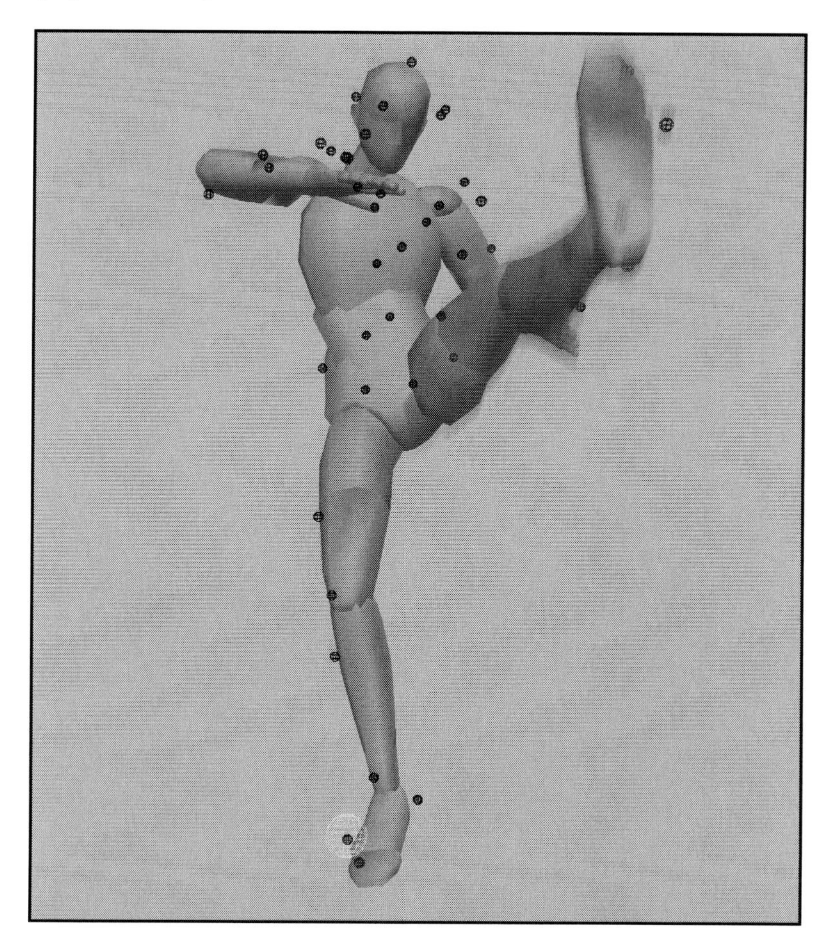

Figure 08_01.

Launch **Motion Builder** and open the "Chapter08"[5] file.

Select the *Import* option from the *File* menu. Navigate to the place where you downloaded the "Chapter08" folder from mocapclub.com. Press *Ctrl + A* to select all the *Trc* files in the folder and click the *Open* button (Figure 08_02).

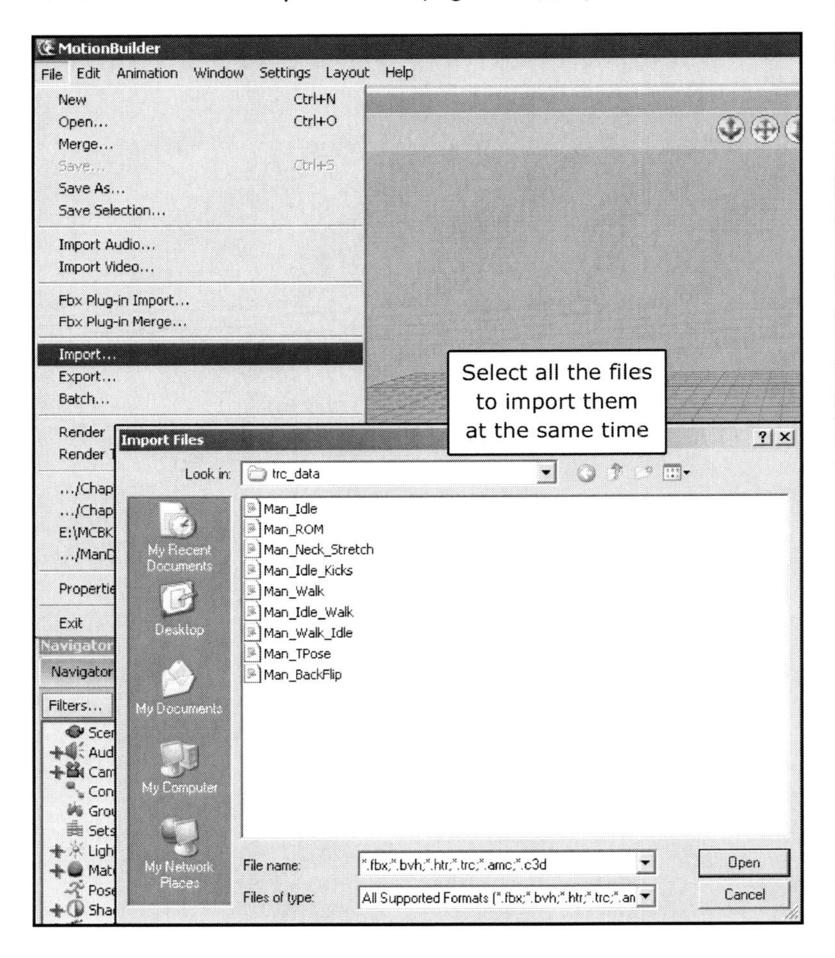

Figure 08_02.

5 Download the files from www.MocapClub.com/TheMocapBook.htm

Click on the *Takes* drop down menu. This menu shows the different motions that have been imported to **Motion Builder**. Select "Man_TPose" from the list and go to frame 50 in the timeline, this is where the pose was best achieved by the performer (Figure 08_03).

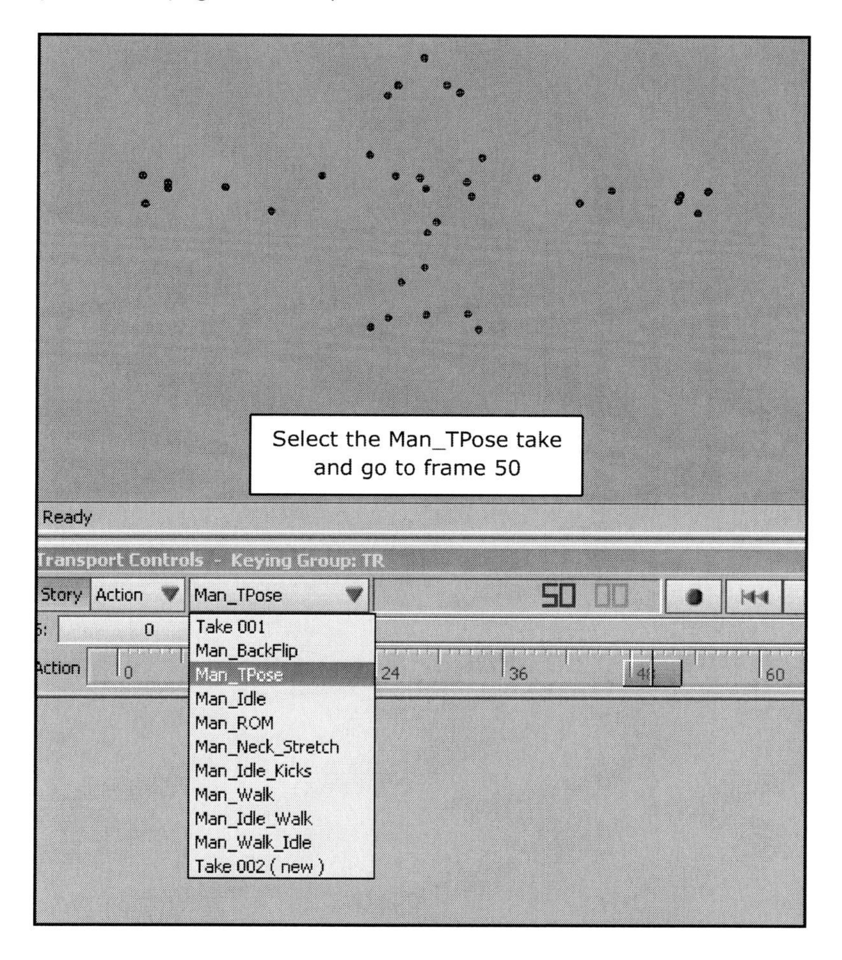

Figure 08_03.

Like it was mentioned in Chapter 05, having a *T-pose* makes the process of transferring the motion from the optical markers to the digital character much easier. Drag an Actor from the *Asset Browser* to the *Perspective View* (Figure 08_04).

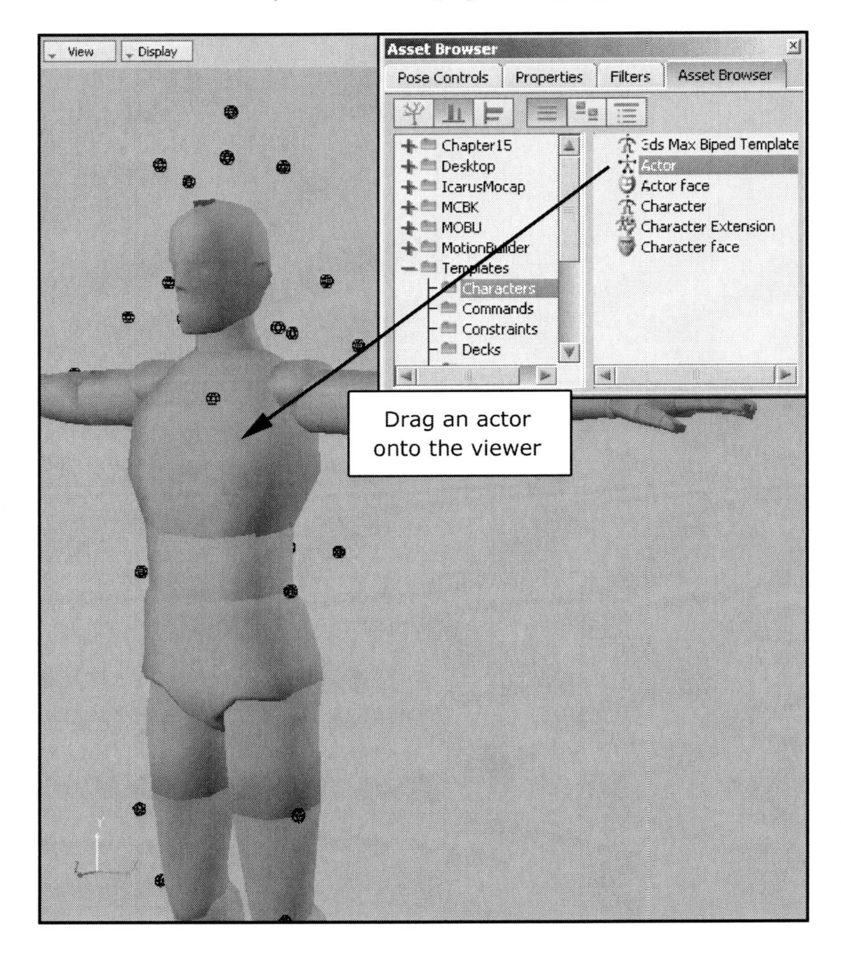

Figure 08_04.

Click *Ctrl + F* to go to the *Front View.* Here we are able to see that the *Actor* is a little small in comparison to the performer that wore the markers. Select all the cells for the body parts of the *Actor* in the *Character Controls* window. Press the *"S"* key to access the scaling mode and in the *Status Bar* area type 1.06 in the X axis value field. The *Actor* now matches the height of the original performer a little better (Figure 08_05).

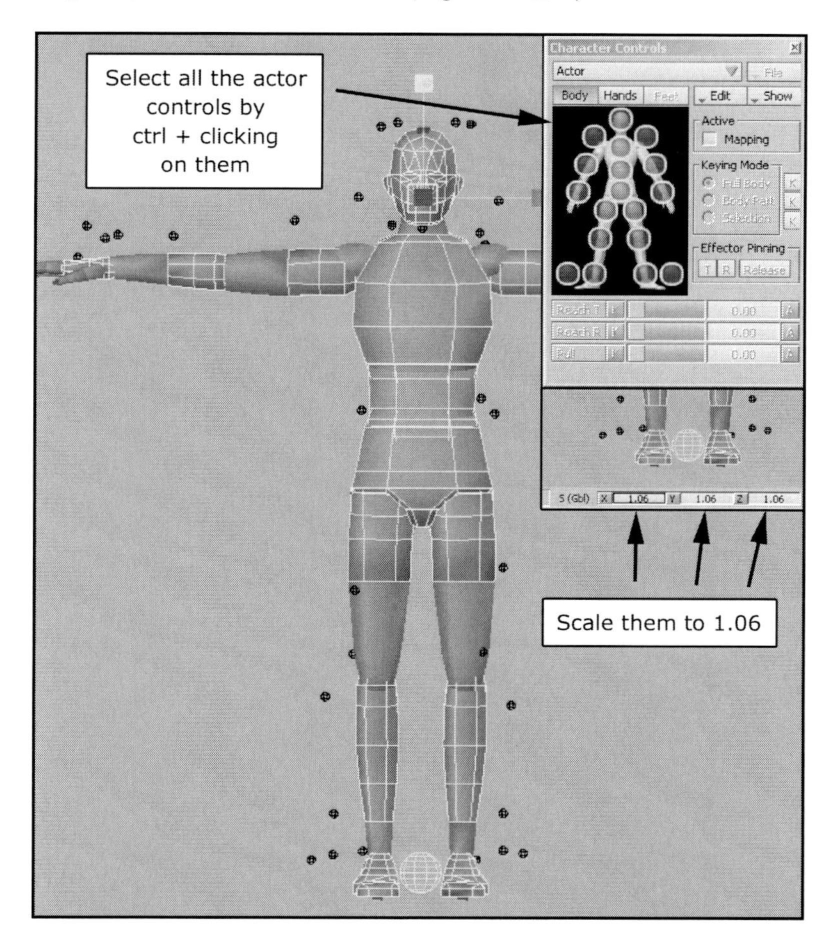

Figure 08_05.

Press *Ctrl + 2* to activate the *Two Panes Viewer Layout* press *Ctrl + R* to turn one of the panes into the *Right View*. Select the hips cell in the *Character Controls* window and use the "T" key to activate the translation mode. Move the *Actor* in X, Y and Z until it resides inside the optical data as figure 08_06 shows (Figure 08_06).

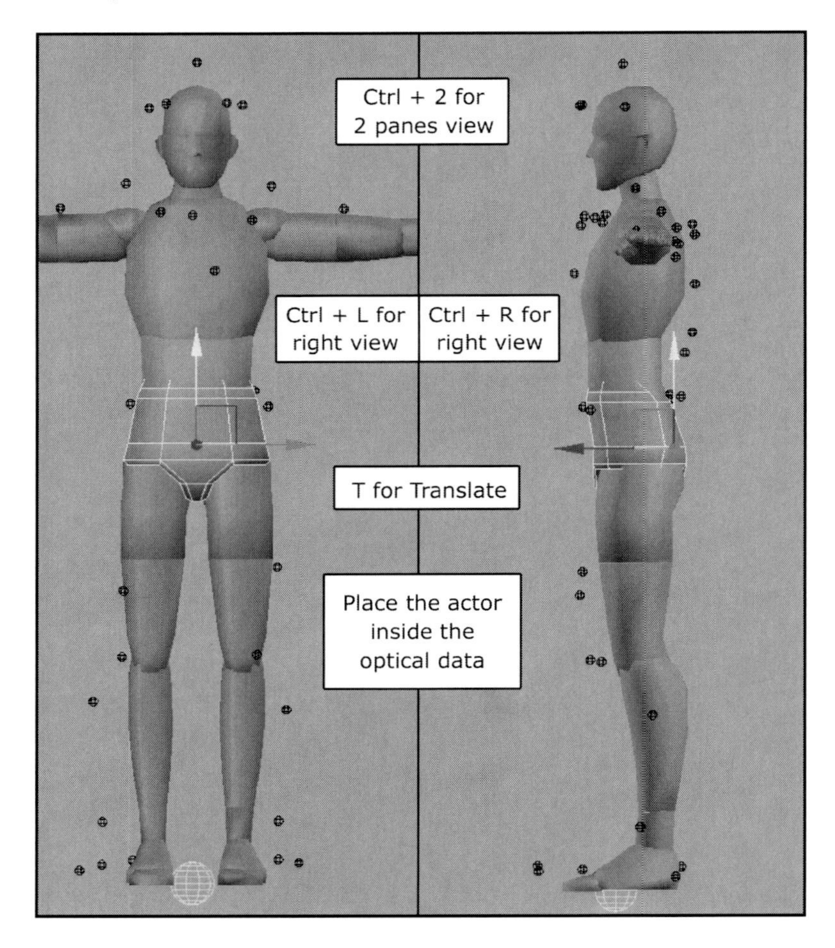

Figure 08_06.

Using the selection cells from the *Character Controls* window and the rotation mode (*"R"* Key), pose the legs like Figure 08_07 illustrates. The feet of the Actor are still too small in comparison with the optical data. Select all the feet and toe cells and scale this body parts to fit the optical data. You will also need to move the feet a little for an optimal fit (Figure 08_07).

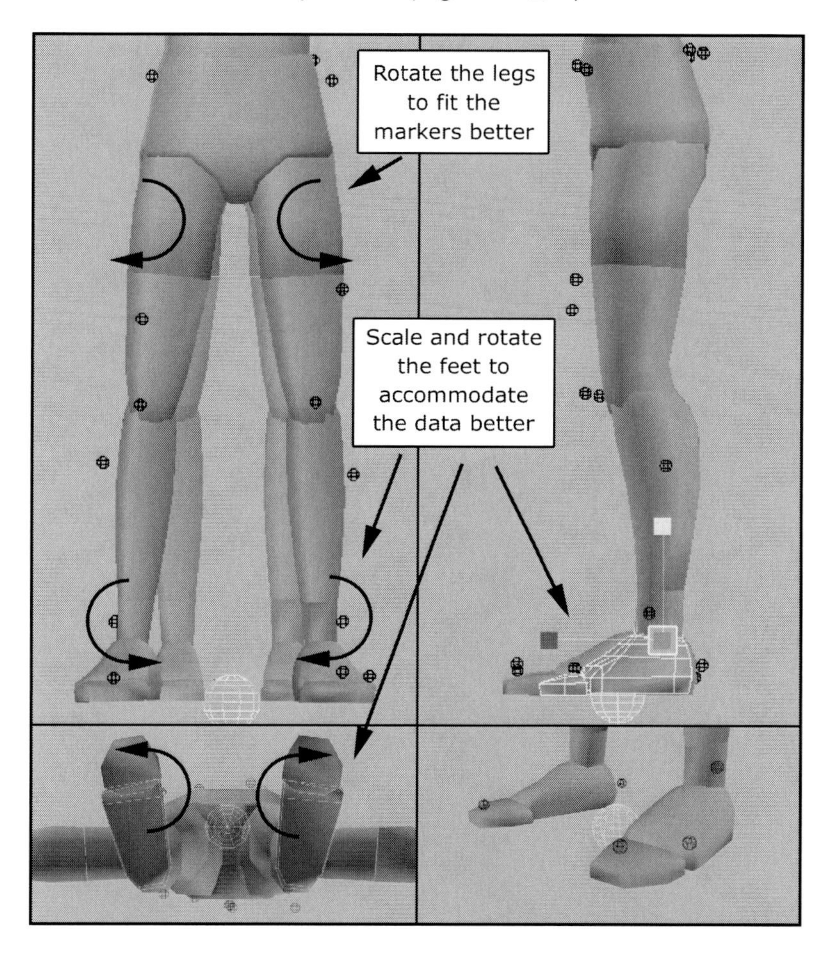

Figure 08_07.

Inside the side view, click on the neck body part (there is no cell for the neck in the *Character Controls* window) and rotate it and scale it to better fit the markers. Also rotate the head to fit the data properly (Figure 08_08).

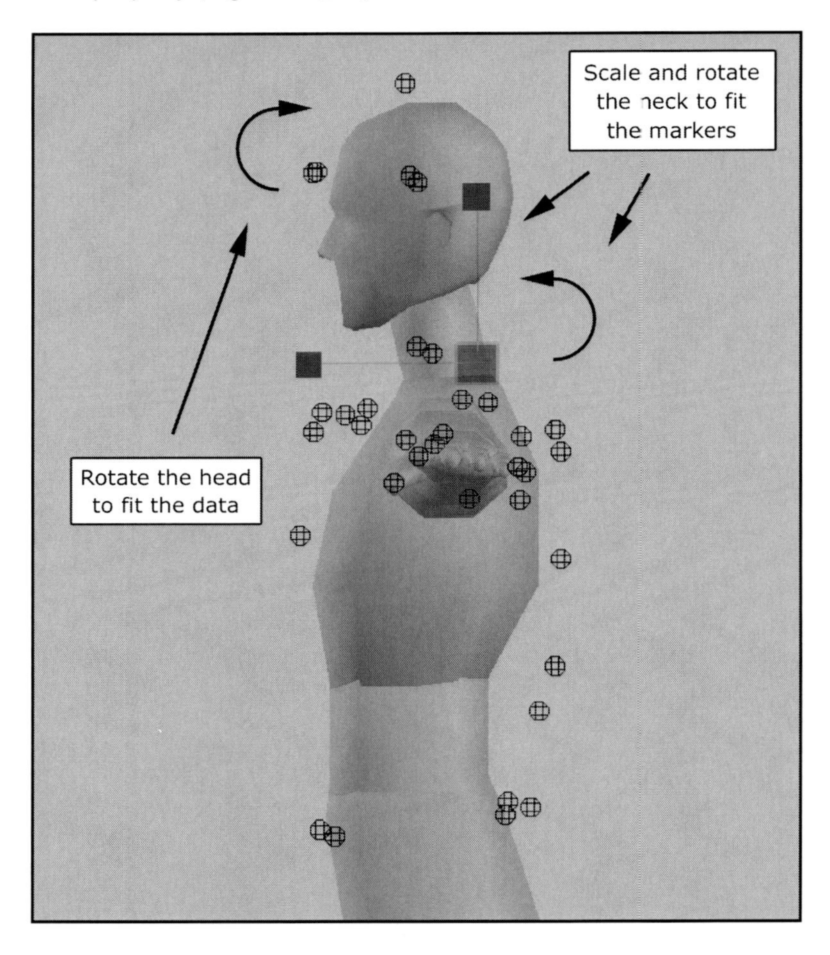

Figure 08_08.

Pose the arms to fit the markers in the front and side views like figure 08_09 indicates (Figure 08_09).

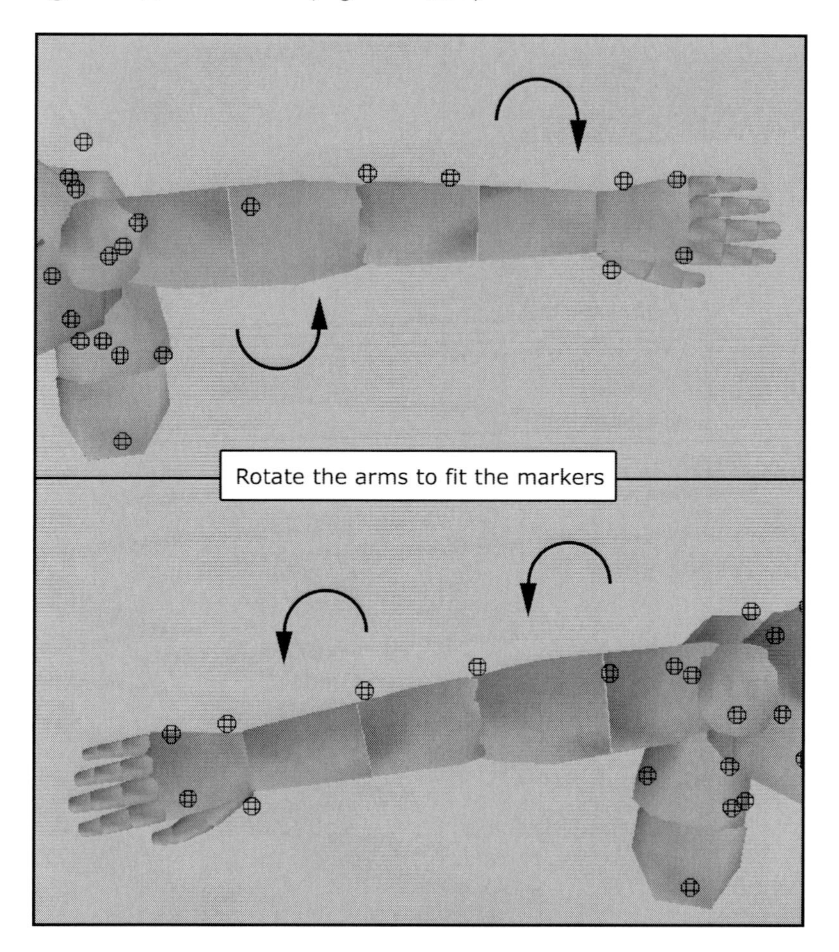

Figure 08_09.

The actor is now accurately fitted inside the optical data.

Note: The more time it is taken in the fitting of the actor to the data, the more accurate the solve will be.

Open the *Actor's* folder in the *Navigator* window and double click on the *Actor* to load its properties. Press the *Marker Set* button in the *Actor Settings* tab and select *Create* from the drop-down list that appears (Figure 08_10).

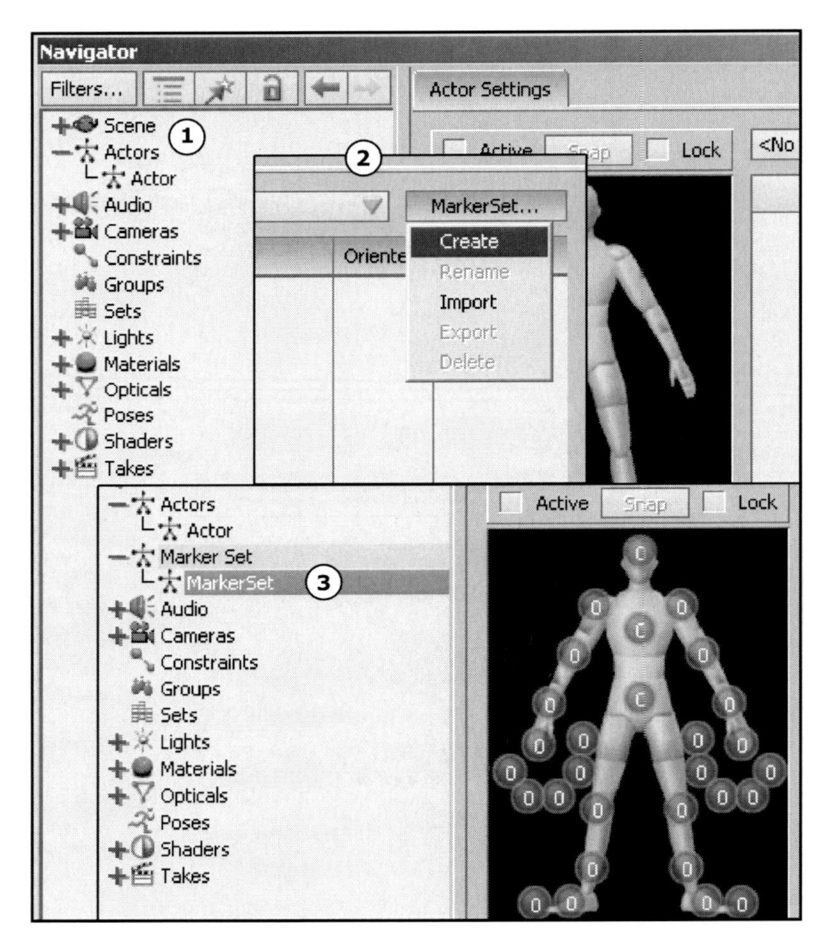

Figure 08_10.

There is now a set of cells in the in the *Actor Representation* section of the *Actor Settings* tab. We will use these tabs to specify what markers we want to be controlling each part of the *Actor*.

Open the *Show* drop-down in the *Character Controls* window and uncheck *Actor (All)*, this will make the selection of the optical markers easier.
Select the 5 markers in the head area by dragging a box around them. Alt + click and drag the markers to the *Head* cell inside the *Actor Settings* tab (Figure 08_11).

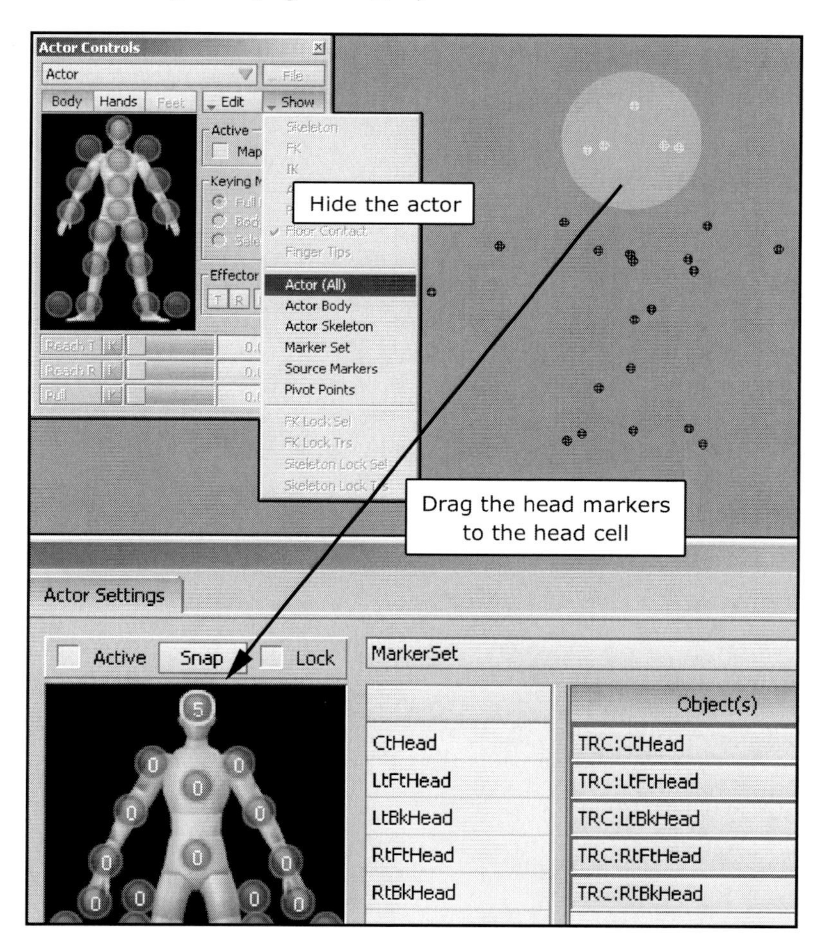

Figure 08_11.

Alt + click and drag the “CtBtChest”, “CtUpChest”, “Spine02” and “Spine03” markers to the *Chest* cell inside the *Actor Settings* tab. Place the “CtBkHip”, “LtBkHip”, “LtFtHip”, “RtBkHip” and “RtFtHip” in the *Hips* cell (Figure 08_12).

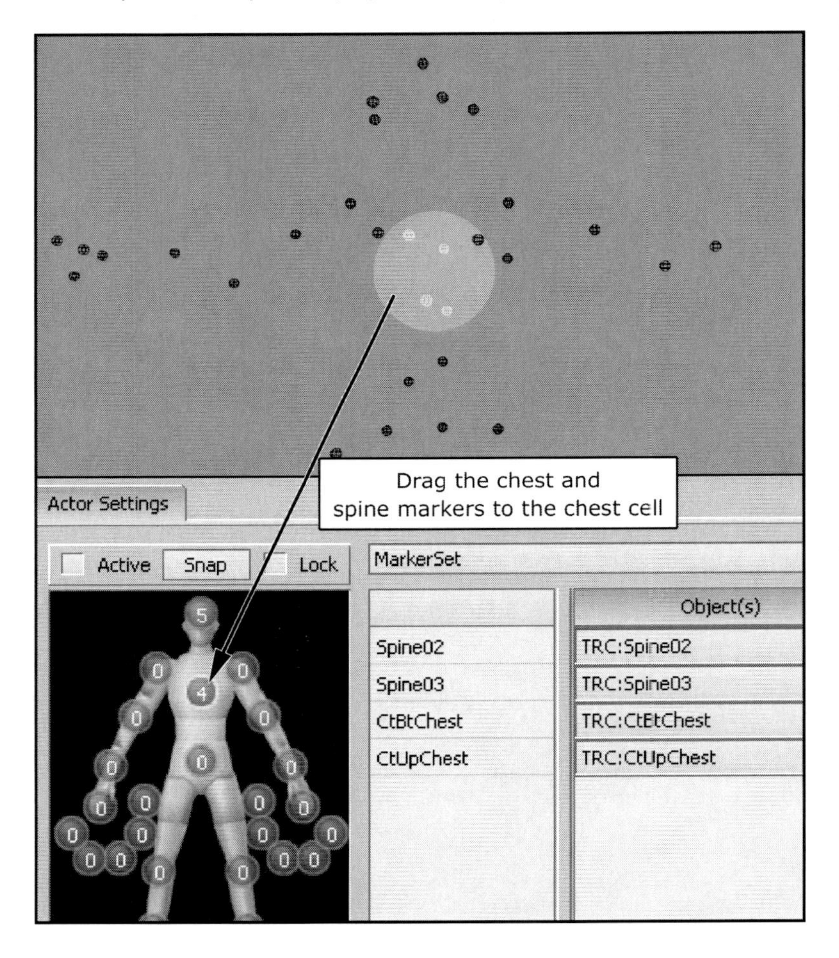

Figure 08_12.

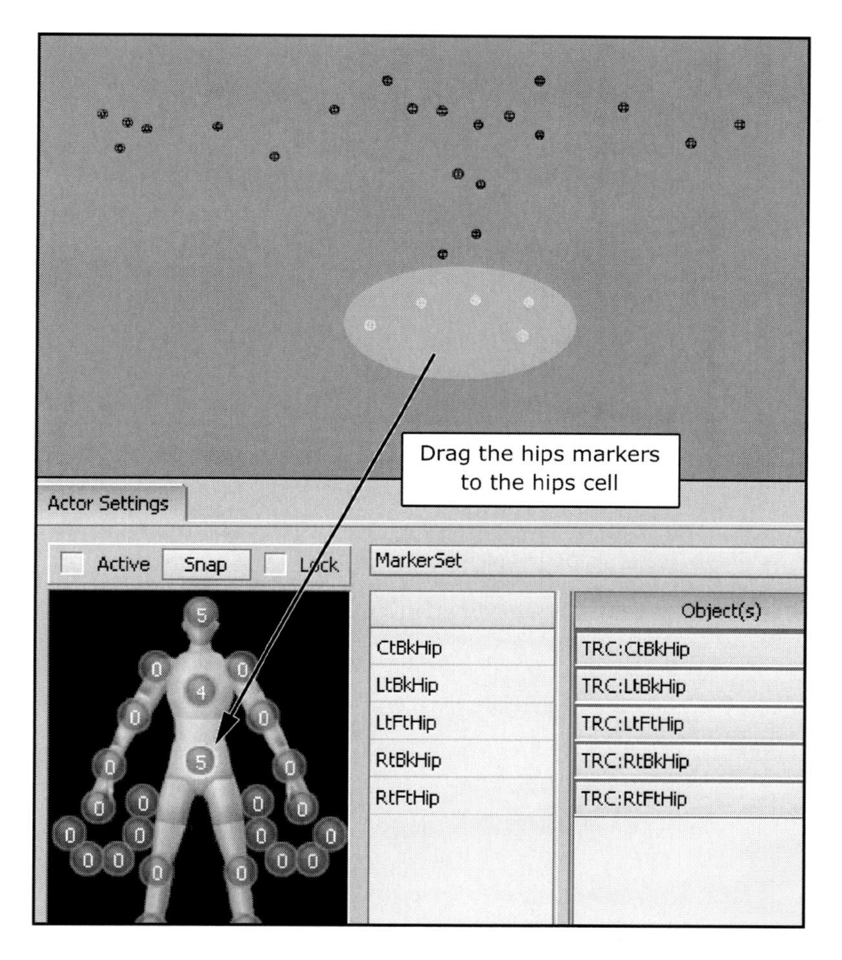
Drag the hips markers to the hips cell
Actor Settings
Active
Snap
Lock
MarkerSet
Object(s)
CtBkHip
LtBkHip
LtFtHip
RtBkHip
RtFtHip
TRC:CtBkHip
TRC:LtBkHip
TRC:LtFtHip
TRC:RtBkHip
TRC:RtFtHip

Figure 08_13

Drop the “LtFtClavicle” and “LtUpClavicle” markers in the *LeftShoulder* cell in the *Actor Settings*. Put the “LtBicep” and “LtElbow” in the *LeftElbow* cell. Place the “LtForeArm” marker in the *LeftForeArm* cell. Drop the “LtOtWrist”, “LtInWrist”, “LtPinky” and “LtPalm” markers into the *LeftWrist* cell. Use the guidelines we have followed so far for the left arm cells to populate the right arm cells (Figure 08_14).

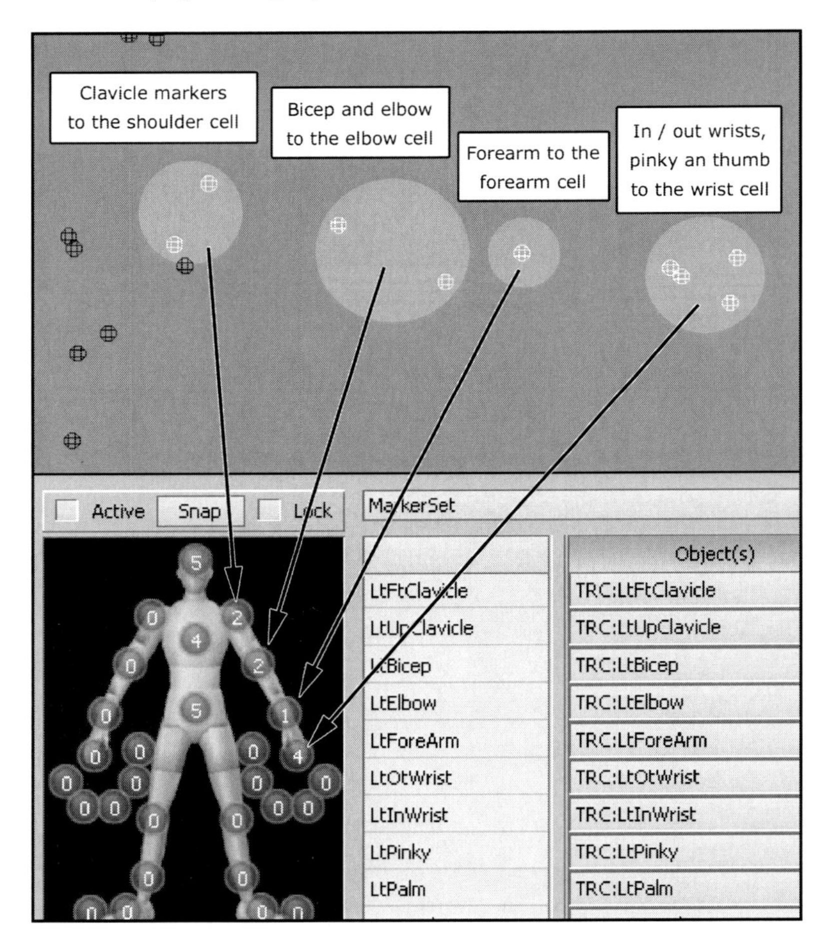

Figure 08_14.

Place the "LtThigh" and "LtKnee" markers in the *Left Knee-Thigh* cell. Put the "LtCalf" marker in the *Left Calf* cell. Drop the "LtAnkle", "LtHeel", "LtBall" and "LtToe" markers to the *Left Ankle* cell. Use the guidelines we have followed so far for the left leg cells to populate the right leg cells (Figure 08_15).

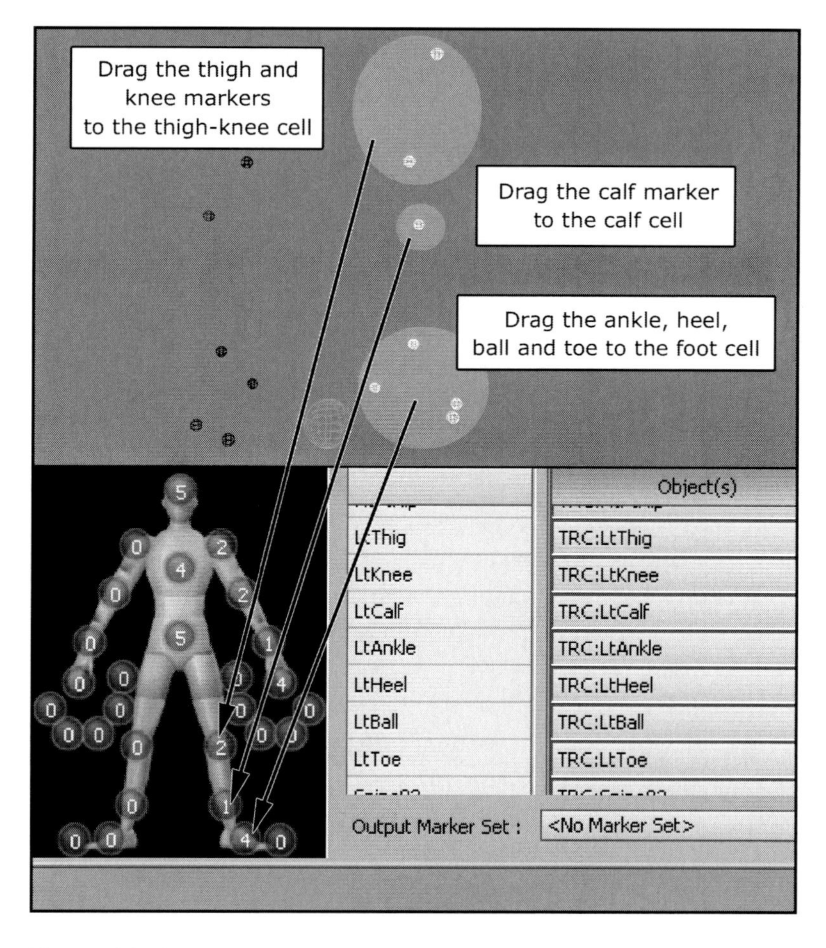

Figure 08_15.

Note: ***Motion Builder*** *will only accept 5 markers per cell in order to drive a body section. Since there is not a universal recipe specifying what markers should be driving which body part, it is up to the user to experiment with different combinations to see what works better for a particular motion.*

All the connections between the markers and the *"Actor's"* parts have been specified; we just need to activate them. In the *Character Controls* window open the *Show* menu and select *Actor Body* to make it visible. Inside the *Actor Settings* tab, check on the *Active* box. The Actor's pose might shift slightly to fit the markers better (Figure 08_16).

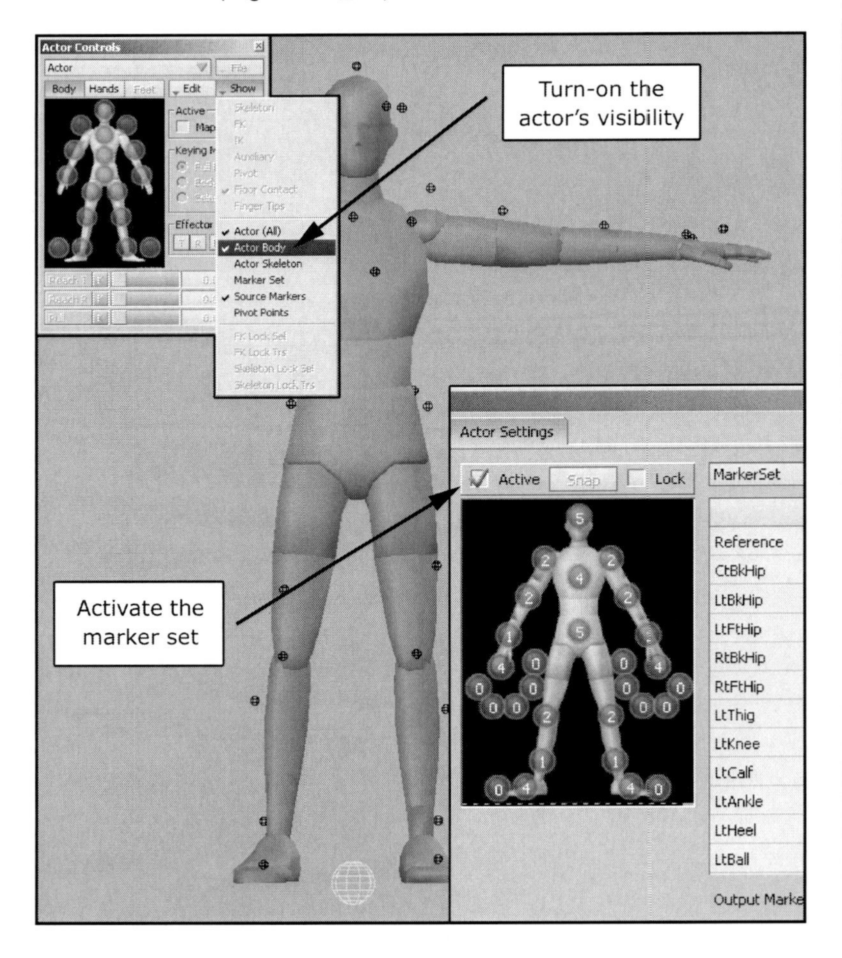

Figure 08_16.

Note: If the Actor's poser shifts drastically, the marker set needs to be deleted and the actor needs to be refitted to the markers more carefully (Figure 08_17).

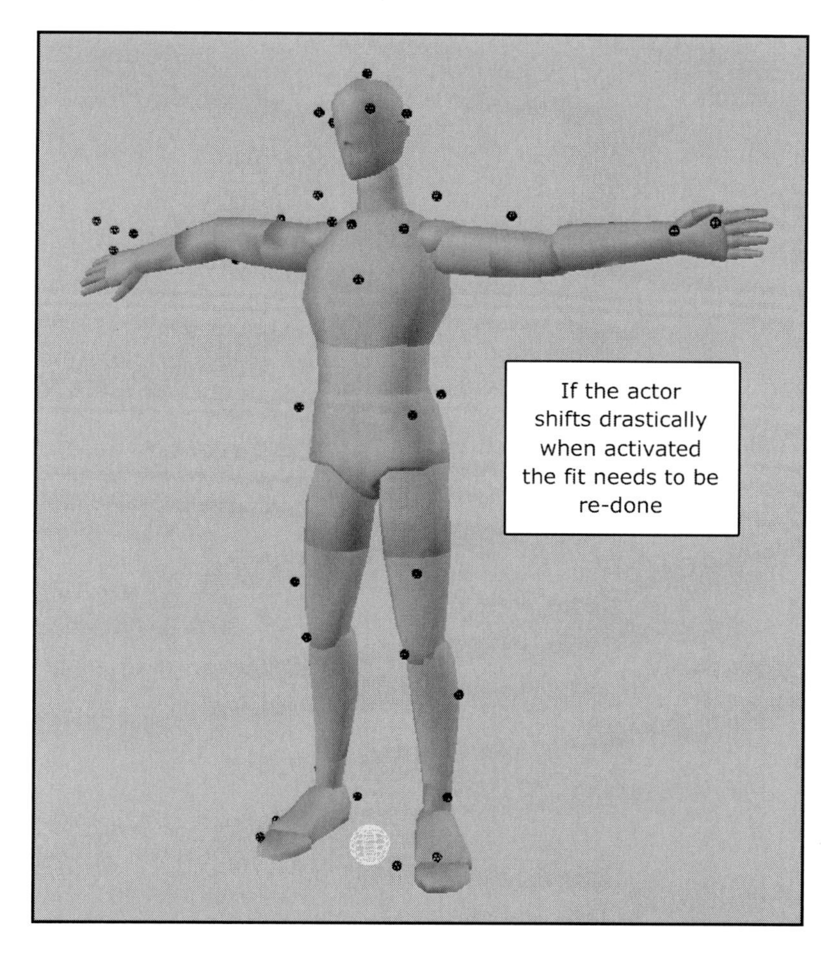

Figure 08_17.

Note: A new actor fitting needs should be done for every capture day as well as every new marker configuration (Fig image of two marker sets we different marker configurations).

The optical data has been properly solved not only for the *T-Pose* take but also for every other take. All the motions in this file are ready to be retargeted to the digital character of your choice. For more information in the retargeting process read chapter 12 (Figure 08_18).

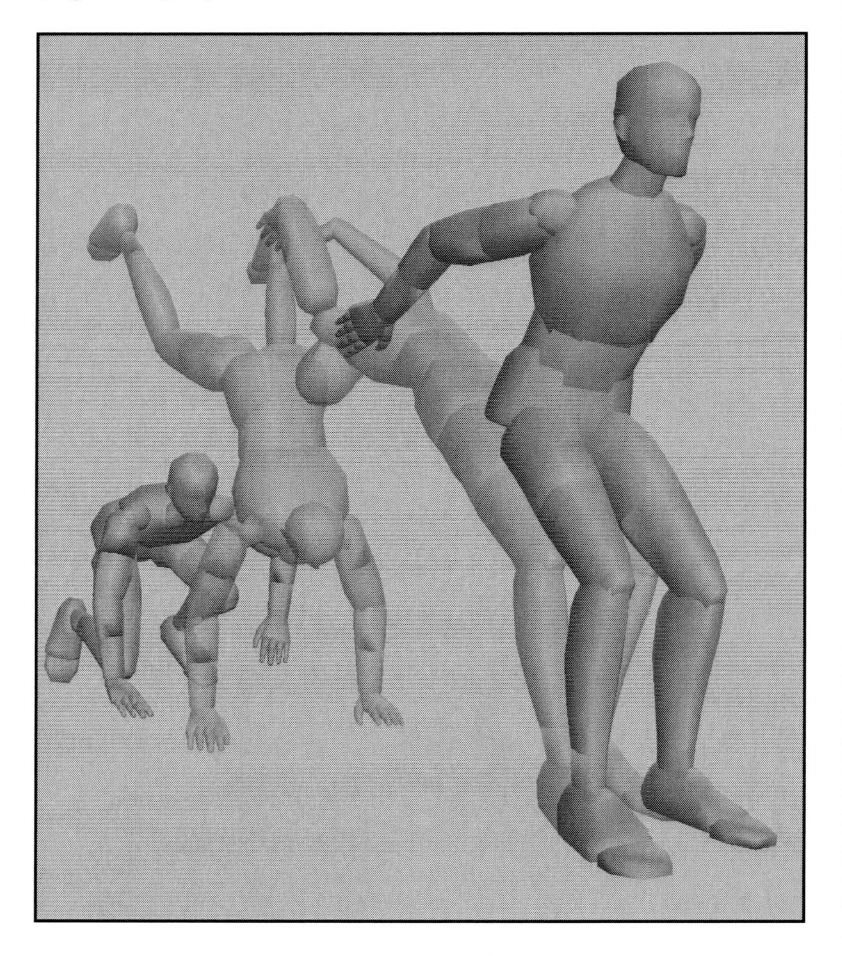

Figure 08_18.

Part 04
Motion Capture Rig

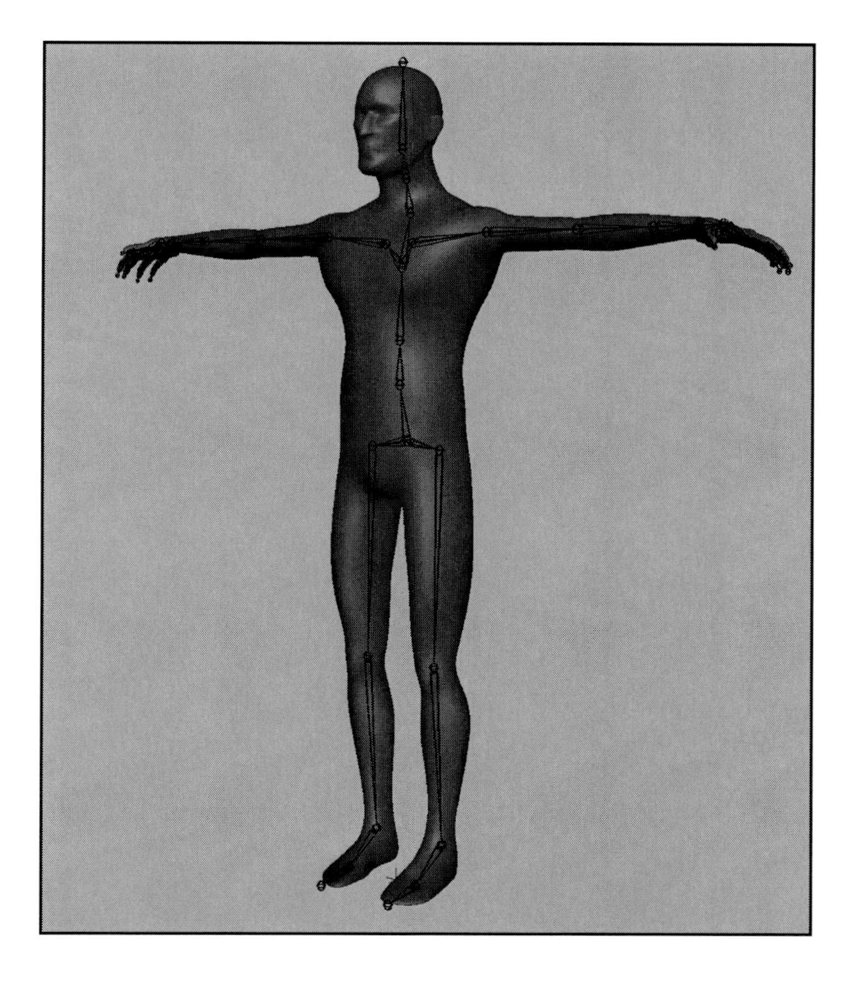

Chapter 09
Rigging For Motion Capture

Rigs for motion capture are mostly simplistic, but like any other part of the CG process complexity can be added to them depending on the needs of the pipeline. In this chapter we are going to focus on creating a simple rig inside **Maya** that can be taken into **Motion Builder** to integrate the solved animation from the actor. This rig will also be able to take motion capture corrections and enhancement to the motion data that gets applied to it. Finally the rig will allow us to transfer animation data from **Motion Builder** back into **Maya** where it was created (Figure 09_01).

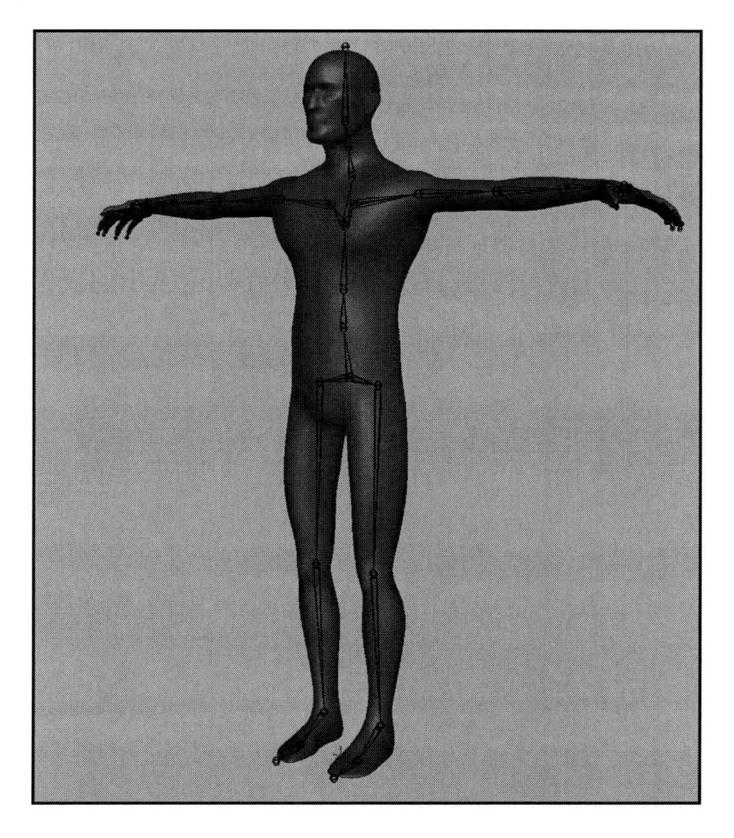

Figure 09_01.

Launch **Maya** and open the "Chapter09"[6] file.

Bone orientation is key to the success of a Rig. To help with this situation we are going to use guides to help with the bone creation process. Lets start with the spine, neck and head. Go to the create menu and select the *EP Curve Tool* option box and check on the *1 Linear* radial button (Figure 09_02).

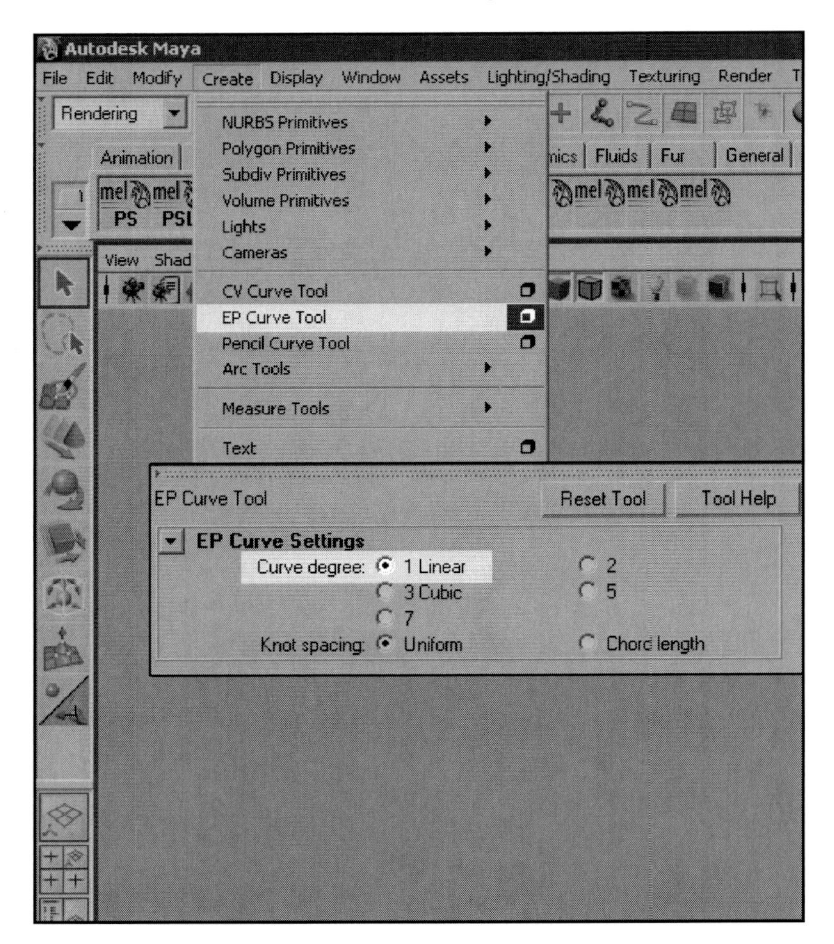

Figure 09_02.

[6] Download the files from www.MocapClub.com/TheMocapBook.htm

On the side view click on the start of the spine, the middle of the lower back, the beginning of the ribcage, the start of the neck, the middle of the neck, the base of the head and the end of the head. Feel free to jump into component mode (*F9*) and change the position of the *Curve Points* to fit the anatomy of the character better (Figure 09_03).

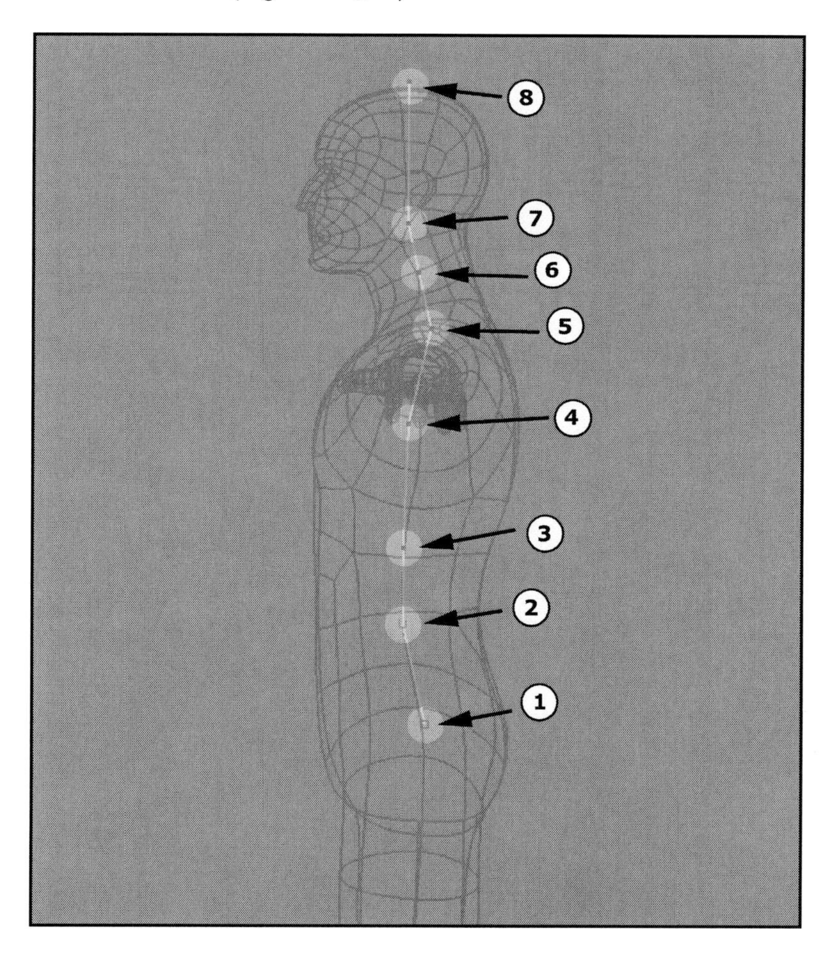

Figure 09_03.

Note: If the spine of your character needs more spine or neck bones feel free to draw an EP Curve with more components.

Template the "MocapGuyGeo" layer (Figure 09_04).
Go to the Skeleton menu and select the *Joint Tool.* With the *"C"* key pressed down click and drag on top of the curve, you would see the joint slide along it. Click and drag on the different sections to create a hierarchy as figure 09_04 shows (Figure 09_04). Select the first bone in the hierarchy and rename it "Hips" in the *Channel Box.* Rename the rest of the bones in the hierarchy "Spine", "Spine1", Spine2", "Neck", "Neck1", "Head" and "Head End" (Figure 09_04).

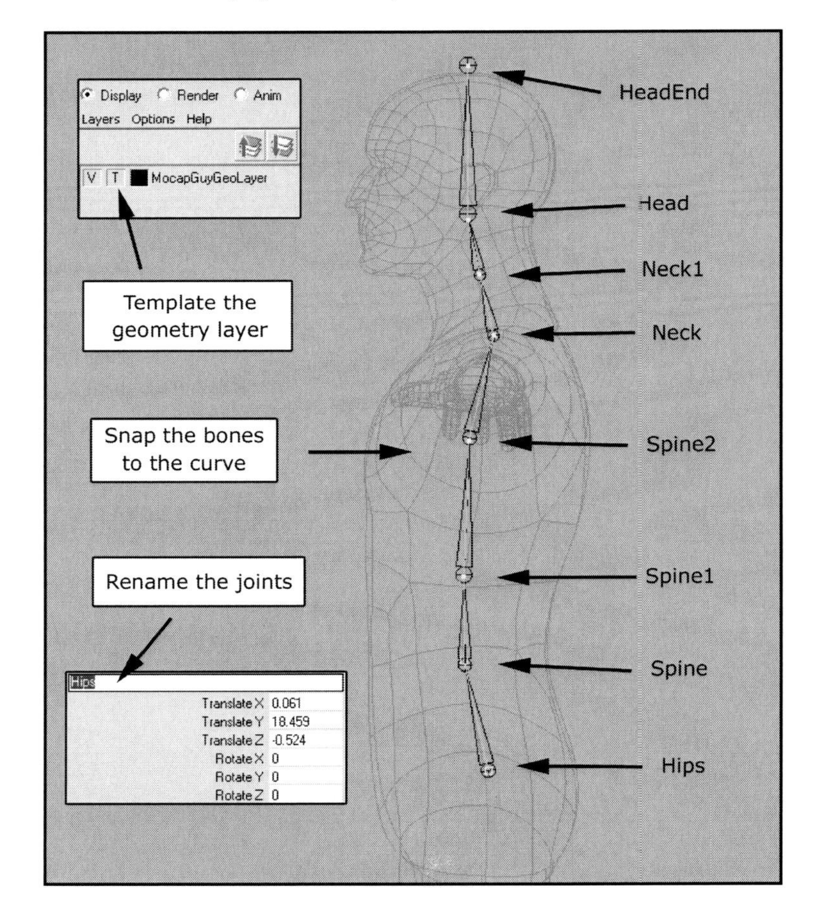

Figure 09_04.

Note: if you have additional bones in the spine, continue naming them spine3, spine4, etc. Apply the same logic for the neck.
Let's now create the left arm. Go to the top view and draw a linear *EP Curve* from the shoulder to the elbow to the palm. In the front view place the curve so it fits inside the arm. Feel free to adjust the components in the top view so they fit the elbow better, however be sure to leave the position of the curve components untouched in the front view. Doing this will assure that the curve and later the bone hierarchy stays coplanar, this means that all the bones in the chain live in the same plain. This will give us optimal results when we add our IK controllers in **Motion Builder** (Figure 09_05).

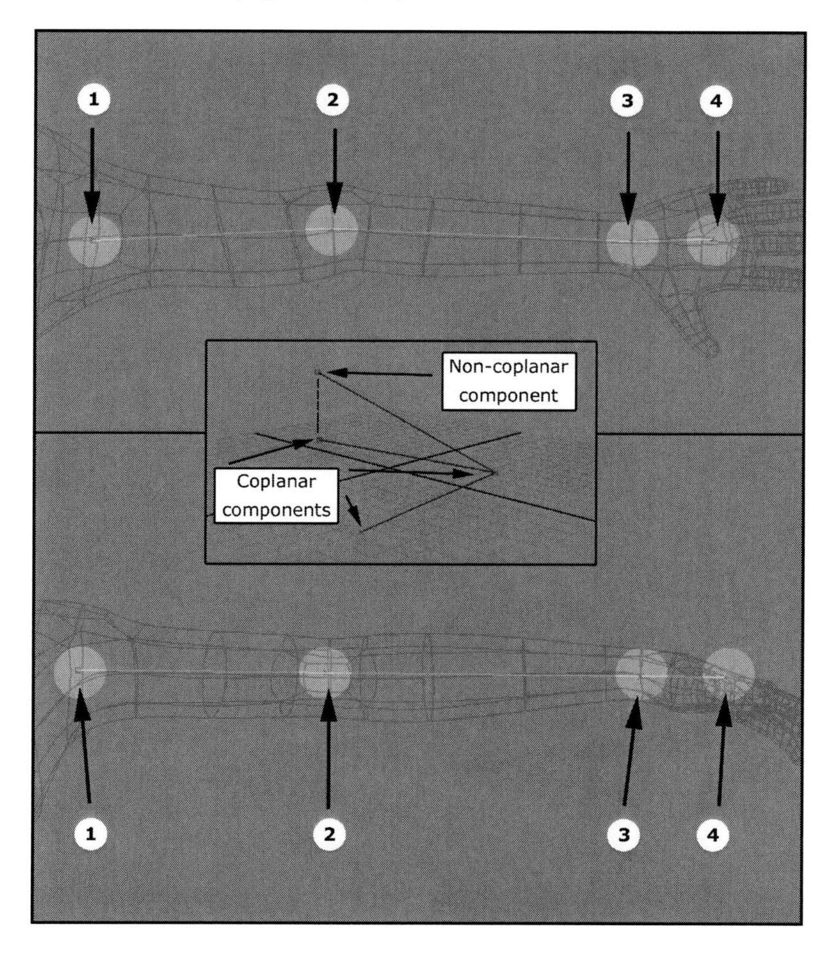

Figure 09_05.

On the top view, draw the arm skeletal hierarchy while holding the *"C"* Key so the bones snap to the guide curve. Rename the bones in the hierarchy "LeftArm", "LeftForeArm", "LeftForeArmRoll", "LeftHand" and "LeftHandEnd" (Figure 09_06).

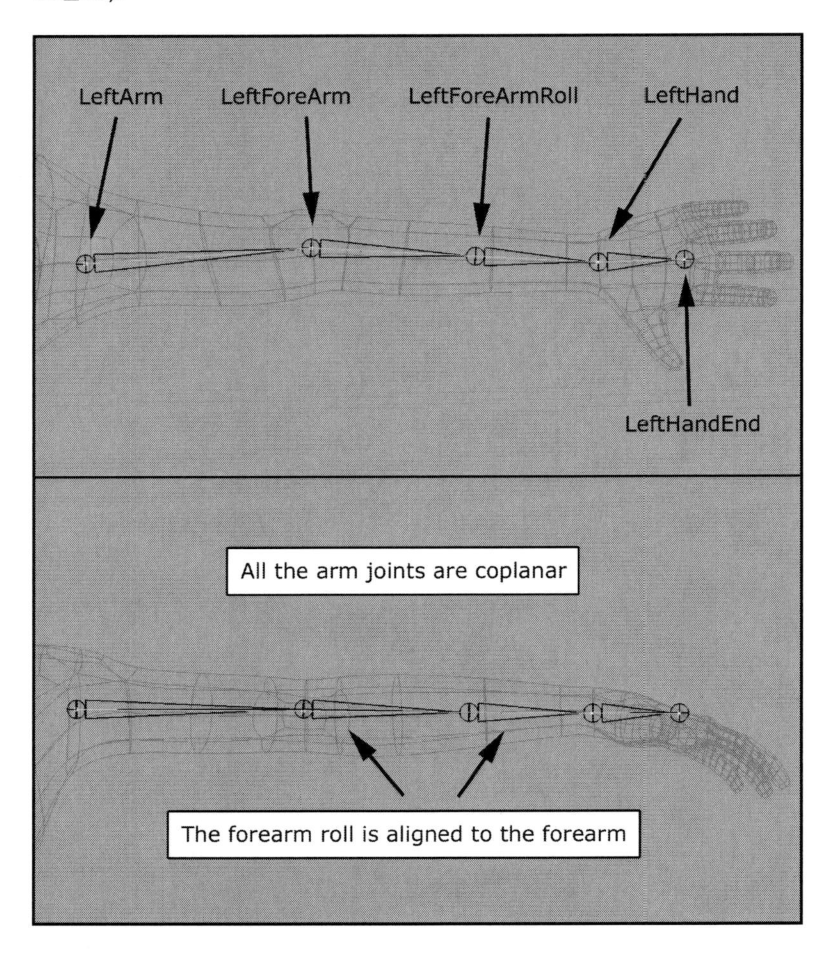

Figure 09_06.

With the left arm done, creating the right one is somewhat automatic, naming conventions included. Select the "LeftArm" bone and under the *Skeleton* menu click the *Mirror Joint* option box. Specify the *Mirror across* option as YZ and under *Mirror function* select *Behavior*. In the *Replacement names of duplicate joints* enter "Left" for *Search for* and "Right" for *Replace with*. Press the *Mirror* button to generate the hierarchy on the opposite side (Figure 09_07).

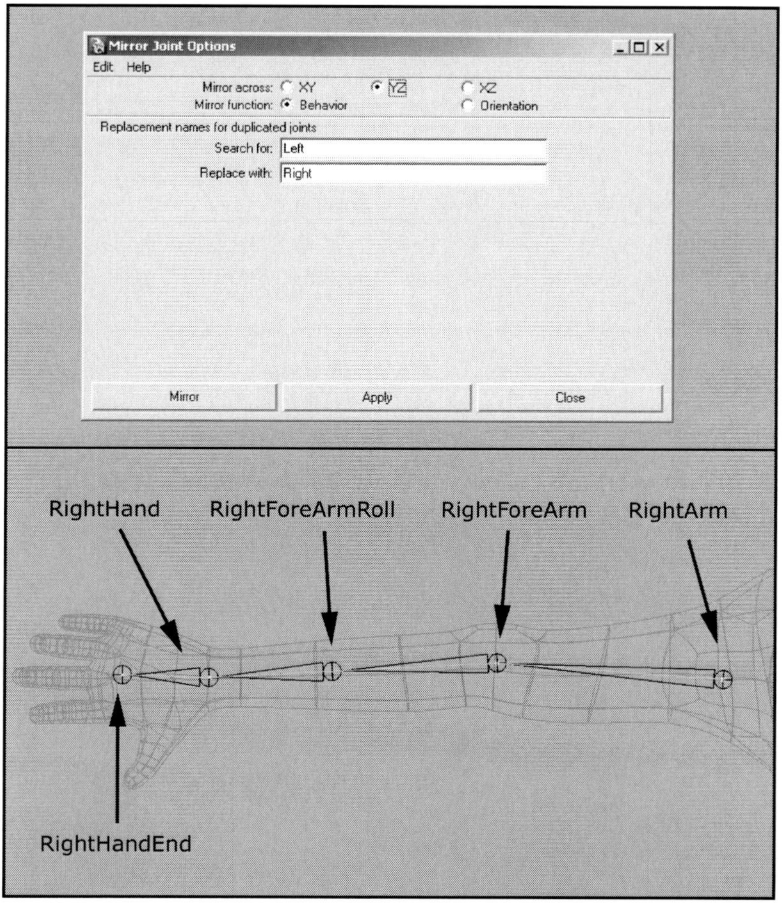

Figure 09_07.

We are now going to create the left leg. The placement of the root of this lower extremity is particularly important for both mechanical as well as deformation purposes. This creates the need of additional guides for the creation of the body part. Go to the side view and create a circle by navigating to the *NURBS Primitives* options under the *Create* menu. Rotate, translate and scale the circle until it encompasses the hips of the model. The center of the circle now indicates where the root of the leg should be placed (Figure 09_08).

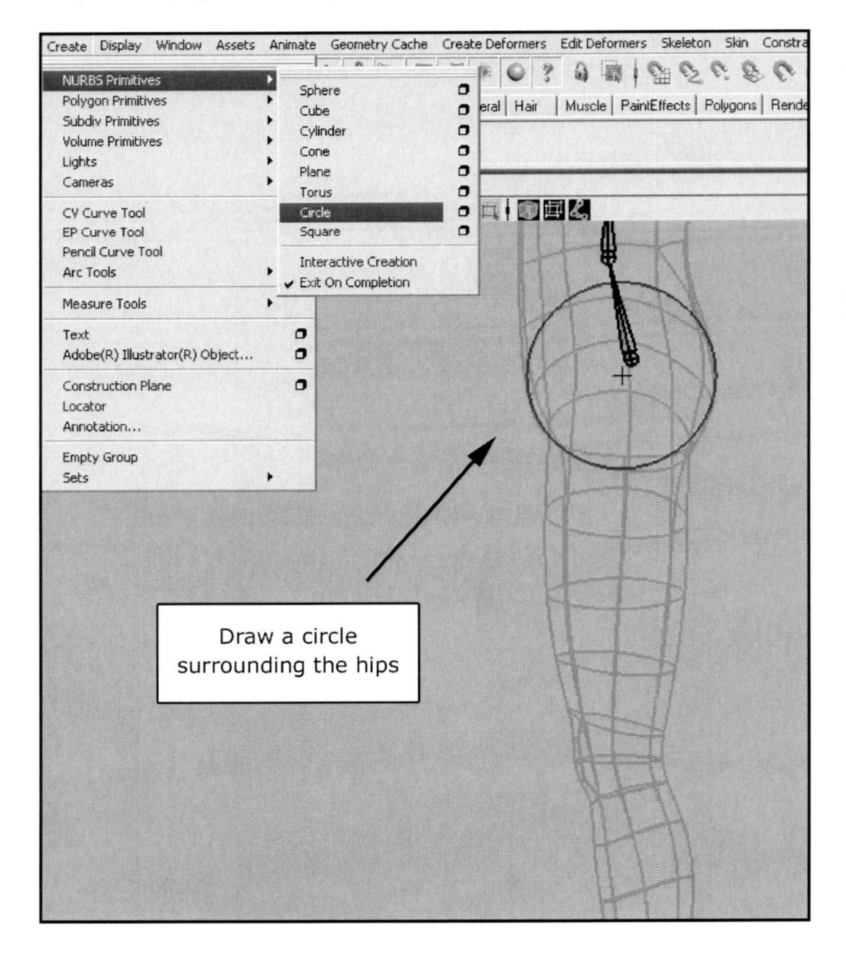

Figure 09_08.

Draw a linear *EP Curve* from the center of the hip circle, to the knee, to the ankle, to the ball off the foot, to the tip of the toe. This is a similar technique to the one illustrators use to draw their figures (Figure 09_09). Draw the leg skeletal hierarchy while holding the *"C"* Key so the bones snap to the guide curve. Rename the bones in the hierarchy "LeftUpLeg", "LeftLeg", "LeftFoot", "LefToeBase", "LeftToeEnd". Mirror the leg to generate the hierarchy on the opposite side (Figure 09_09).

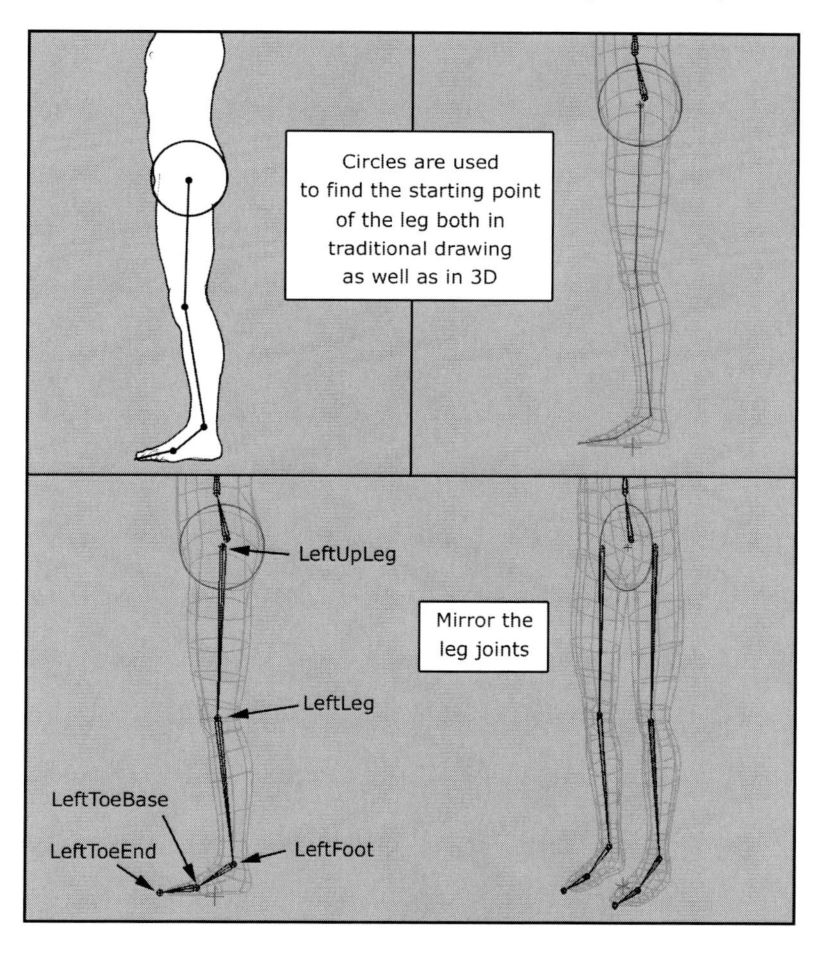

Figure 09_09.

Lets now take care of the shoulder. We will not need any guides for this hierarchy since it is a simple 2-bone chain. With the *Join Tool* active click a few inches below and to the left of the "Neck" bone. While pressing and holding the *"V"* key click on top of the "Left Arm" bone (this will snap the position of the end of the shoulder hierarchy to the beginning of the arm chain). Select the end joint of the shoulder skeleton and delete it. Rename the shoulder bone "LeftShoulder". Mirror the shoulder to generate the opposite shoulder (Figure 09_10).

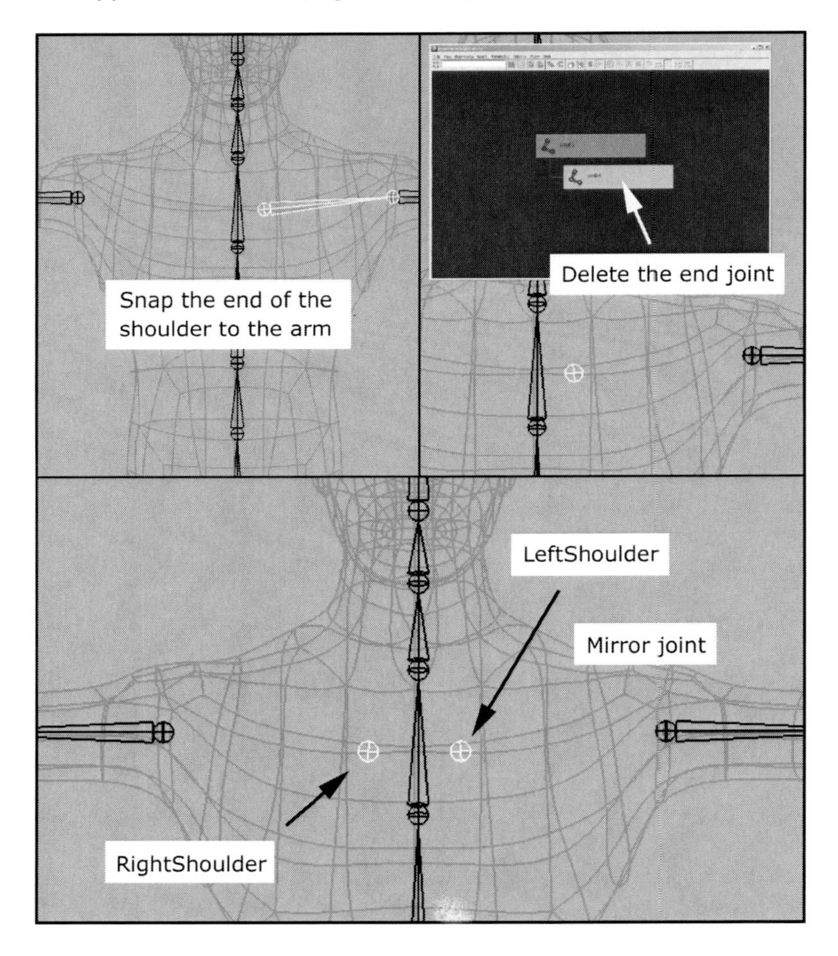

Figure 09_10.

We are now going to draw the fingers. Go to the top view and draw a linear *EP Curve* running through the index with components at every bending section of the finger. In the perspective view translate the components in the Y-axis so they fit inside the geometry of the finger (Figure 09_11).

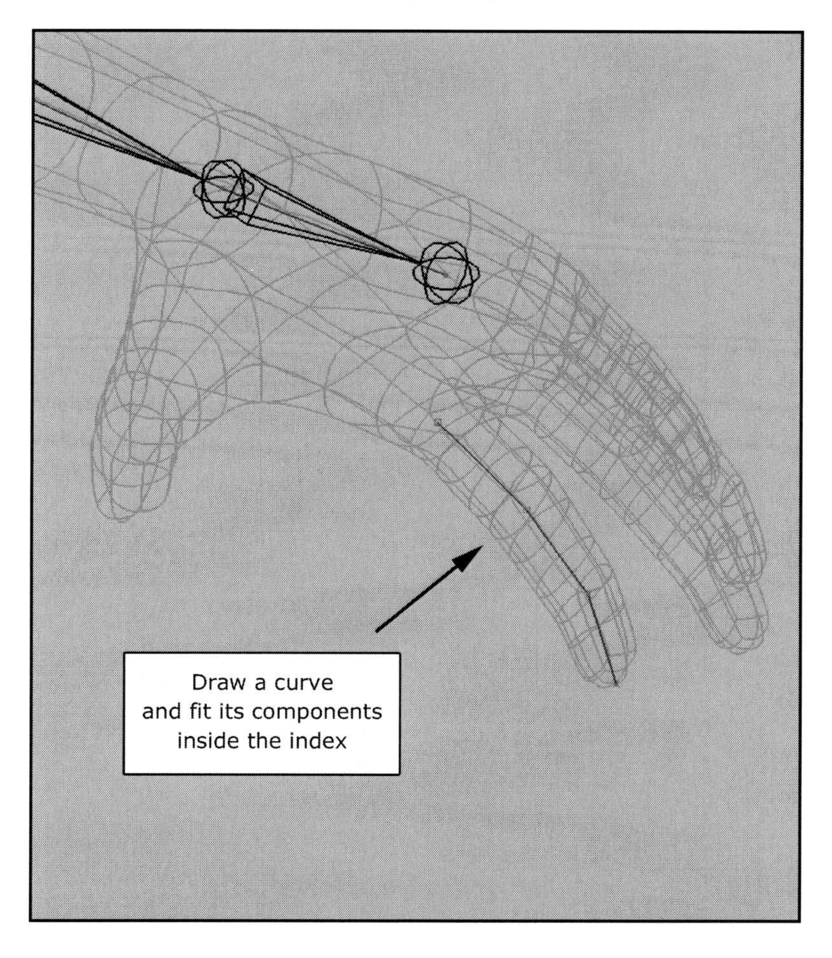

Figure 09_11.

Draw the index skeletal hierarchy while holding the *"C"* Key so the bones snap to the guide curve. Rename the bones in the hierarchy "LeftHandIndex1", "LeftHandIndex2", "LeftHandIndex3" and "LeftHandIndexEnd". Mirror the index to generate the hierarchy on the opposite side (Figure 09_12).

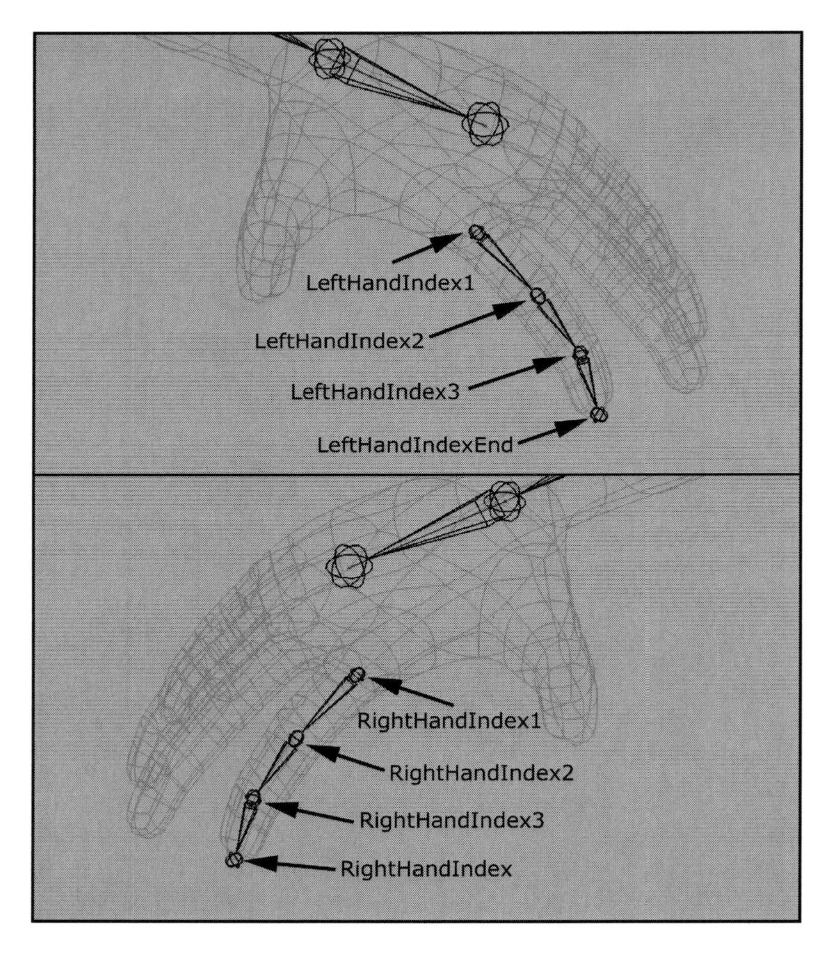

Figure 09_12.

Repeat the process for the middle finger as well as the ring finger. Rename the hierarchies middle finger hierarchy "LeftHandMiddle1", "LeftHandMiddle2", "LeftHandMiddle3" and "LeftHandMiddleEnd". Rename the ring finger hierarchy "LeftHandRing1", "LeftHandRing2", "LeftHandRing3", "LeftHandRingEnd". Mirror the middle and ring fingers to generate the hierarchy on the opposite side (Figure 09_13).

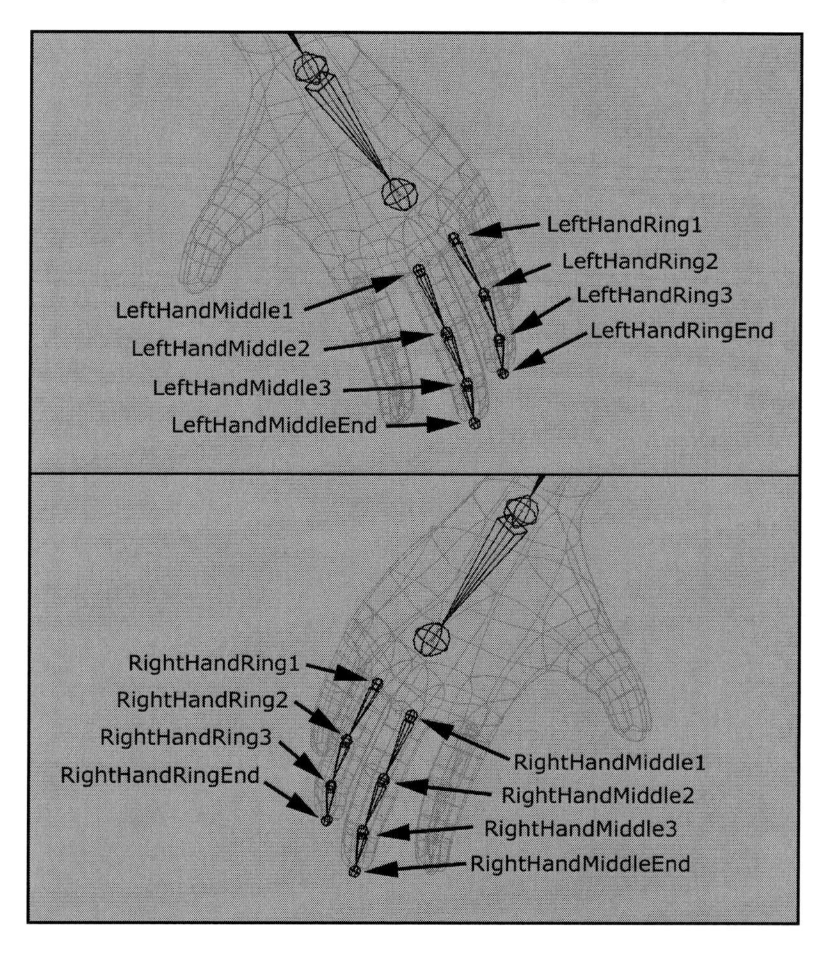

Figure 09_13.

The pinky is a little different that the fingers that we have drawn so far. From the top view draw a linear *EP Curve* running through the pinky but this time make sure to start it at the palm. Place components at every bending section of the finger. Draw the pinky skeletal hierarchy while holding the *"C"* Key so the bones snap to the guide curve. Rename the bones in the hierarchy "LeftHandPinkyPalm", "LeftHandPinky1", "LeftHandPinky2", and "LeftHandPinky3", "LeftHandPinkyEnd". Mirror the pinky to generate the hierarchy on the opposite side (Figure 09_14).

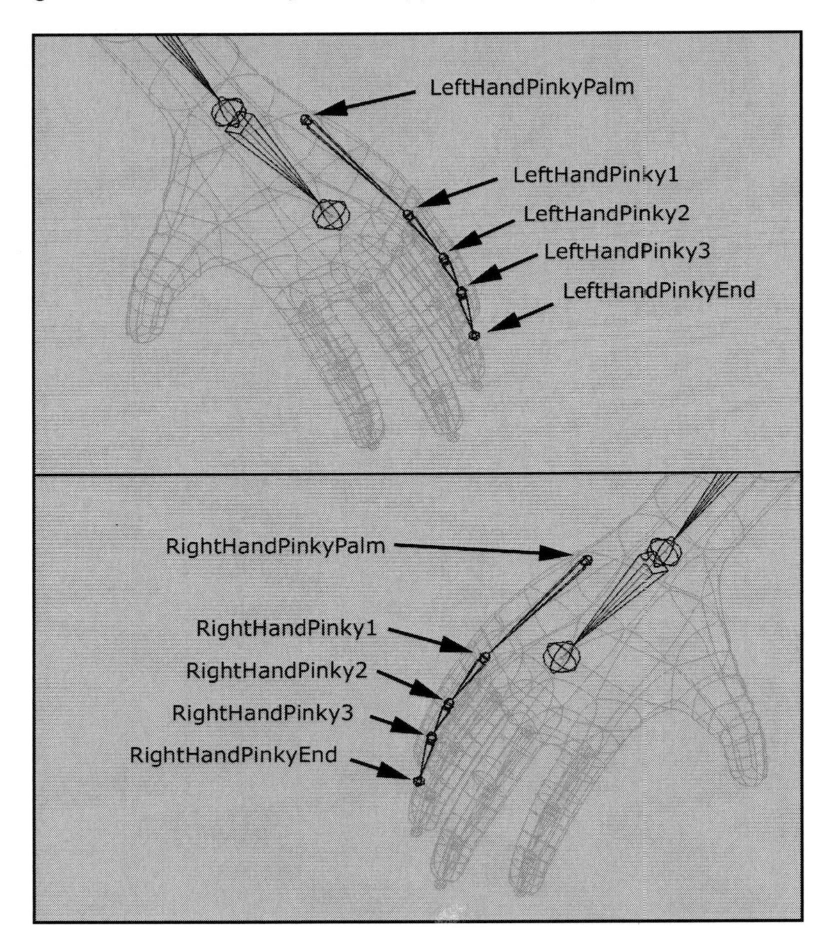

Figure 09_14.

We now move on to the thumb. This finger like the pinky also has specific needs. The main thing to keep in mind with the thumb is that it does not rotate in the same direction as the others. Draw a linear *EP Curve* running through the thumb making sure to start at the palm. Draw the Thumb skeletal hierarchy while holding the *"C"* Key so the bones snap to the guide curve. Rename the bones in the hierarchy "LeftHandThumb1", "LeftHandThumb2", "LeftHandThumb3", and "LeftHandThumbEnd" (Figure 09_15).

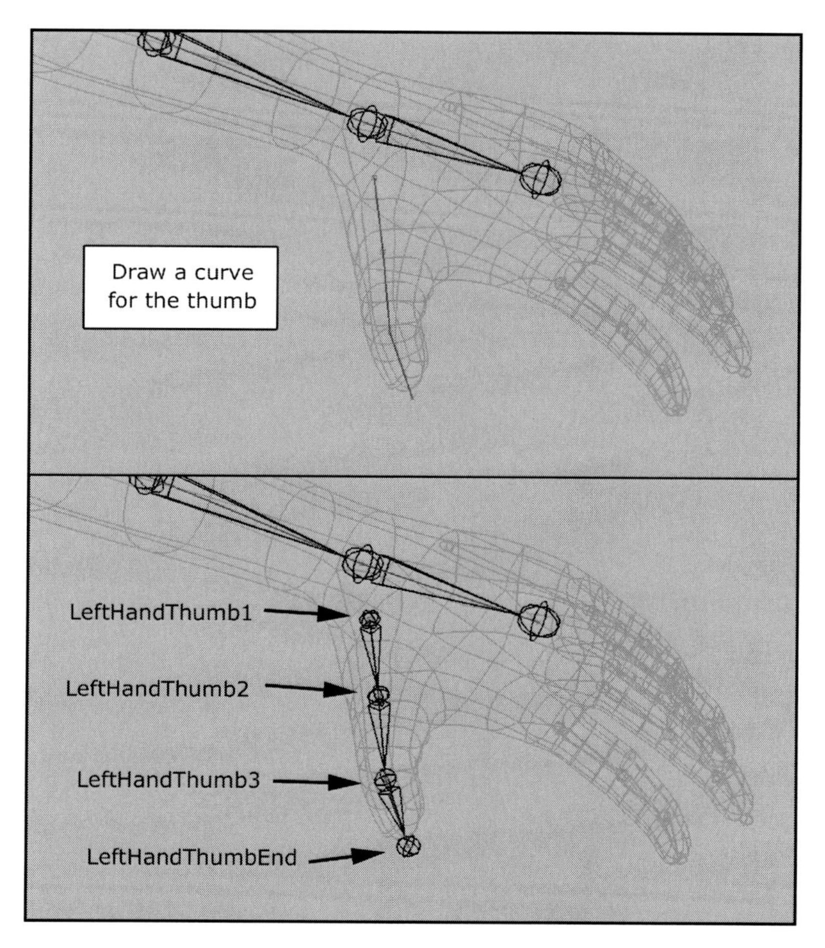

Figure 09_15.

Although the thumb has been created, its mechanics are not accurate yet. With the "LeftHandThumb1" bone selected press the *"F9"* key to access component mode. Select the axis of every thumb finger, in the *Command Line* type "rotate –r –os 90 0 0;" and hit the *Enter* key to execute. All the thumb bones should now be properly oriented and the finger should rotate in the proper direction (Figure 09_16).

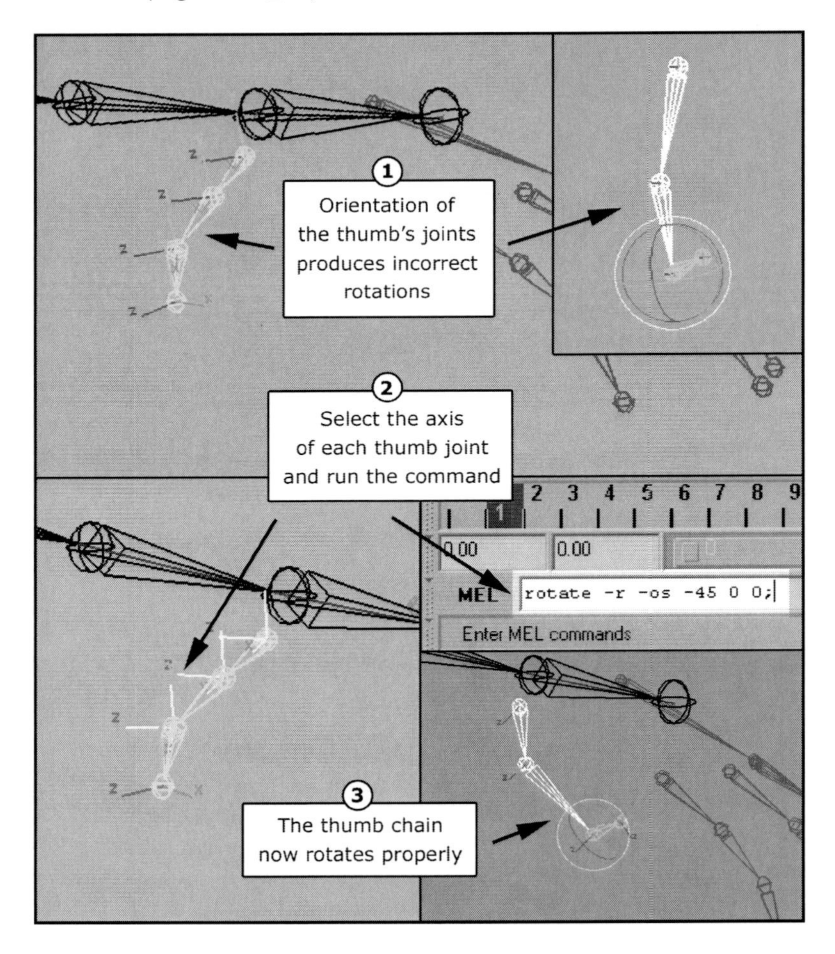

Figure 09_16.

Now that we are done with the creation of the bones, lets put the hierarchy together. Open the *Hypergraph* by selecting its title from the *Window* menu. Select the "LeftUpLeg" node inside the *Hypergraph* window and middle click and drag it to the "Hips" node. Middle click and drag the "RightUpLeg" to the "Hips", the "LeftShoulder" to "Spine2", the "RightShoulder" to "Spine2", "LeftArm" to "LeftShoulder" and "RightArm" to "RightShoulder". Drag the root of every finger in the left side to the "LeftHand" and every finger root in the right side to the "RightHand". Your hierarchy should look something like figure 09_17 (Figure 09_17).

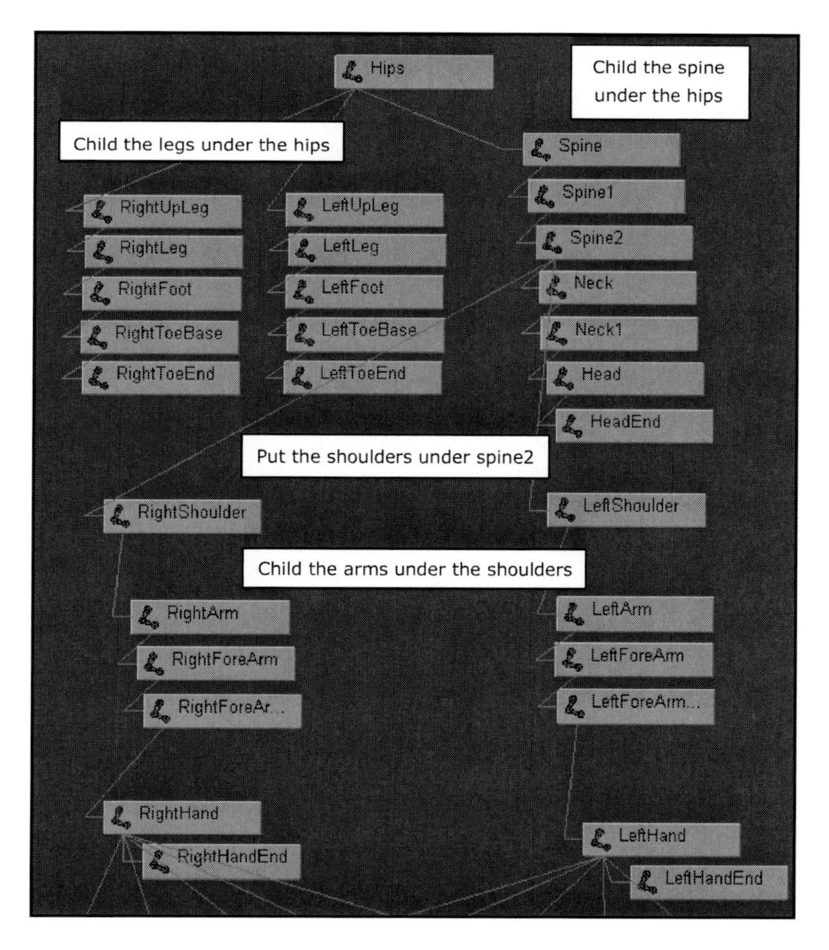

Figure 09_17.

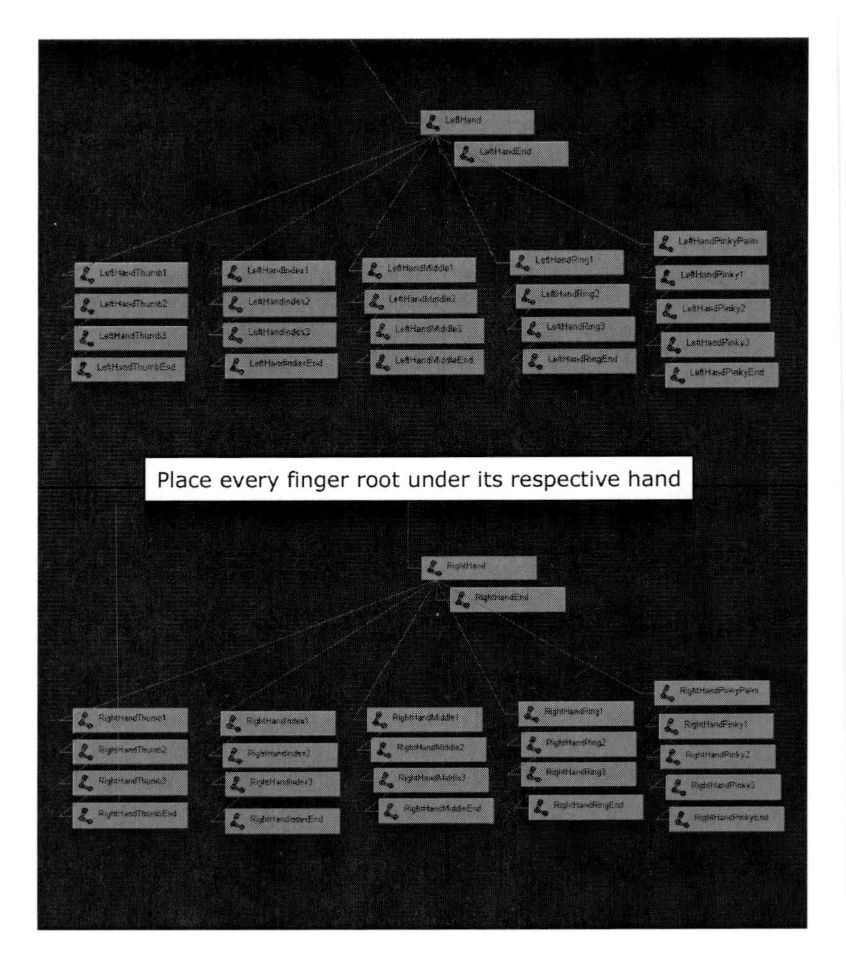
LeftHand
LeftHandEnd
LeftHandThumb1
LeftHandThumb2
LeftHandThumb3
LeftHandThumbEnd
LeftHandIndex1
LeftHandIndex2
LeftHandIndex3
LeftHandIndexEnd
LeftHandMiddle1
LeftHandMiddle2
LeftHandMiddle3
LeftHandMiddleEnd
LeftHandRing1
LeftHandRing2
LeftHandRing3
LeftHandRingEnd
LeftHandPinkyPalm
LeftHandPinky1
LeftHandPinky2
LeftHandPinky3
LeftHandPinkyEnd
Place every finger root under its respective hand
RightHand
RightHandEnd
RightHandThumb1
RightHandThumb2
RightHandThumb3
RightHandThumbEnd
RightHandIndex1
RightHandIndex2
RightHandIndex3
RightHandIndexEnd
RightHandMiddle1
RightHandMiddle2
RightHandMiddle3
RightHandMiddleEnd
RightHandRing1
RightHandRing2
RightHandRing3
RightHandRingEnd
RightHandPinkyPalm
RightHandPinky1
RightHandPinky2
RightHandPinky3
RightHandPinkyEnd

Figure 09_18.

Hierarchy wise we only need one additional node to serve as reference for the rig so we can scale the entire thing up or down depending on the size of the scene. From the *Create* menu, select a *Locator*. Rename this new node "MocapGuy_Reference". Inside the *Hypergraph* middle click and drag it to the "Hips" node to the "MocapGuy_Reference". The hierarchy is complete and should look something like Figure 09_19 (Figure 09_19).

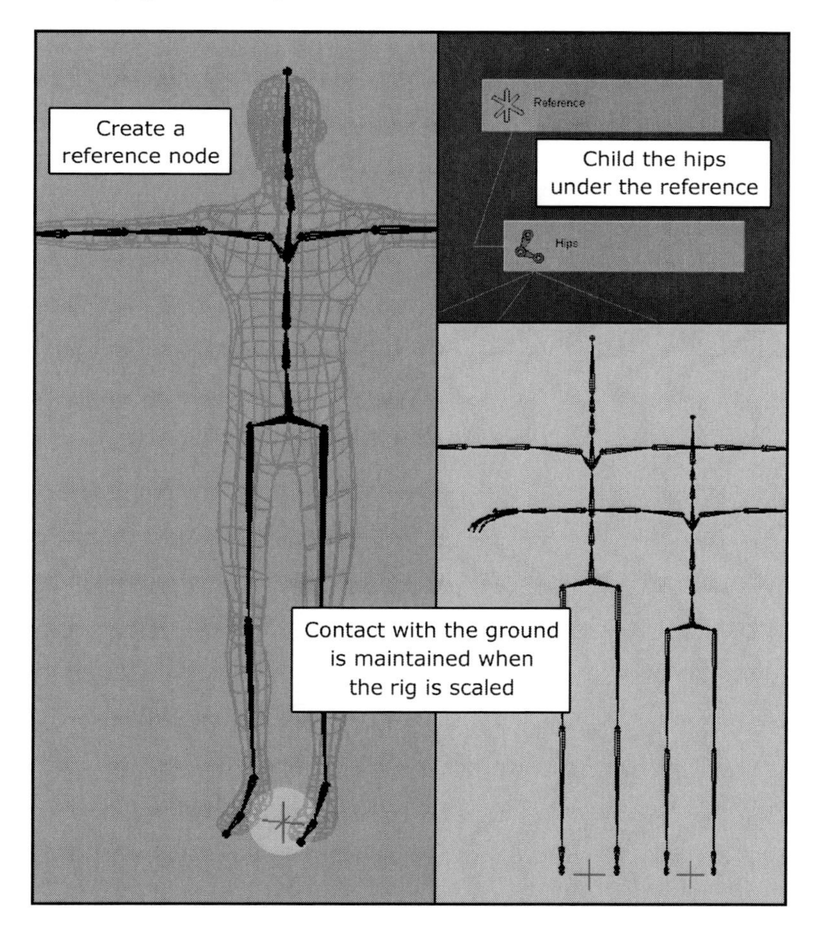

Figure 09_19.

Note: by leaving the reference in at feet level, step contact is maintained regardless of scale.

We will not create any controllers for this rig inside of **Maya** since we are going to be generating all the controllers that we need to deal with motion capture inside **Motion Builder**.
Let's move on then to the skinning of the model to the skeleton. Select the "MocapGuy_Geo" node and *Shift* + select the "Hips" node. Go to the *Skin* menu and select the *Smooth Bind* option box under the *Bind Skin* submenu. Under the *Bind to* option choose *Joint hierarchy*, on *Bind method* select *Closest in hierarchy*, on *Max influences* select 3, check *Maintain max influences* off and set the *Dropoff rate* to 7 (Figure 09_20).

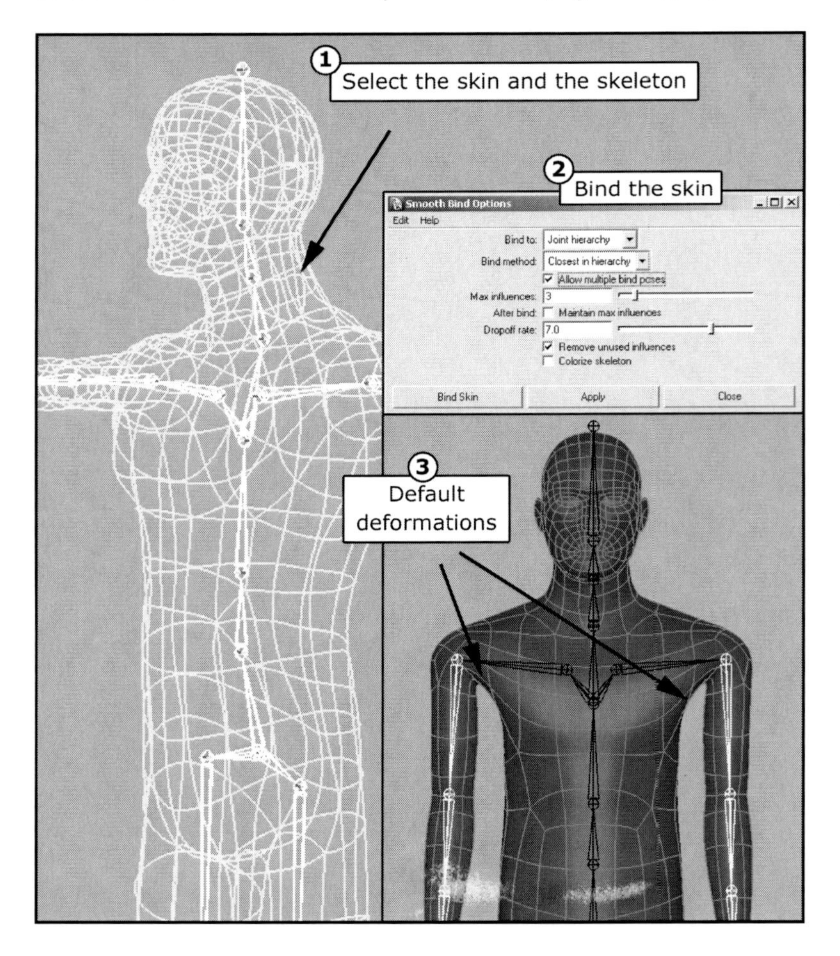

Figure 09_20.

The next step would be to weight the skin of the model so we get better deformations, however, there is no real different between weighting a mesh to a motion capture rig and weighting a model to a hand animation rig. For this exercise we are going to import some pre-weighted maps so we can continue exploring motion capture. In the *Skin* menu select the *Import Skin Weights Maps* option inside the *Edit Smooth Skin* submenu. Use the pop-up window that appears to navigate to the "MocapGuy_Weights" folder that you downloaded from "mocapclub.com".
Select the "MocapGuy_WeighMaps.weightmap" file and press import. Accept the warning dialog that appears. The skin of MocapGuy should look much better when you rotate the bones (Figure 09_21).

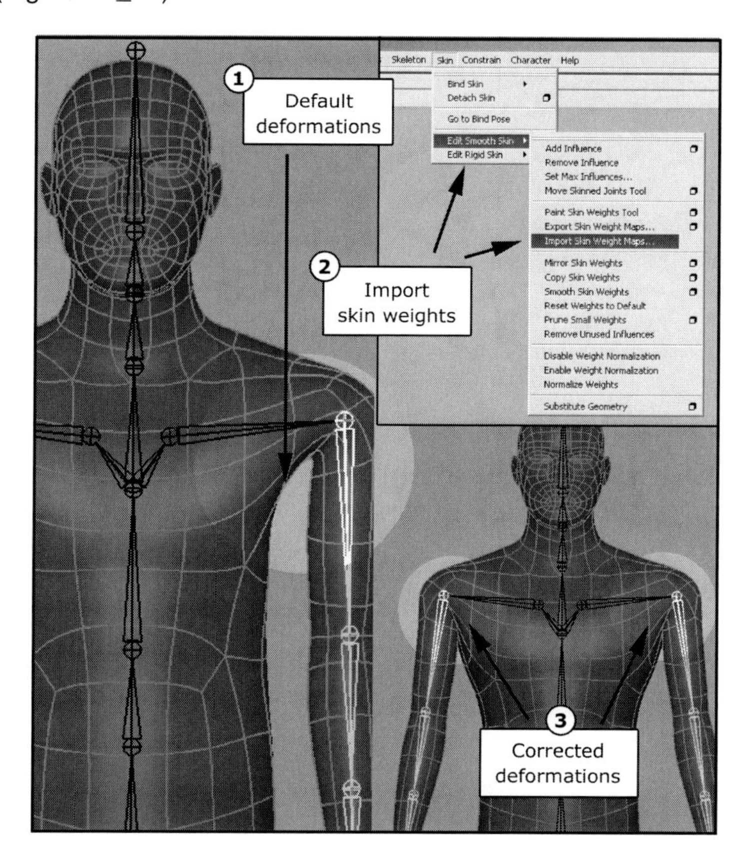

Figure 09_21.

Note: For more information on the weighting process refer to the Maya help docs (F1 Key), Chris Maraffi's "Mel Scripting a Rig in Maya" book or the "Hyper-Real" series from Autodesk (formerly Alias).

The last thing that we need to do is export the rig so it can be opened in **Motion Builder**. Go to the *File* menu and select the *Export All* option. Navigate to the place where you want to save the file. Under the *File name* section type "MocapGuy", select *Fbx* from the *Files of type* option and press the *Export* button. Accept the new *FBX Exporter* window that pops up as the default settings will work for now. Close the *Warnings and Errors* window that appears. The file has been exported as an fbx and can now be opened in **Motion Builder** (Figure 09_22).

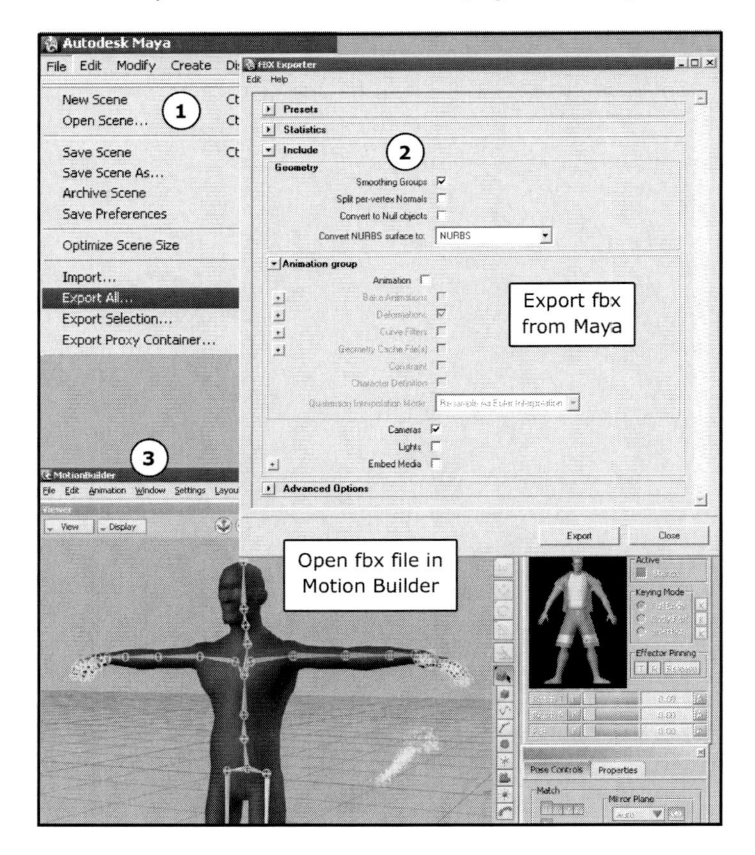

Figure 09_22.

Part 05
Back In Motion Builder

Chapter 10
Automatic Characterization

The *Characterization* is the tool that allows **Motion Builder** to retarget animation from the *Actor* tool to a rig or from rig to rig. *Characterizing* is one of the software's most important building blocks allowing platforms like the *Story Tool*, *Pose Controls* and much more. When you characterize, you are informing **Motion Builder** on how the rig that you created relates to the joint hierarchy that it expects from a Biped or a Quadruped (Figure 10_01).

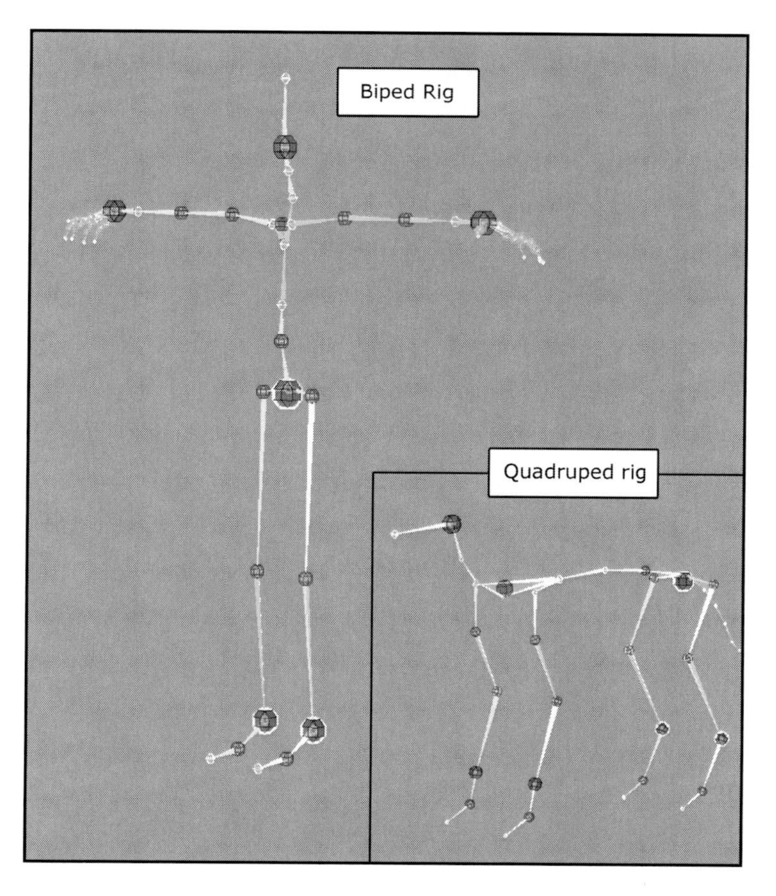

Figure 10_01.

Because of the naming conventions that we followed on chapter 09, the characterization process will be somewhat automatic (Figure 10_02).

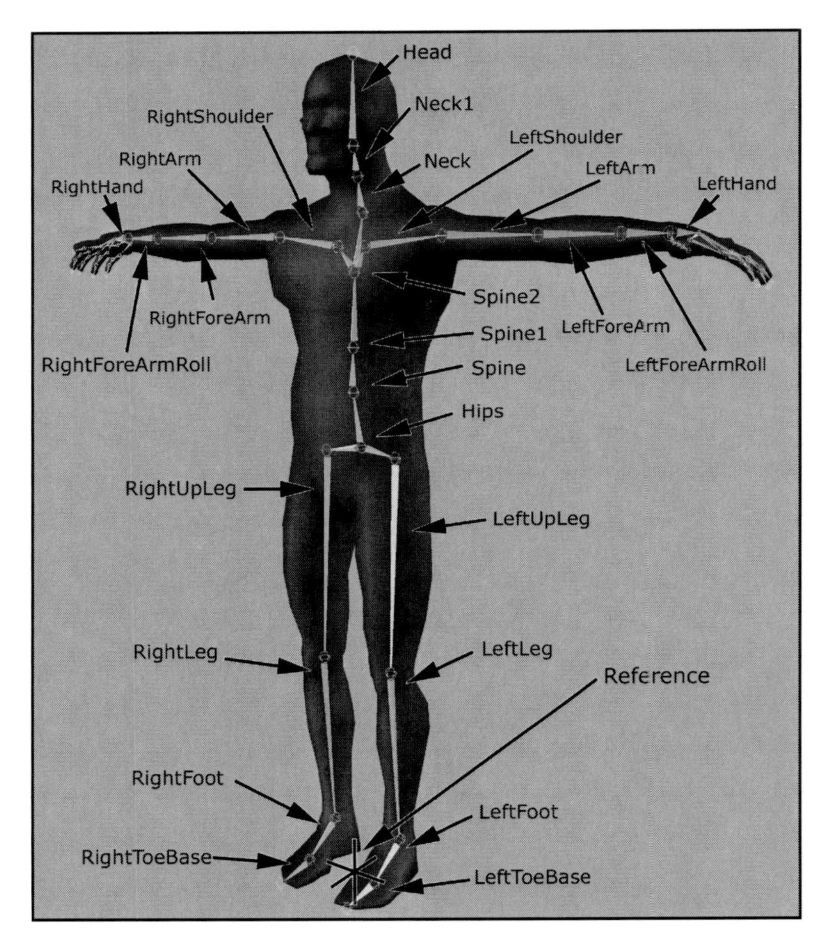

Figure 10_02.

Launch **Motion Builder** and open the "Chapter10"[7] file.

In order to have a successful Characterization, the rig needs to be in full *T-Pose* and facing the positive Z axis. Even if the character is rigged and modeled in a different stance in needs to be taken into *T-Pose* to be characterized (Figure 10_03)

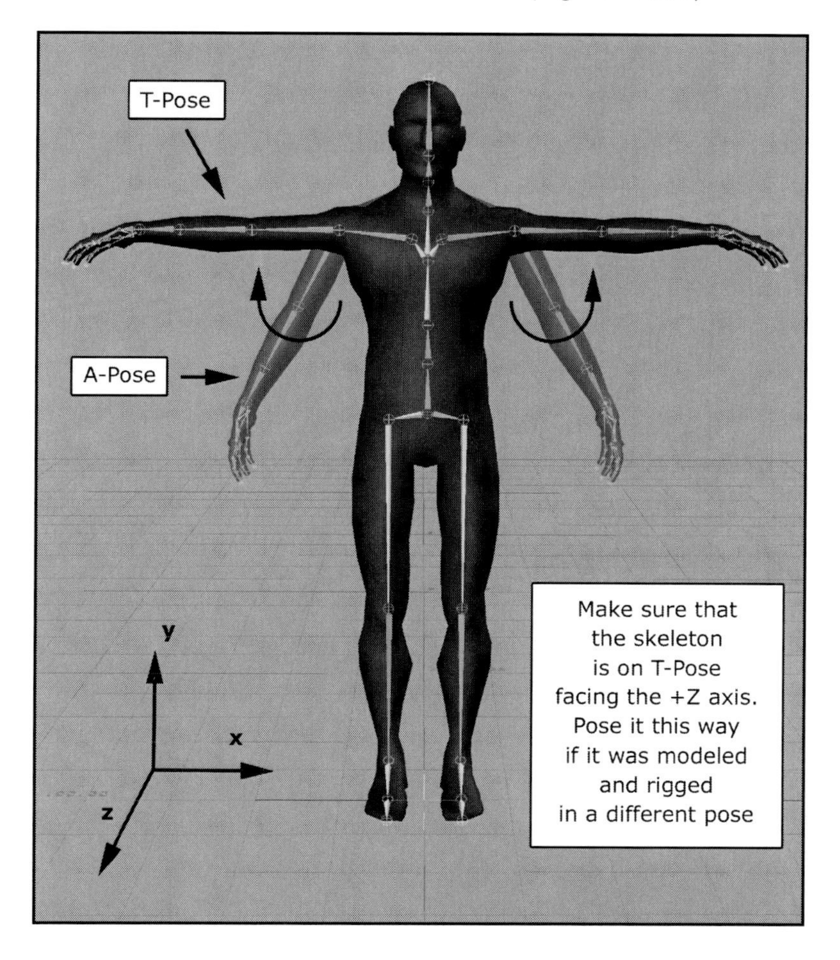

Figure 10_03.

[7] Download the files from www.MocapClub.com/TheMocapBook.htm

Go to the Asset Browser window, from the *Characters* section inside the *Templates* folder drag a *Character* icon to any joint of your hierarchy. Select *Characterize* from the pop-up option that appears. **Motion Builder** will ask you if you want to characterize the rig as a *Biped*, or a *Quadruped*, since we are using a human model for this tutorial click on the *Biped* button. (Figure 10_04).

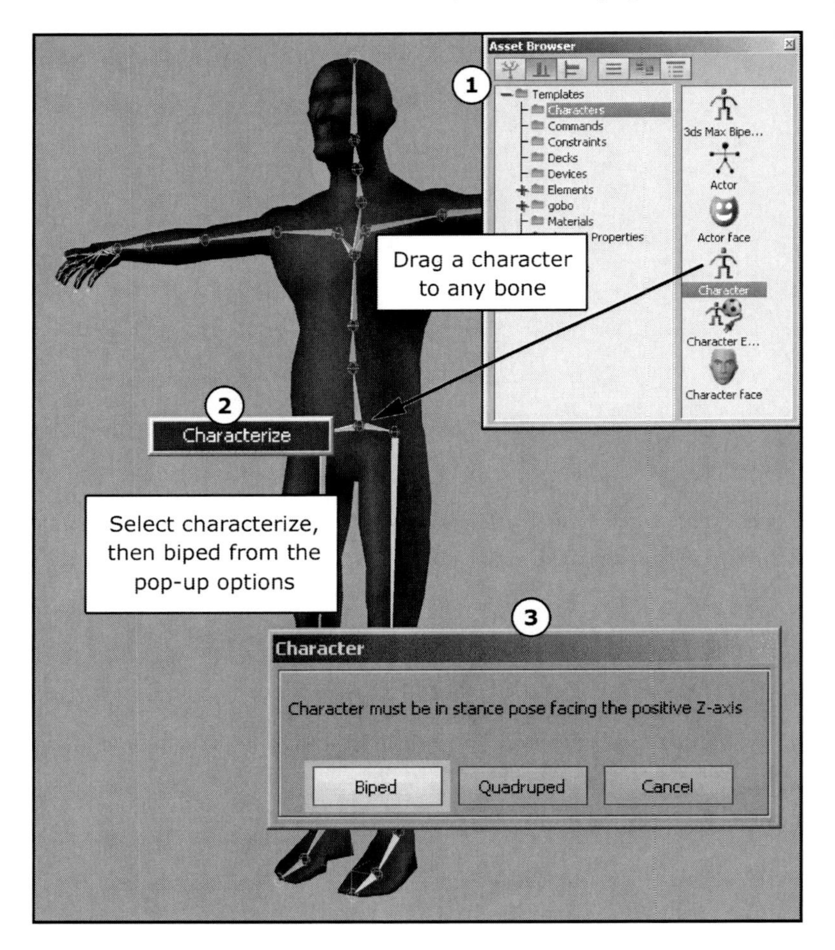

Figure 10_04.

A *Characters* section appears under the navigator window, open it up and rename Character to "MocapGuy_Character" by right clicking on it and choosing *Rename* from the pop-up menu. (Figure 10_05).

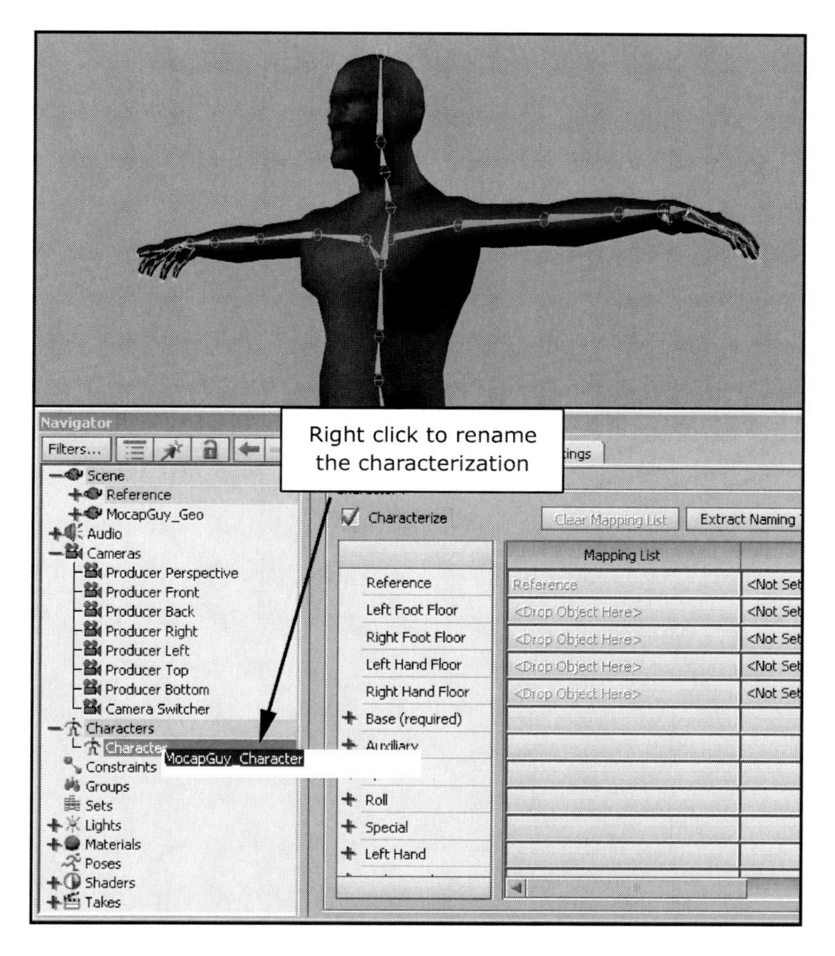

Figure 10_05.

If you get close to the feet or the hands of the character, you will see some green and purple boxes surrounding those body parts. These are called *Floor Contacts* and they will later help the rig recognize the grid or a selected piece of geometry as the ground preventing those body parts from going bellow it. Fit the contact planes of the character by placing the green and purple boxes around the feet and hands to the appropriate places. Heel, ball and toe tip for the feet and wrist, finger base and finger tip for the hands. Use figure 10_06 to guide you on the placement of the *Floor Contact Components* (Figure 10_06).

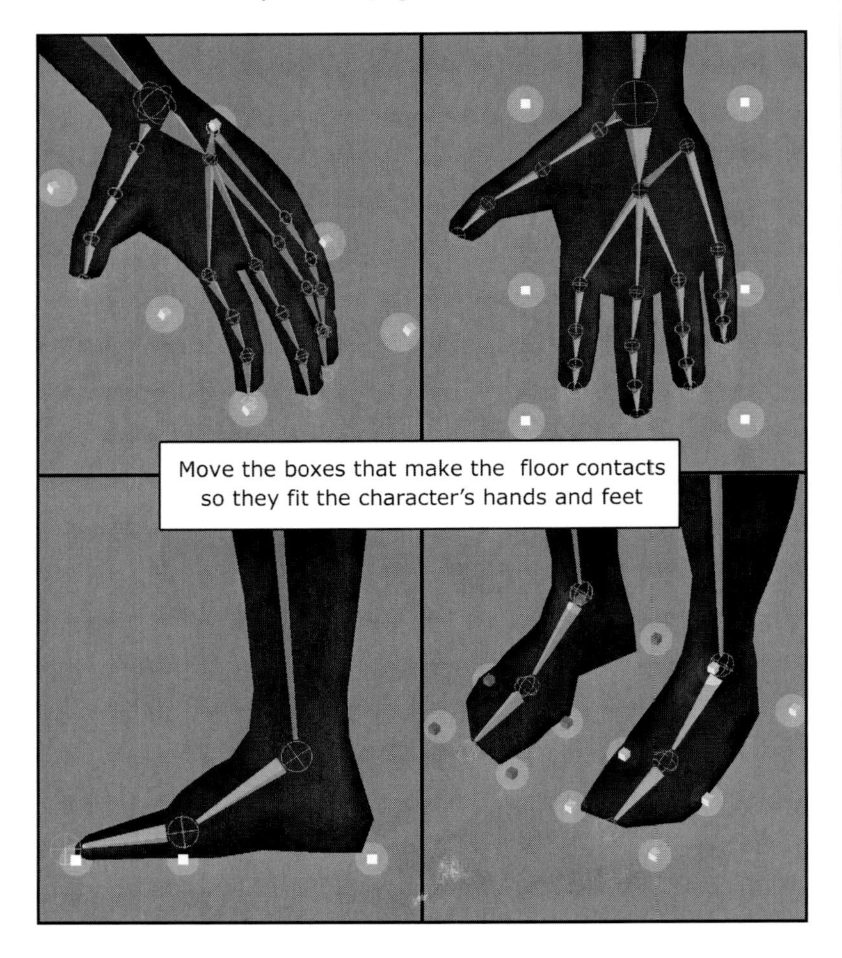

Figure 10_06.

The rig has been properly *Characterized* and has the ability to take motion capture from an *Actor* tool as well as animation from another rig. We will cover these two operations in depth in the *Retargeting* chapter (Figure 10_07).

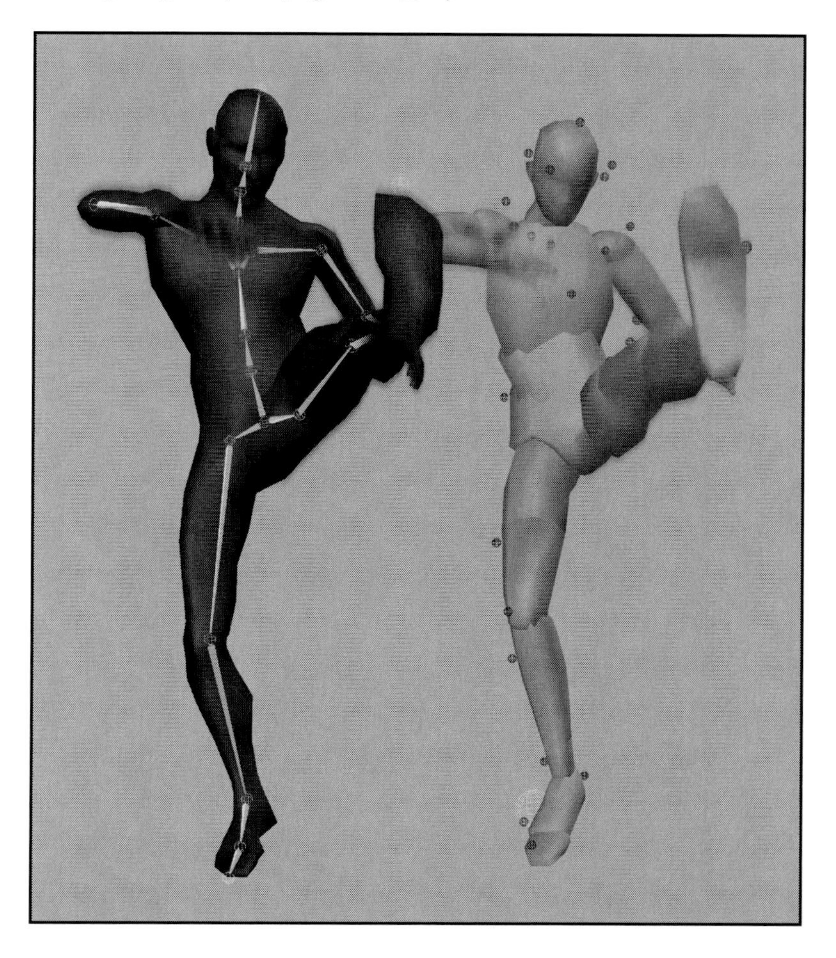

Figure 10_07.

Chapter 11
Manual Characterization

Some production pipelines do not allow for automatic *Characterization* because they require very strict naming conventions for their rigs. In this case you need to *Characterize* the rig manually (Figure 11_01).

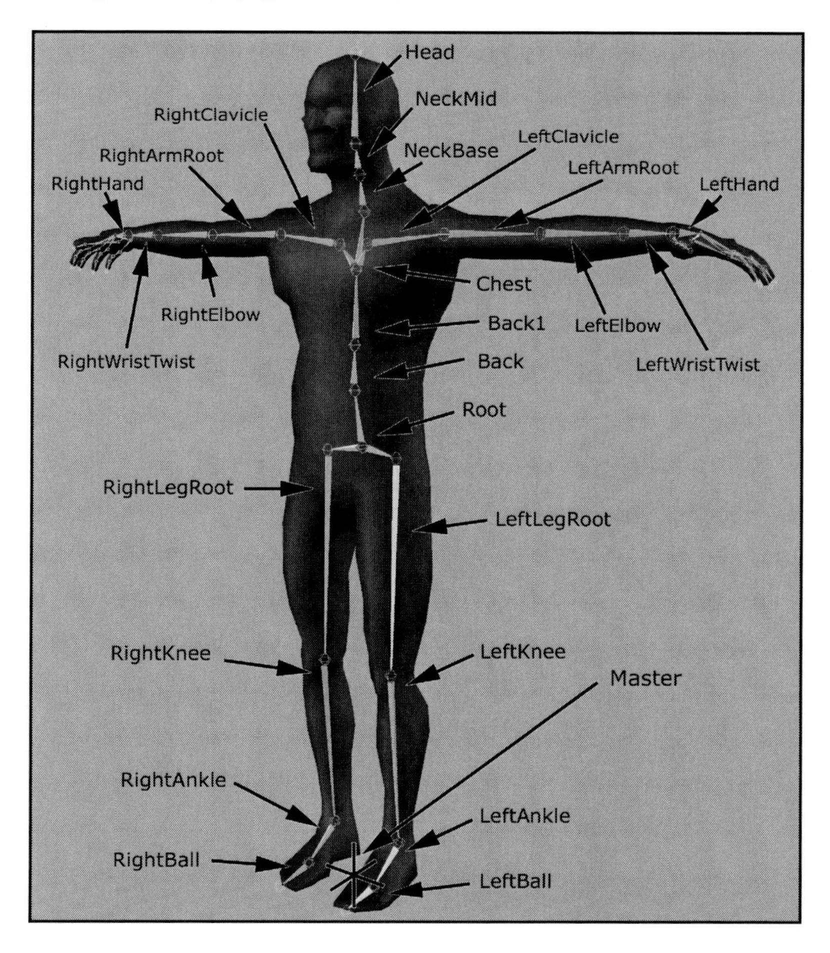

Figure 11_01.

Launch **Motion Builder** and open the "Chapter11"[8] file.

From the Asset Browser drag a *Character* icon to any joint of your hierarchy. Select *Characterize* from the pop-up option that appears. **Motion Builder** will give you an error window stating the failure of the *Characterization*. Press the *OK* button and accept the following window that lists the nodes that you are missing for the *Character Definition* (Figure 11_02).

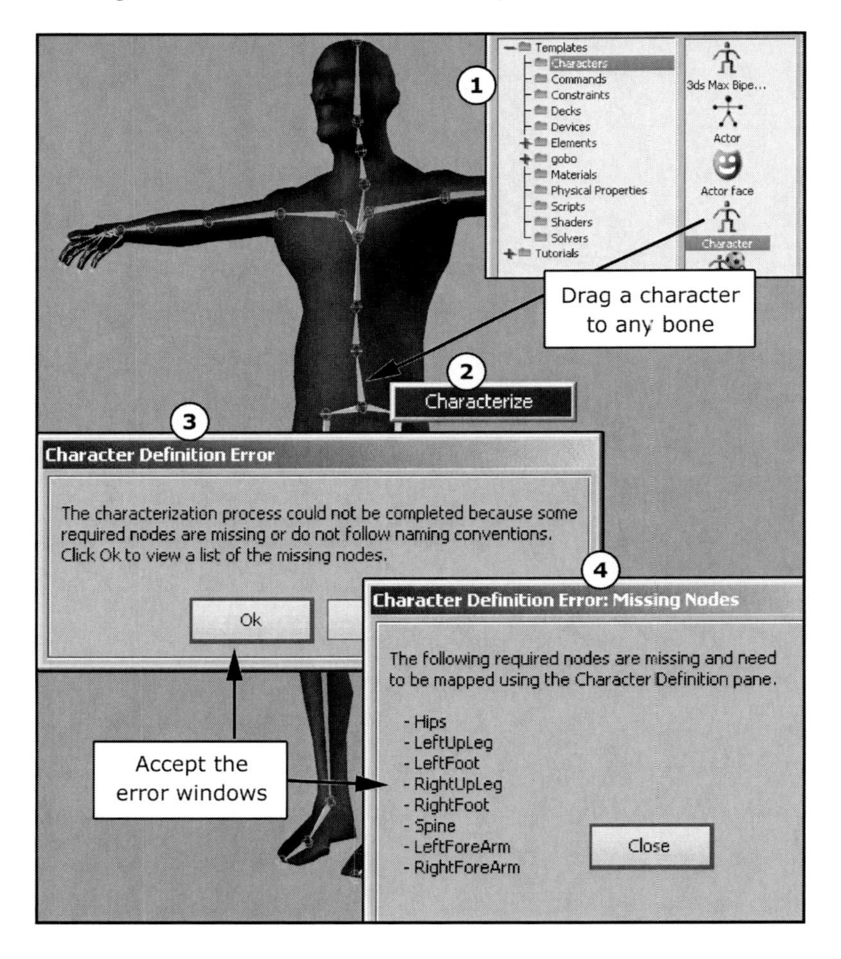

Figure 11_02.

---

[8] Download the files from www.MocapClub.com/TheMocapBook.htm

If you look under the navigator window, a new *Character* does appear in the *Navigator* window, it just lacks the proper information to be connected with your skeletal hierarchy. Start by renaming this new *Character* to "MocapGuy_Character". Double click on it to load its properties. Open the *Base* section of the *Character* tool and *Alt* + click + drag the "Root" bone to the "Hips" opening (Figure 11_03).

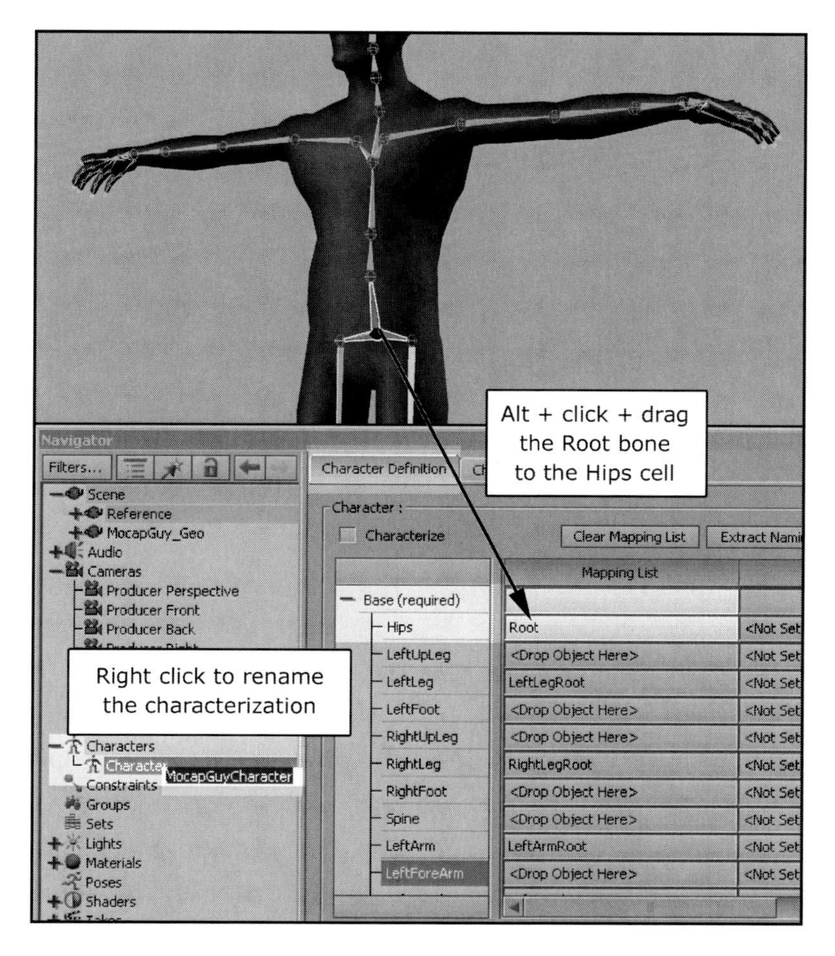

Figure 11_03

Alt + click + drag the “LeftLegRoot” to the “LeftUpLeg” Section. Inside the “LeftLeg” spot you will now see the “LeftLegRoot” repeated from the preceding slot. Basically **Motion Builder** tried to fill the *Characterization* list as best as it could, however, we will have to override a few definitions. To do this, just drag the correct node on top of the improper definition. *Alt* + click + drag the “Left Knee” to the “LeftLeg” slot (Figure 11_04).

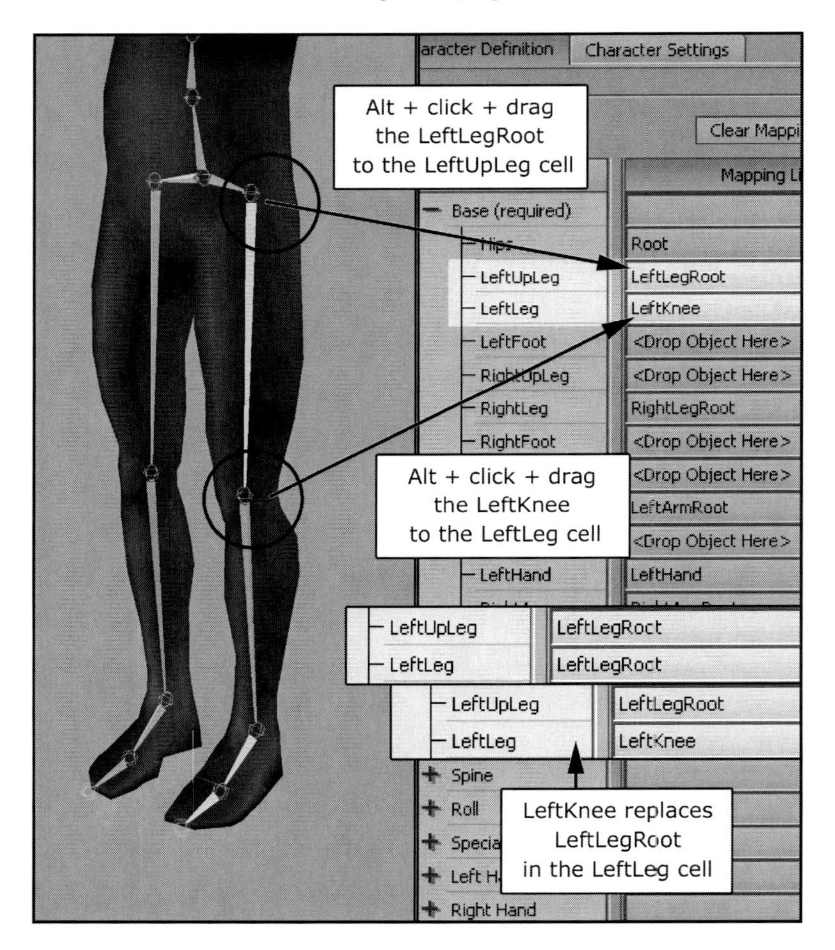

Figure 11_04.

Continue by *Alt* + clicking and dragging the "LeftAnkle" to the "Left Foot" spot. Use the guidelines we have followed so far for the left leg to populate the right leg (Figure 11_05).

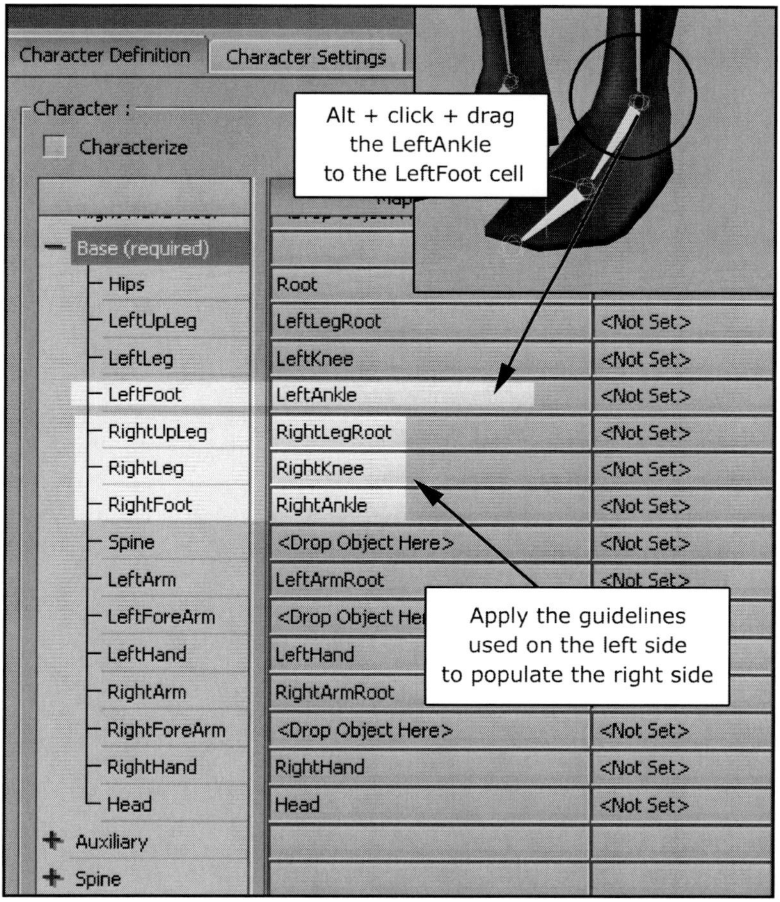

Figure 11_05.

Alt + click+ drag the "Back" joint to the "Spine" slot. The next spot to fill would be the "LeftArm", as you will be able to see from the list this specific node has already been correctly populated. Move on to the "LeftForeArm" and drag the "LeftElbow" to its definition section. Like the "LeftArm", the "LeftHand" and "Head" are already correctly defined. Use the guidelines we have followed for the left arm to populate the right arm (Figure 11_06).

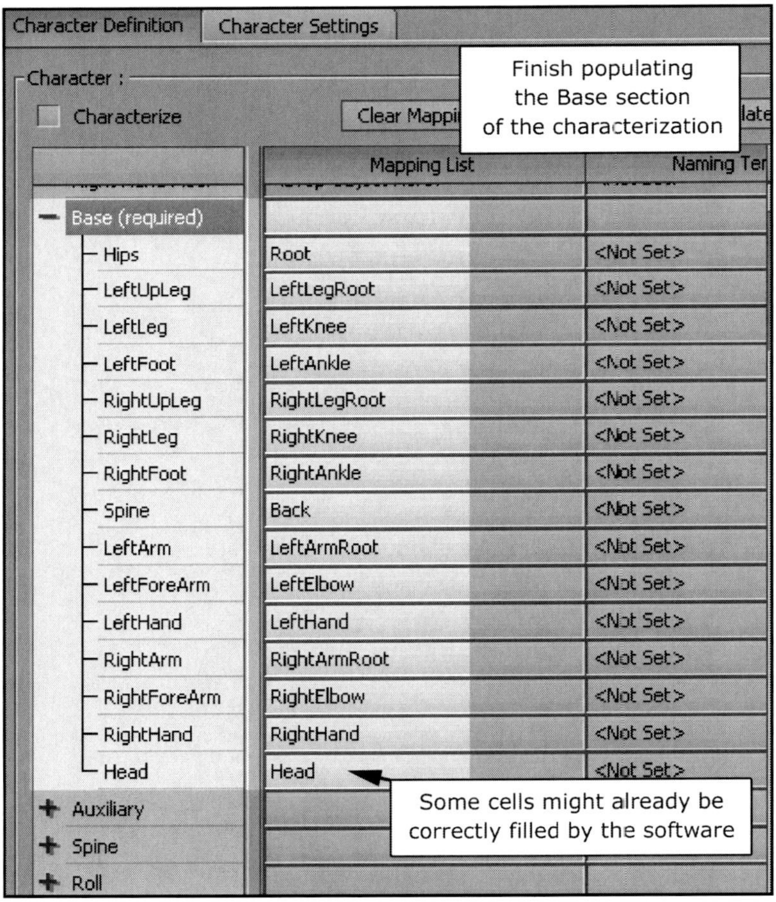

Character Definition | Character Settings

Character :

Characterize | Clear Mappi…

Finish populating the Base section of the characterization

Base (required)	Mapping List	Naming Ter…
Hips	Root	<Not Set>
LeftUpLeg	LeftLegRoot	<Not Set>
LeftLeg	LeftKnee	<Not Set>
LeftFoot	LeftAnkle	<Not Set>
RightUpLeg	RightLegRoot	<Not Set>
RightLeg	RightKnee	<Not Set>
RightFoot	RightAnkle	<Not Set>
Spine	Back	<Not Set>
LeftArm	LeftArmRoot	<Not Set>
LeftForeArm	LeftElbow	<Not Set>
LeftHand	LeftHand	<Not Set>
RightArm	RightArmRoot	<Not Set>
RightForeArm	RightElbow	<Not Set>
RightHand	RightHand	<Not Set>
Head	Head	<Not Set>

+ Auxiliary
+ Spine
+ Roll

Some cells might already be correctly filled by the software

Figure 11_06.

Now that we are done with the *Base* section of the *Character* definition, lets continue with the *Auxiliary* section. *Alt* + click + drag the following nodes to the following sections: “LeftBall” to “LeftToeBase”, “LeftClavicle” to “LeftShoulder” and “NeckBase” to “Neck”. Use the guidelines we have followed for the left side to populate the right side (Figure 11_07).

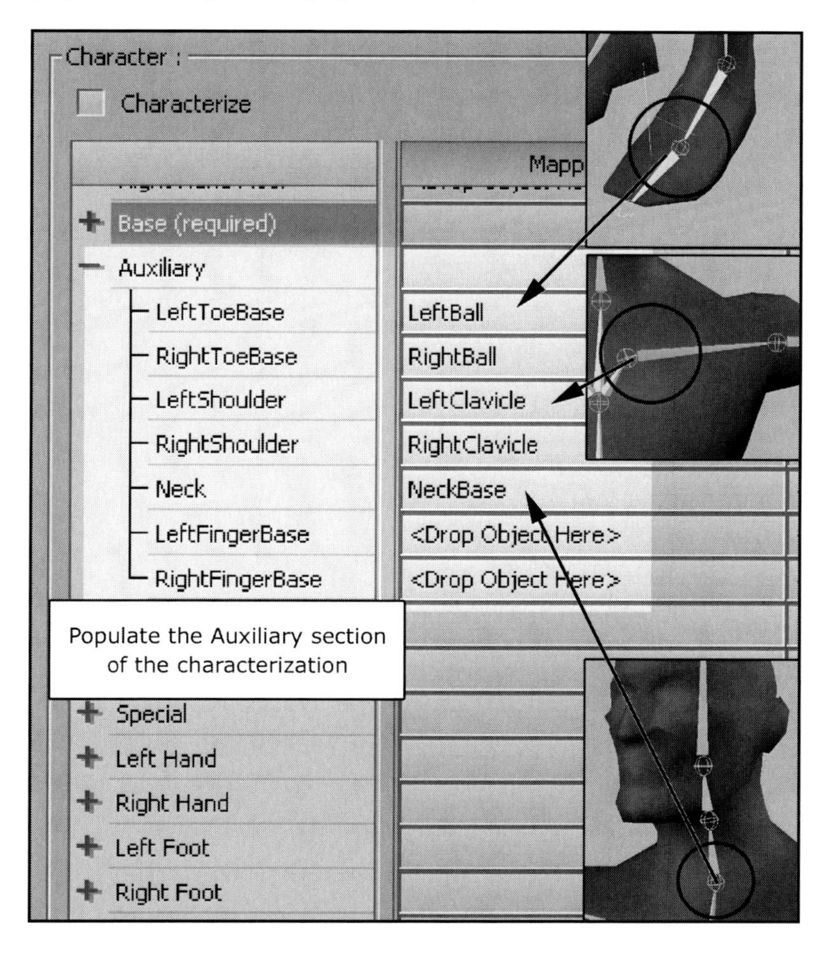

Figure 11_07.

Lets now populate the *Spine* section. *Alt* + click + drag the "Back1" bone to the "Spine1" section and the "Chest" bone to the "Spine2" section (Figure 11_08).

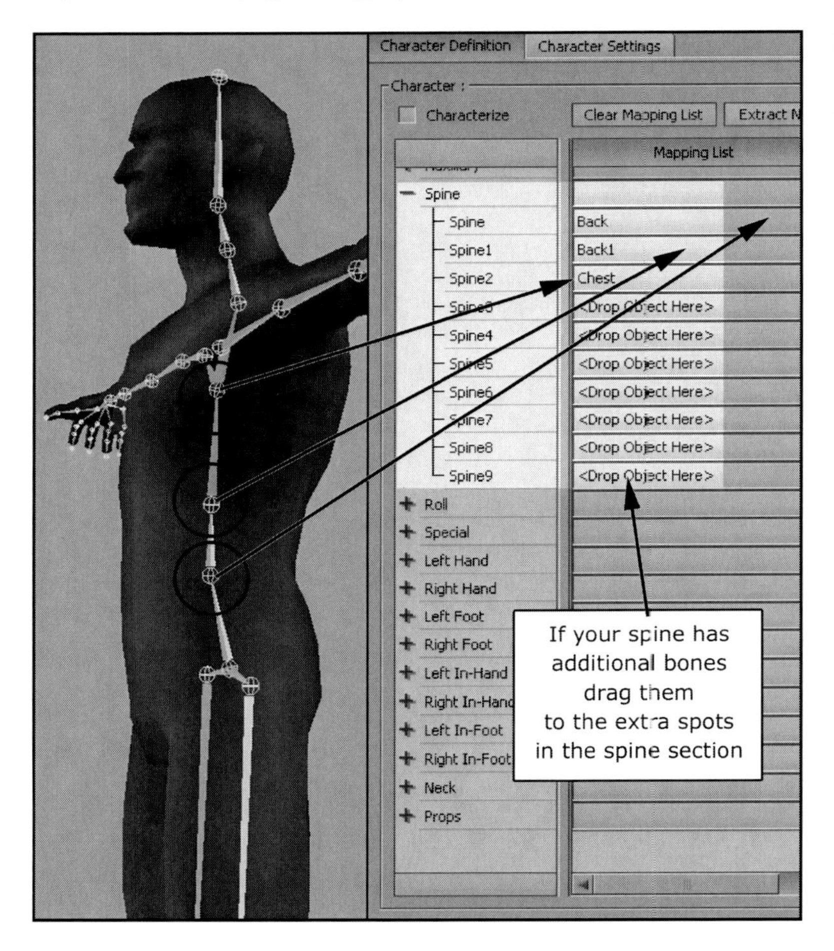

Figure 11_08.

In the *Roll* section of the *Character* definition, *Alt* + click + drag the “LeftWristTwist” bone to the “LeftForeArmRoll”. Follow the same guideline for the right side. In the *Neck* Section you will be able to see that the “NeckBase” is already correctly assigned to the “Neck” cell. *Alt* + click + drag the “NeckMid” bone to the “Neck1” spot. (Figure 11_09).

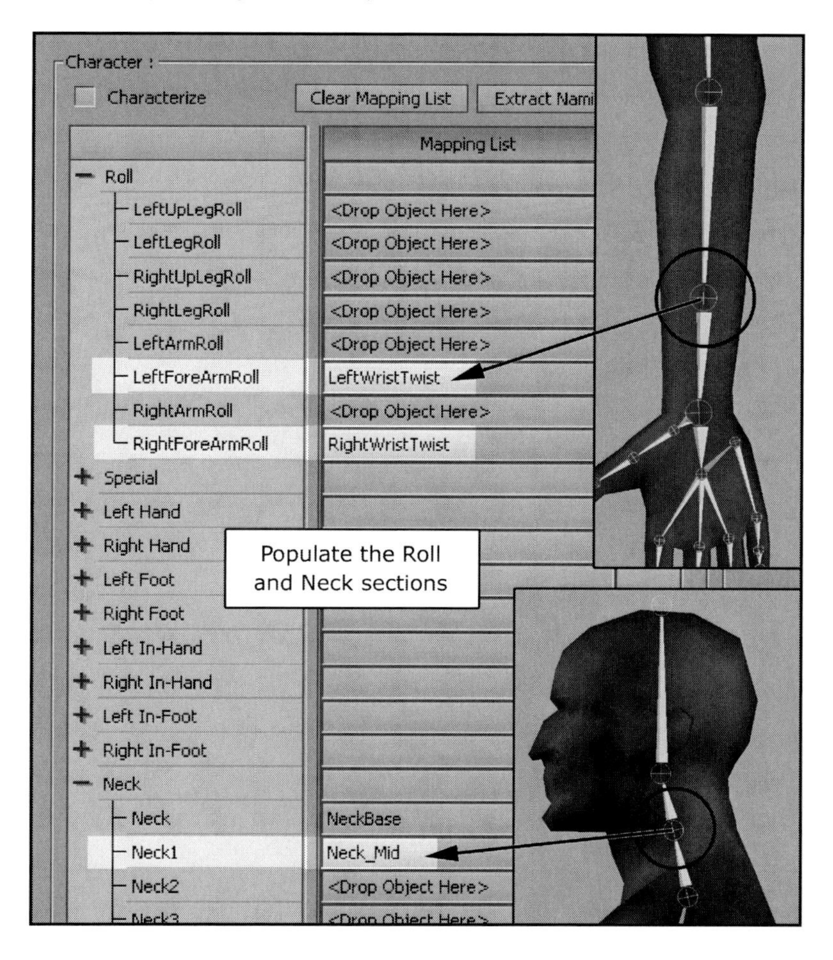

Figure 11_09.

Use figure 11_10 to guide you on how to populate the *Left Hand* section of the *Character* definition. Follow the same guidelines for the “Right Hand” (Figure 11_10, 11_11).

Populate the Left Hand section of the characterization

Left Hand		
LeftHandThumb1	LeftThumb1	Se
LeftHandThumb2	LeftThumb2	Se
LeftHandThumb3	LeftThumb3	Se
LeftHandThumb4	<Drop Object Here>	Se
LeftHandIndex1	LeftIndex1	<Not Se
LeftHandIndex2	LeftIndex2	<Not Se
LeftHandIndex3	LeftIndex3	<Not Se
LeftHandIndex4	<Drop Object Here>	<Not Se
LeftHandMiddle1	LeftMiddle1	<Not Se
LeftHandMiddle2	LeftMiddle2	<Not Se
LeftHandMiddle3	LeftMiddle3	<Not Se
LeftHandMiddle4	<Drop Object Here>	<Not Se
LeftHandRing1	LeftRing1	<Not Se
LeftHandRing2	LeftRing2	<Not Se
LeftHandRing3	LeftRing3	<Not Se
LeftHandRing4	<Drop Object Here>	<Not Se
LeftHandPinky1	LeftPinky1	<Not Se
LeftHandPinky2	LeftPinky2	<Not Se
LeftHandPinky3	LeftPinky3	<Not Se
LeftHandPinky4	<Drop Object Here>	<Not Se
LeftHandExtraFin...	<Drop Object Here>	<Not Se

Figure 11_10.

Right Hand		
RightHandThumb1	RightThumb1	
RightHandThumb2	RightThumb2	
RightHandThumb3	RightThumb3	
RightHandThumb4	<Drop Object Here>	
RightHandIndex1	RightIndex1	<Not S
RightHandIndex2	RightIndex2	<Not S
RightHandIndex3	RightIndex3	<Not S
RightHandIndex4	<Drop Object Here>	<Not S
RightHandMiddle1	RightMiddle1	<Not S
RightHandMiddle2	RightMiddle2	<Not S
RightHandMiddle3	RightMiddle3	<Not S
RightHandMiddle4	<Drop Object Here>	<Not S
RightHandRing1	RightRing1	<Not S
RightHandRing2	RightRing2	<Not S
RightHandRing3	RightRing3	<Not S
RightHandRing4	<Drop Object Here>	<Not S
RightHandPinky1	RightPinky1	<Not S
RightHandPinky2	RightPinky2	<Not S
RightHandPinky3	RightPinky3	<Not S
RightHandPinky4	<Drop Object Here>	<Not S
RightHandExtraFin...	<Drop Object Here>	<Not S

Populate the Right Hand section of the characterization

Figure 11_11.

Finally, *Alt* + click + drag the "MocapGuy_Master" node to the *Reference* section. Check on the *Characterize* box and select *Biped* from the pop-up window. The rig has been properly *Characterized*, even though its bones used a naming convention unknown to **Motion Builder**. The rig can now take motion capture from an *Actor* tool as well as animation from another rig (Figure 11_12).

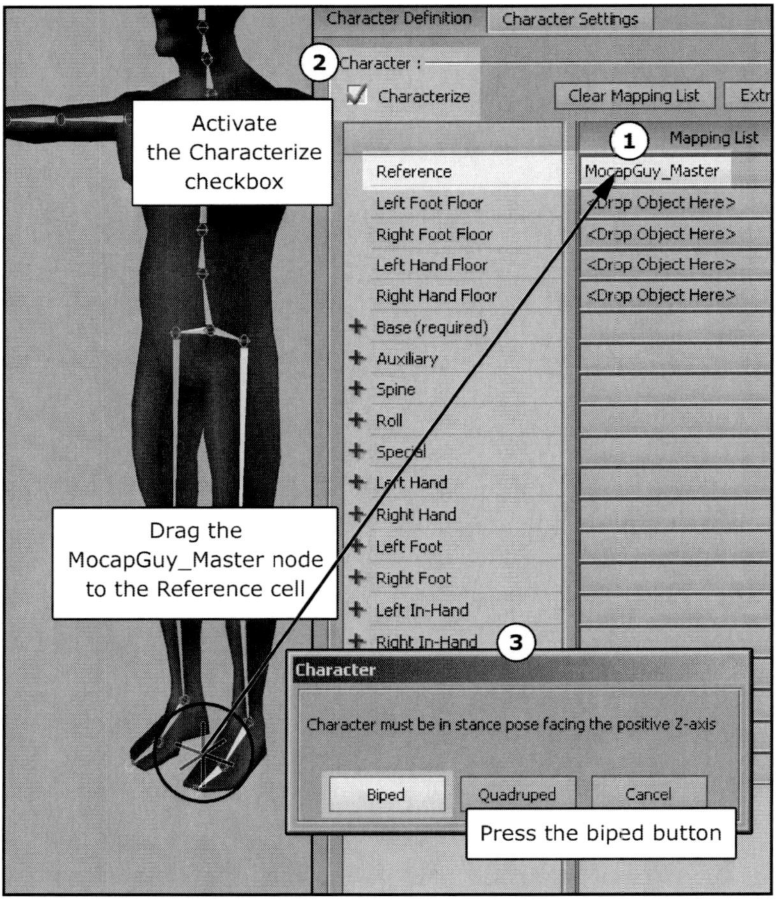

Figure 11_12.

Chapter 12
Retargeting

Retargeting is the process of transferring animation from one skeletal hierarchy to another. This process is extremely important in motion capture because there are very few cases in which the physical performer completely matches the digital one (Figure 12_01).

Figure 12_01.

Launch **Motion Builder** and open the "Chapter12" file.
This file contains an already *Characterized* MocapGuy, very similar to the one we ended up with at the end of Chapter 10.
Go to the *File* menu and select the *Merge* option. Navigate to the place where you downloaded the "Chapter12"[9] folder from the mocapclub.com site. Select the "SolvedMotions" file and press the Open button (Figure 12_02).

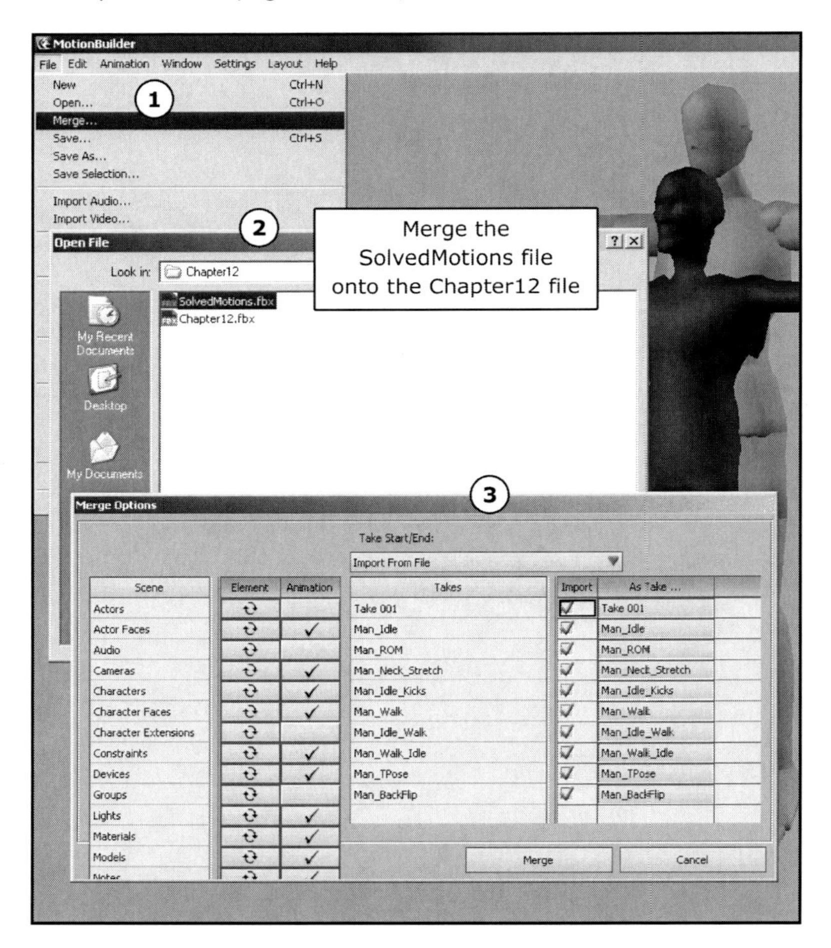

Figure 12_02.

---

[9] Download the files from www.MocapClub.com/TheMocapBook.htm

In the *Character Controls* window, make sure that MocapGuy is selected in the *Current Character* section. Go to the *Edit* drop-down and select the *Actor* from the list that appears. MocapGuy will now follow the motions of the *Actor* in every take (Figure 12_03).

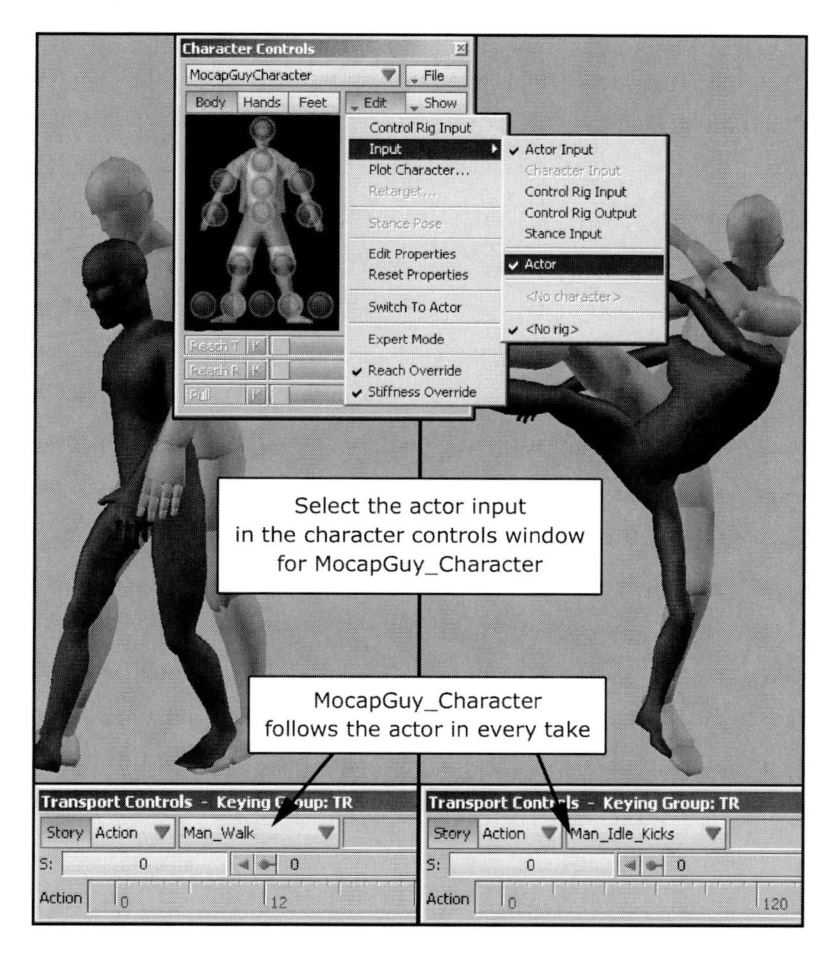

Figure 12_03.

At the moment it is a little difficult to evaluate how the motions are being transferred between the *Actor* and “MocapGuy”. To correct this, select “MocapGuy’s” *Reference* cell in the *Character Controls* window and scale it to 2.85. The *Actor* and “MocapGuy” are now roughly the same size and we can see the motions are transferring very well in every take. There are some minor discrepancies here and there, but we will be able to correct those with the options in the *Character Settings* tab and later with the use of layered animation. For more information on layered animation see chapter 14 (Figure 12_04).

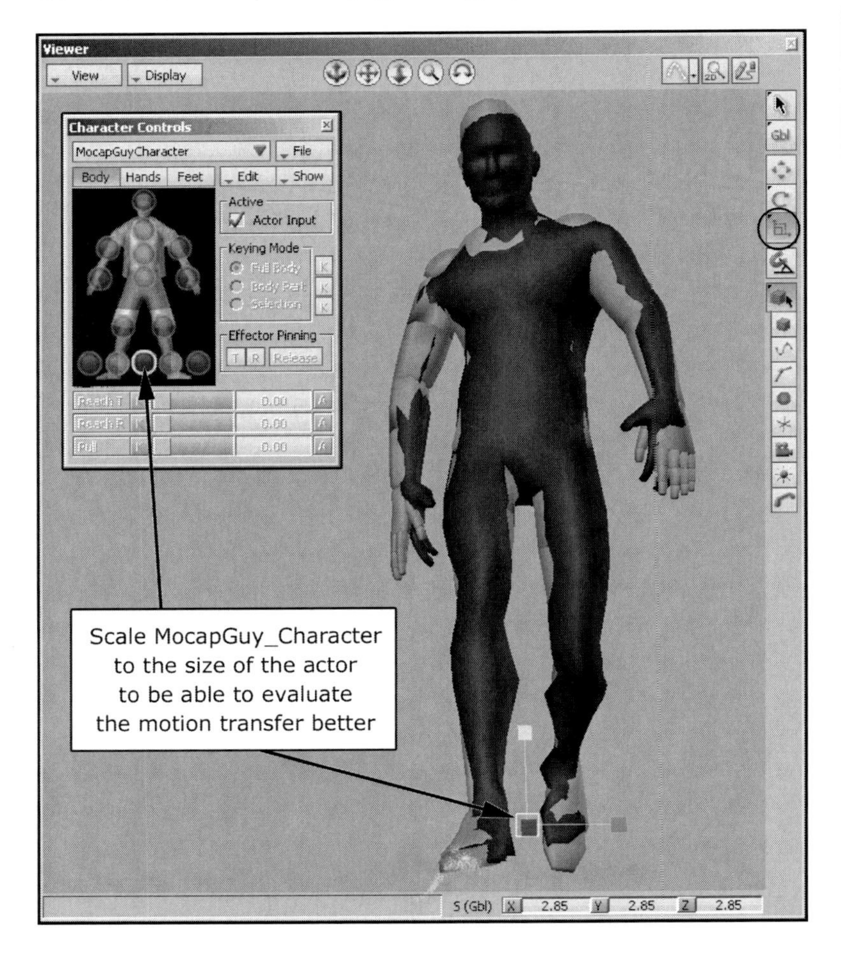

Figure 12_04.

From this point on we can hide the *Actor* to reduce visual clutter in the scene. Go to the *Character Controls* window select the *Actor* for the *Current Character* section and uncheck *Actor All* from the *Show* drop-down (Figure 12_05).

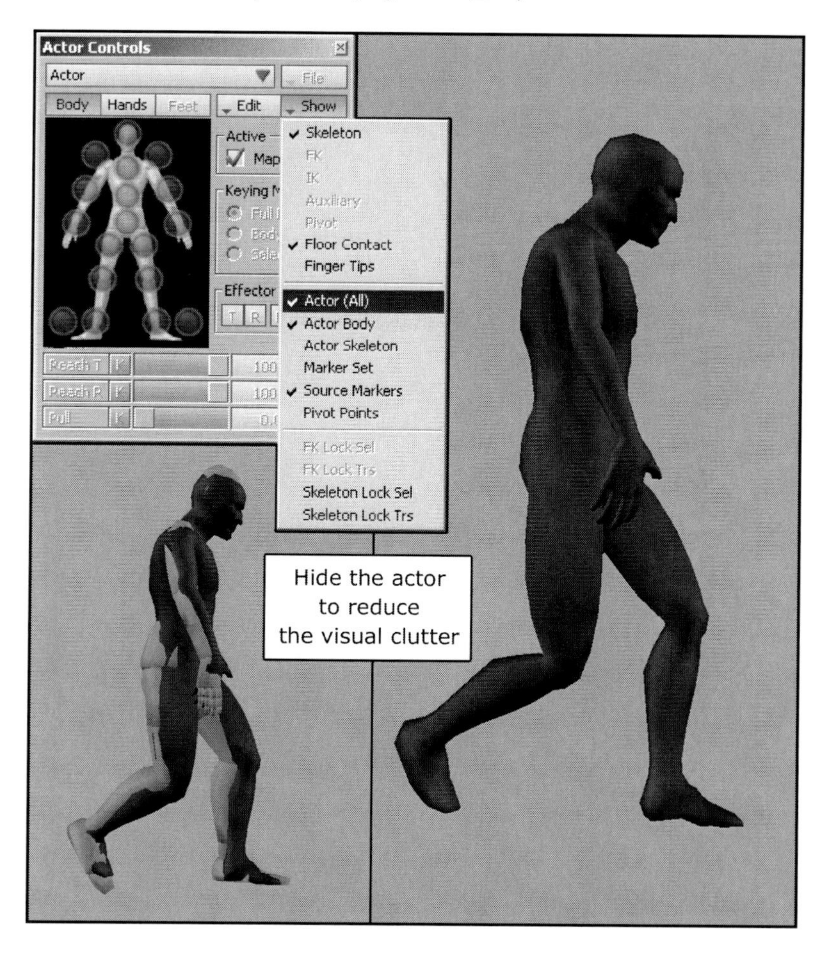

Figure 12_05.

We are now going to concentrate on polishing the transfer of the "Man_BackFlip" motion by using the *Character Settings* options. Select "Man_BackFlip" under the *Takes* drop-down. Even though most of the motion is being transferred properly, the palms do not make adequate contact with the ground in the middle of the flip. There is also an issue with the right hand contact with the ground at the end of the flip (Figure 12_06).

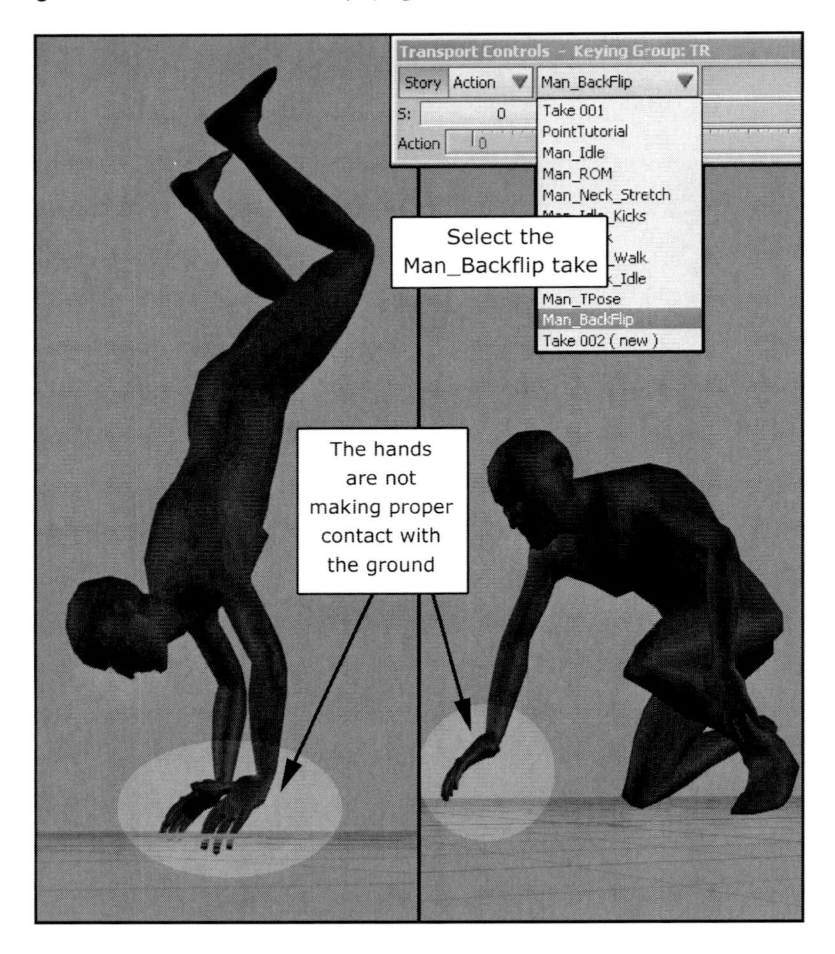

Figure 12_06.

Make sure that "MocapGuyCharacter" is selected in the *Current Character* section of the *Character Controls* window as well as the *Navigator*. Inside the *Character Settings* open the *Retargeting*, then the *Reach* and finally the *Left Arm*, *Right Arm* and *Spine* drop-downs (Figure 12_07).

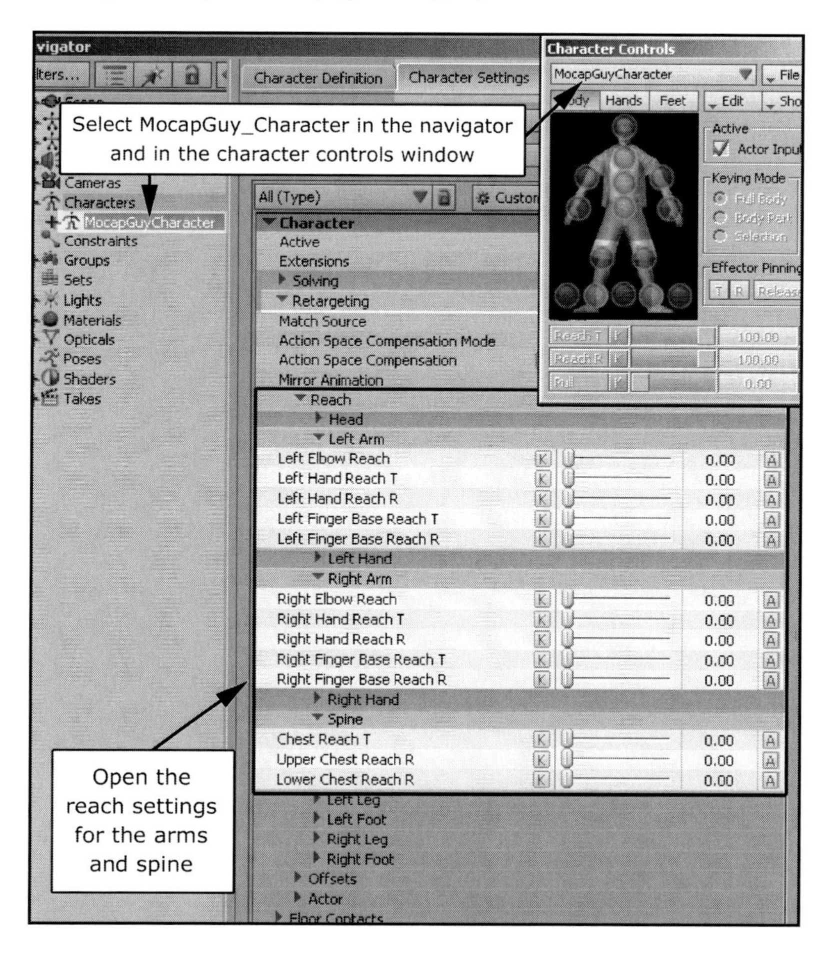

Figure 12_07.

Inside the *Left Arm* section turn the *Left Elbow Reach*, *Left Hand Reach T* and *Left Hand Reach R* to a 100. Do the same for the *Right Arm*. Also turn the Lower Chest Reach to a 100 inside the *Spine* section.
When the default retargeting between the actor and the character is made, **Motion Builder** creates a few offsets between the two hierarchies based on the differences in their proportions. Most times these offsets are desirable; such is the case of every other *Take* in this file. However, in the case of the "Man_Backflip" motion, activating the *Reach* values for the arms forces these body parts to try and match the position of the *Actor's* corresponding parts (Figure 12_08).

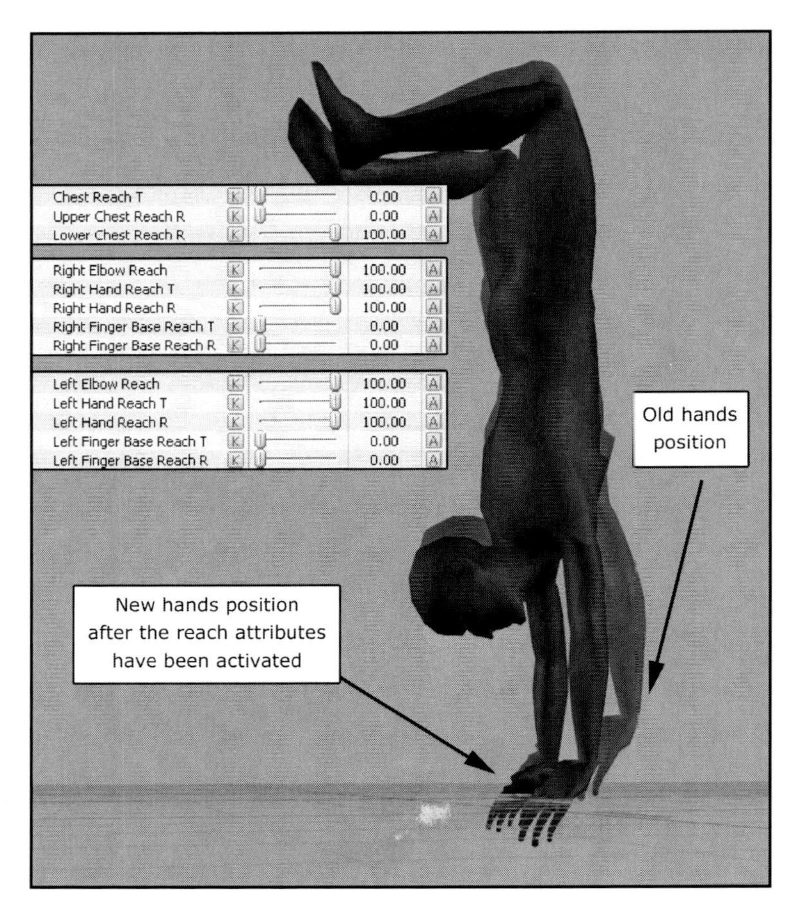

Figure 12_08.

The motion transfer will look a little better if the *Reach* values for the arms and the spine would not stay on for the entire motion. Fortunately these values can be animated using the *"K"* button besides the attribute slider. Go to frame 302 and key the arms and spine *Reach* with a value of 0. At frame 311 key the arms and spine *Reach* at a value of 100. At frame 314 key the arms and spine at a 100. At frame 319 key the arms and spine at 0. At frame 325 key the arms and spine at a 100. At frame 337 key the arms and spine at a 100. At frame 365 key the arms and spine at 0 (Figure 12_09).

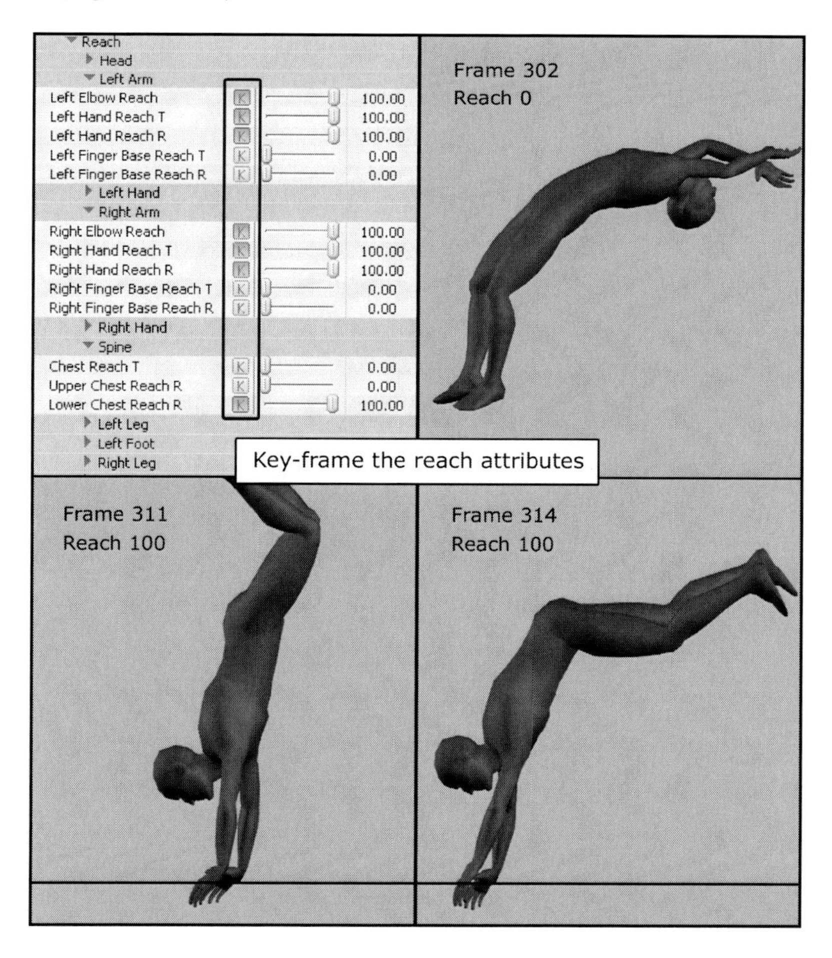

Figure 12_09.

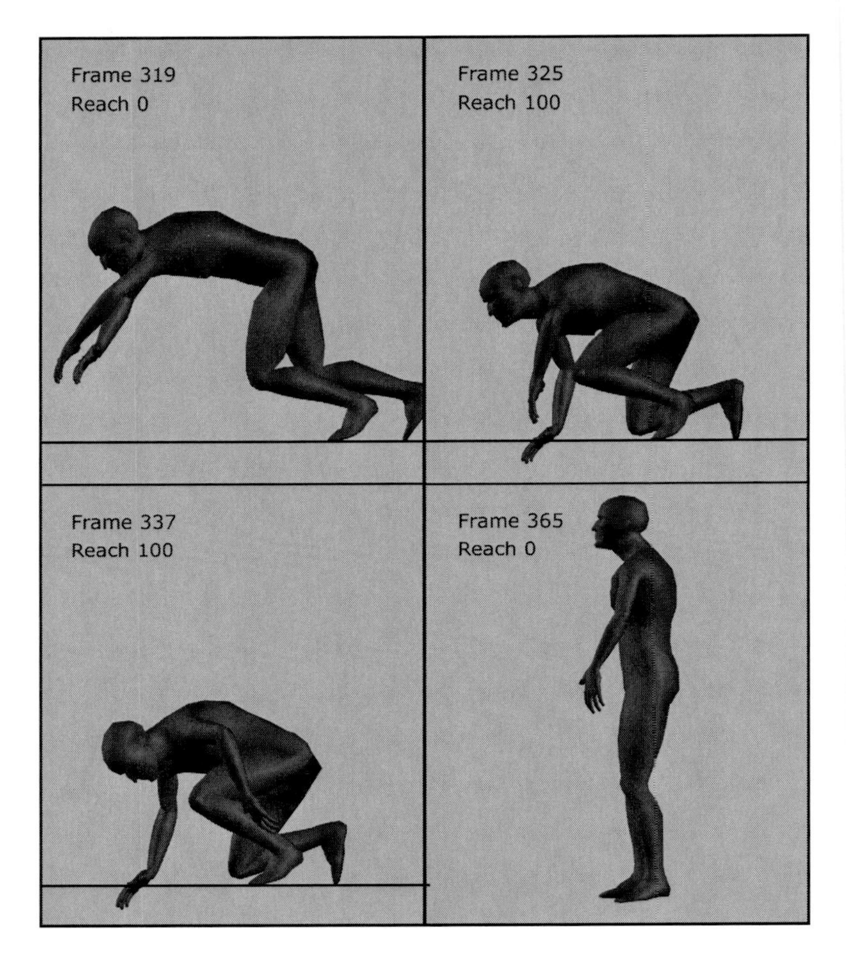
Frame 319
Reach 0
Frame 325
Reach 100
Frame 337
Reach 100
Frame 365
Reach 0

Figure 12_10.

Like it was mentioned in Chapter10, the *Floor Contacts* help the rig recognize the grid or as the ground and they prevent the hands and the feet from going bellow it.
Activate all the boxes inside the *Floor Contacts* drop-down. Doing this prevents the hands and feet from going through the ground (Figure 12_11).

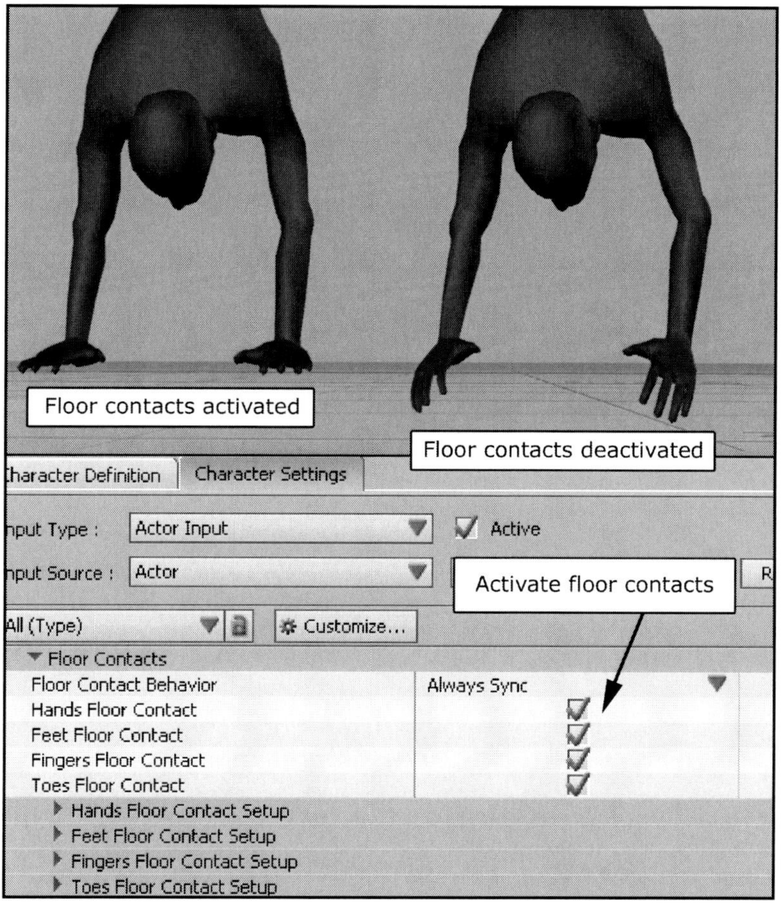

Figure 12_11.

The transfer between the actor and "MocapGuyCharacter" is looking much better. However, what we have right now is just a live connection between these two hierarchies. We need to plot the animation to the *Character* in order to have animation curves in our rig instead of a data stream if we later want to do hand key changes to the motion.
Making sure that "MocapGuyCharacter" is selected in the *Character Controls* window, open the *Edit* drop-down and select the *Plot Character* option. Press the *Control Rig* button in the *Character* popup window that appears and select *FK/IK* from the *Create Control Rig* window (Figure 12_12).

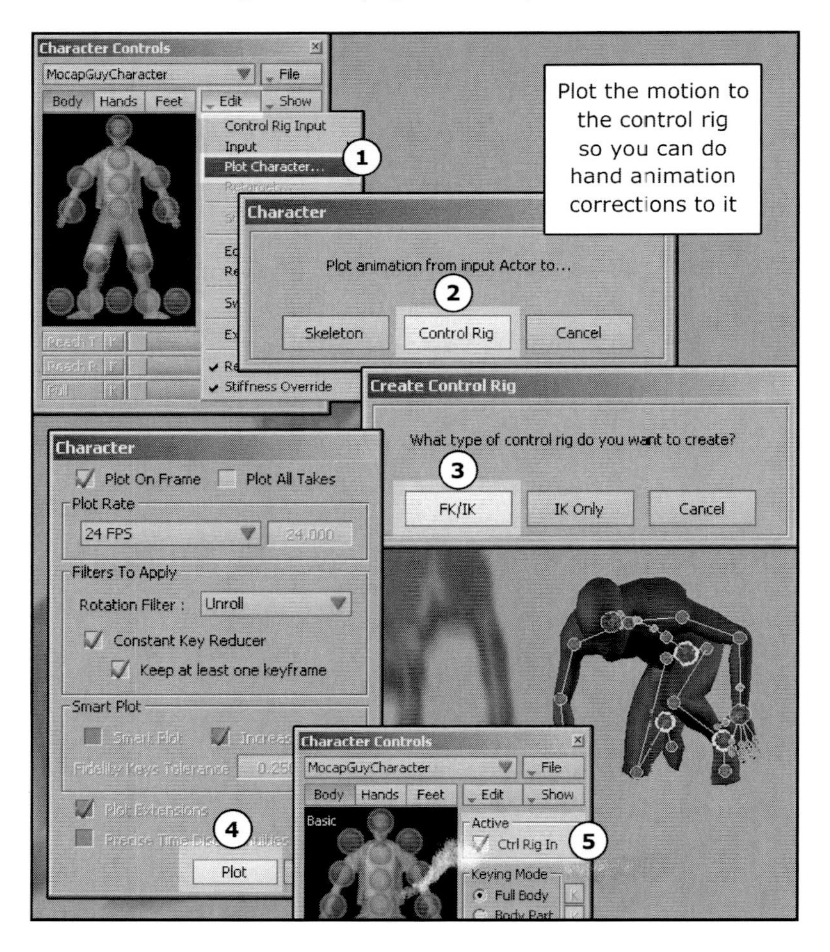

Figure 12_12.

The “Man_BackFlip” motion has accurately been *Retargeted* to the “MocapGuyCharacter”. Since the motion has been plotted to a *Control Rig* the motion can be further polished by the use of layered animation. The next chapter will expand upon the *Control Rig* subject.

Chapter 13
Understanding The Control Rig

The *Control Rig* is one of **Motion Builder's** most powerful animation tools. With the click of a button you can get a set of highly advanced *FK* and *IK* controllers that work seamlessly in with each other. This allows the user to quickly and intuitively create animation from the ground up, reuse pre-made animation or enhance existing motions.
In the previous chapter we created a *Control Rig* at the end of the retargeting process. However, it is important to note that the *Control Rig* can be created right after the *Characterization* process (Figure 13_01).

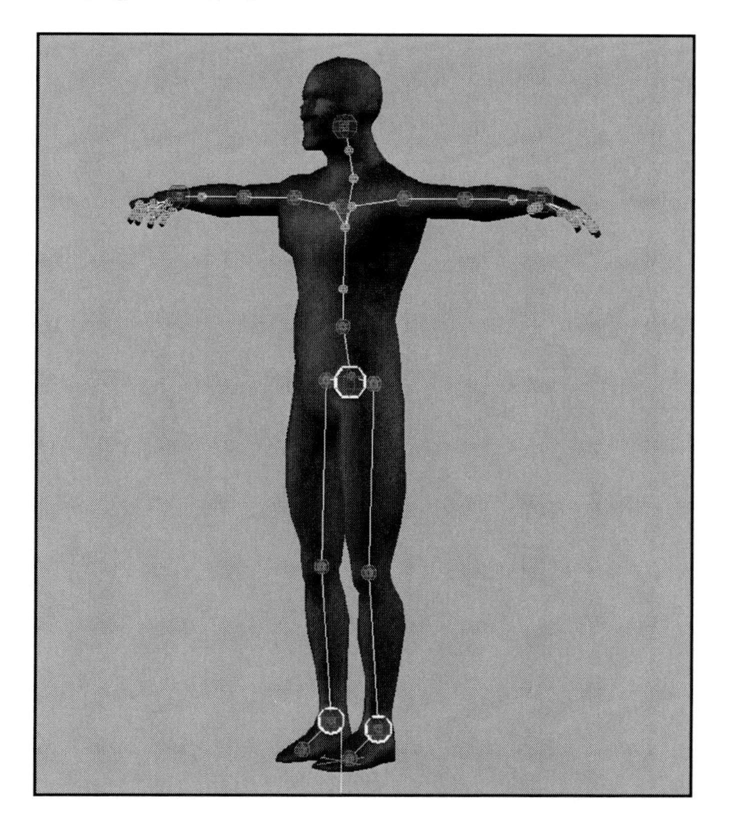

Figure 13_01.

Launch **Motion Builder** and open the "Chapter13"[10] file.

Double click on "MocapGuyCharacter" inside the *Navigator* window. Select the *Character Definition* tab and press the *Create* button below the *Control Rig* section. Select *FK/IK* from the *Create Control Rig* window (Figure 13_02).

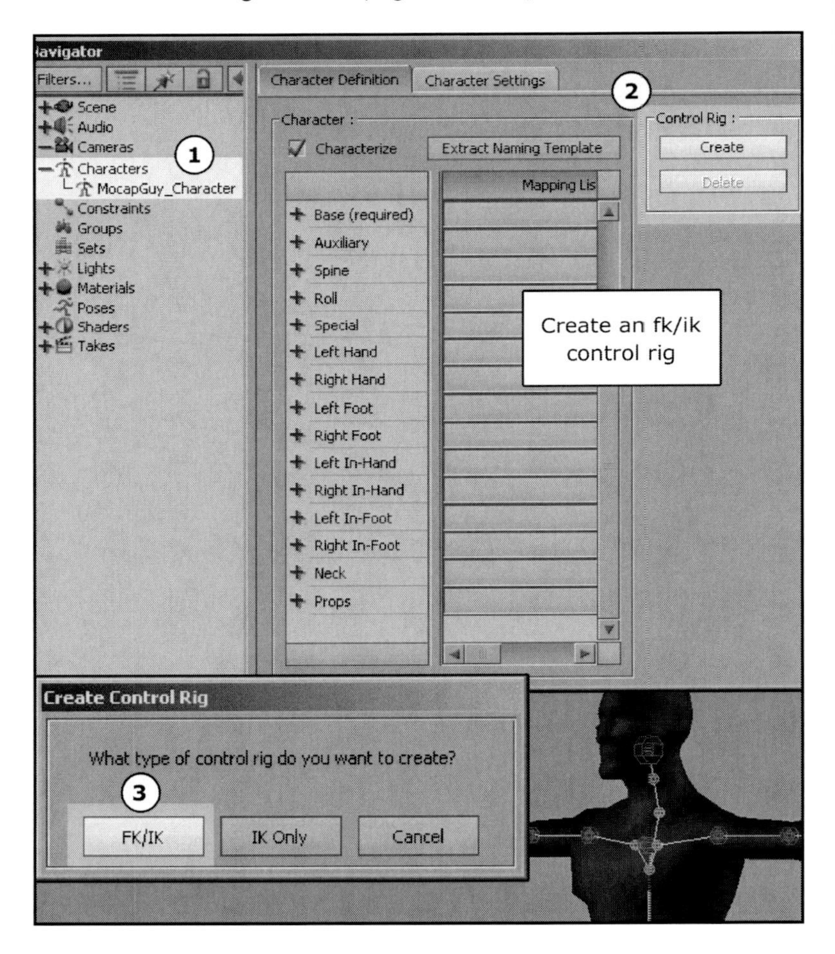

Figure 13_02.

---

[10] Download the files from www.MocapClub.com/TheMocapBook.htm

An *IK/FK Control Rig* has successful being created. This means that the rig is composed out of two sets of controllers that are driving your original skeletal hierarchy. There is the skeleton that you created and bound your geometry to, a *Forwards Kinematics* skeleton and the *Inverse Kinematics* set of controllers (Figure 13_03).

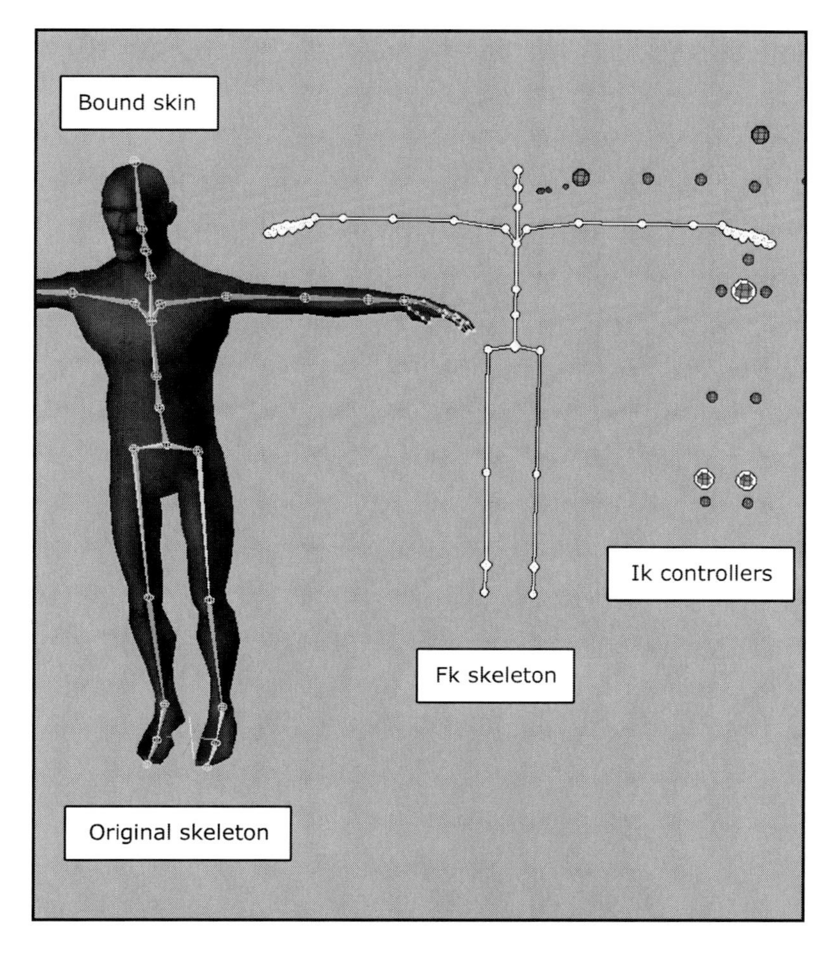

Figure 13_03.

Like most *IK/FK* setups on the planet there are switches in the *Control Rig* that allow you to decide which one of the rig elements (*IK* or *FK*) is going to be driving your bound skeleton. The *Reach Translation* and *Reach Rotation* are the attributes that switch between controllers (*IK* and *FK*) for any selected element. You can find the *Reach T* and *Reach R* attributes for the *Control Rig effectors* in the *Character Controls* window. (Figure 13_04).

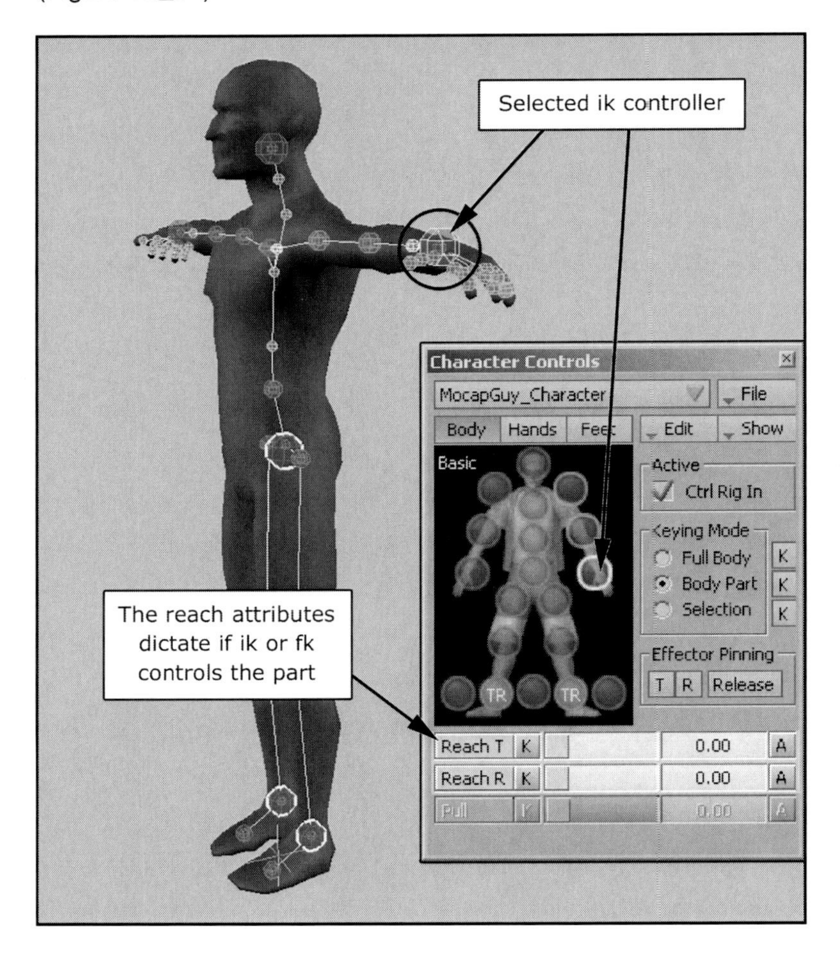

Figure 13_04.

When the *Reach T* and *Reach R* of an *Effector* are at 100 it means the section is being controlled by *Inverse Kinematics*. If the *Reach* values are at 0 the section is being controlled by *Forward Kinematics* (Figure 13_05).

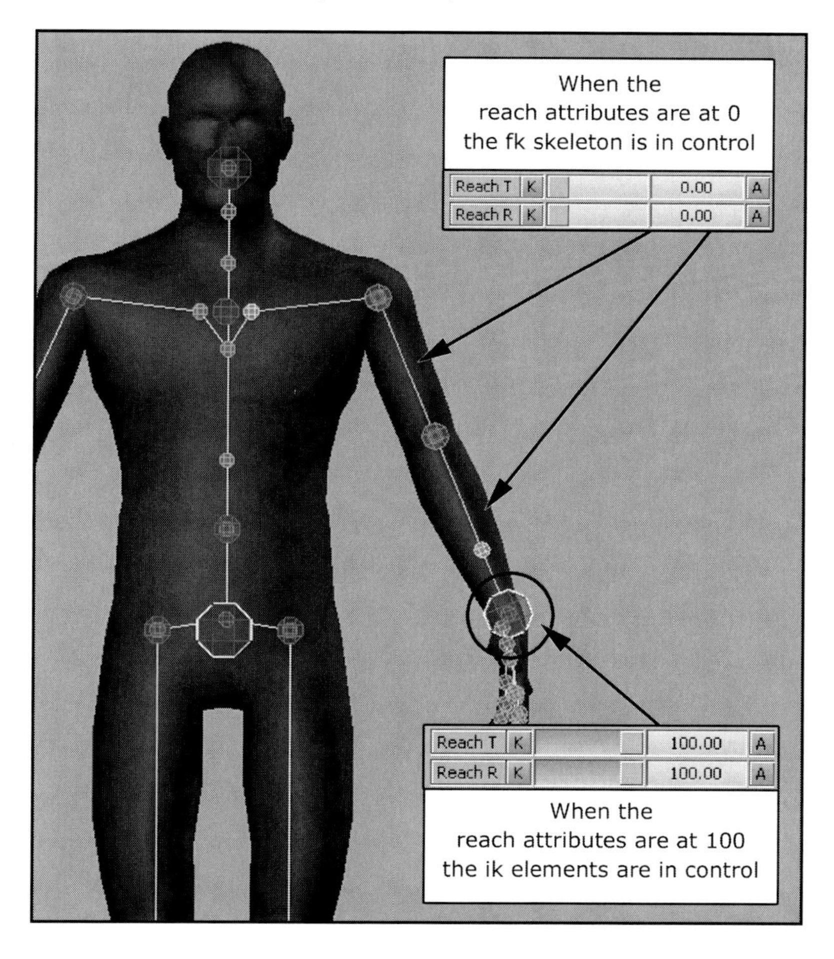

Figure 13_05.

There are three keying modes in **Motion Builder** which are, *Full Body*, *Body Part* and *Selection*. Depending on which-one you have selected you might be keying both set of controllers, a section of the controllers or one single element (Figure 13_06).

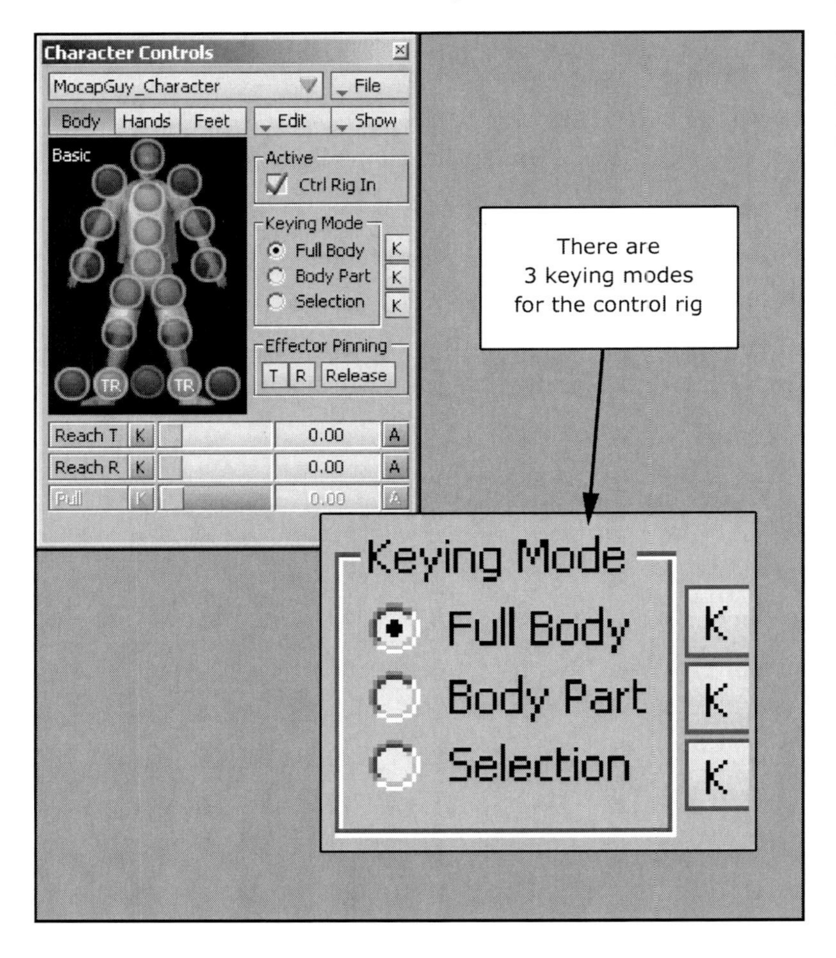

Figure 13_06.

When *Full Body* mode is selected you are keying every element in the entire *Control Rig* (both *IK* and *FK*). The keys from *Full Body* mode display red in the *Action Timeline*.
When *Body Part* is selected a section of controllers gets keyed on the *FK* level as well as the *IK* level. Keys from *Body Part* display green in the *Action Timeline* (Figure 13_07).

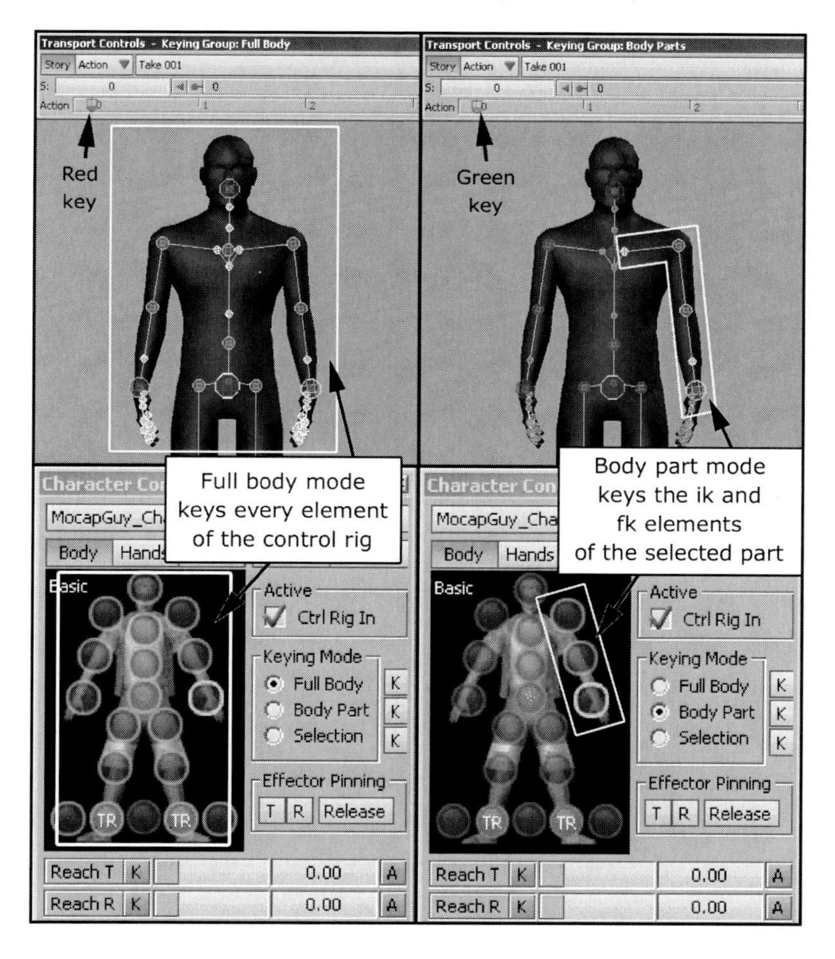

Figure 13_07.

There are seven sections of the basic *Control Rig* in *Body Part* mode. These are *Hips*, *Spine*, *Head*, *Left Lower Extremity*, *Right Lower Extremity*, *Left Upper Extremity*, *Right Upper Extremity*, *Left Fingers* and *Right Fingers* (Figure 13_08).

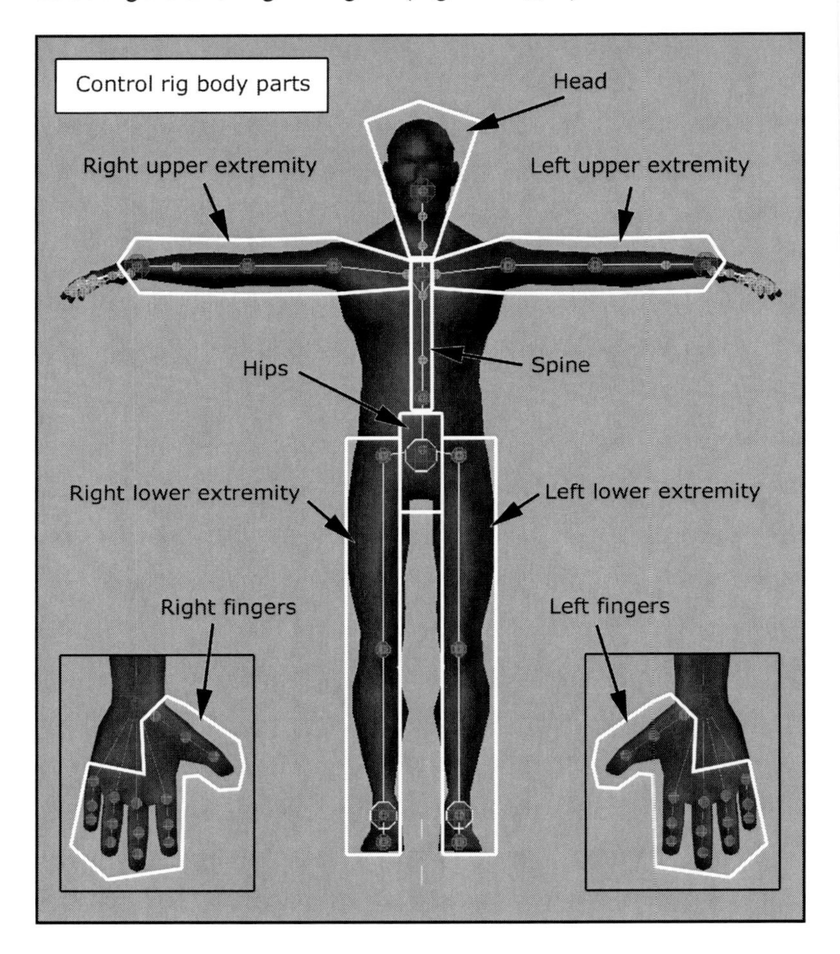

Figure 13_08.

The last keying mode is *Selection*. When using this mode, any key that you set affects only the element selected and nothing else. Keys set on *Selection* mode display transparent on the *Action Timeline* (Figure 13_09).

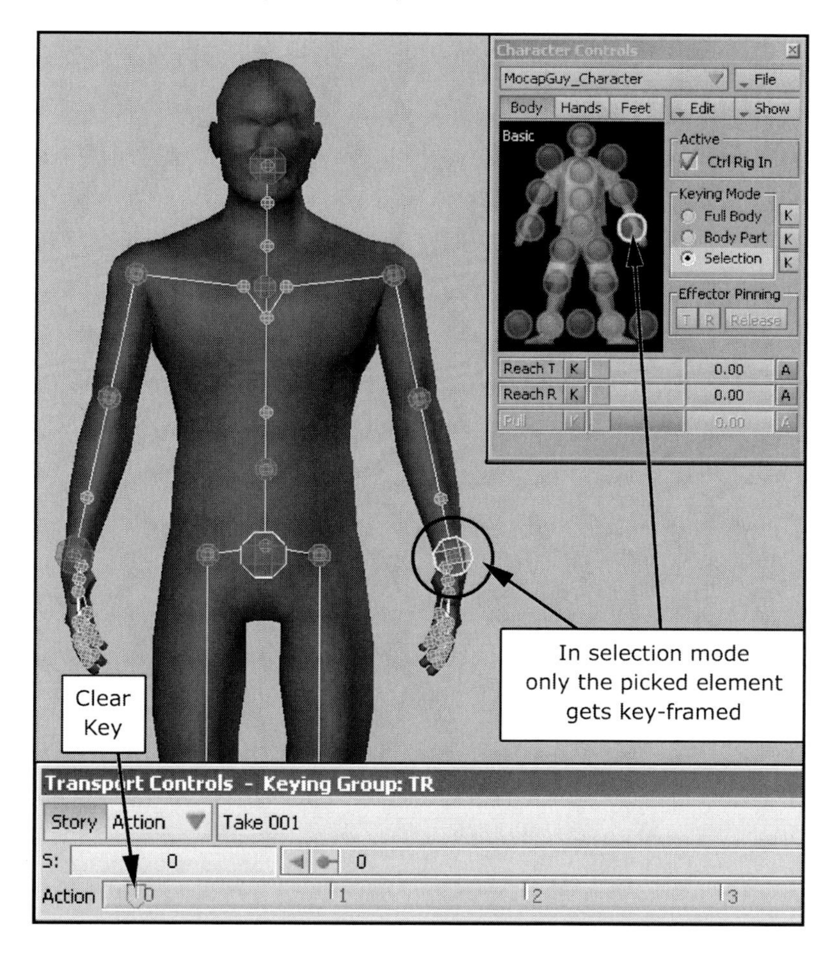

Figure 13_09.

When keying on *Full Body* mode and *Body Part* mode, **Motion Builder** keeps the *Control Rig* synchronized. This means that *FK* and *IK* controllers move together so you can blend seamlessly between them.
However when you key on *Selection* mode the *Control Rig* can get un-synchronized quickly. To illustrate this point make sure that *Body Part* is selected in the *Character Controls* window. Select the *Left Arm FK* controller and key it at frame 0. Go to frame 24 rotate the controller 70 degrees on the local Y axis and key frame (Figure 13_10).

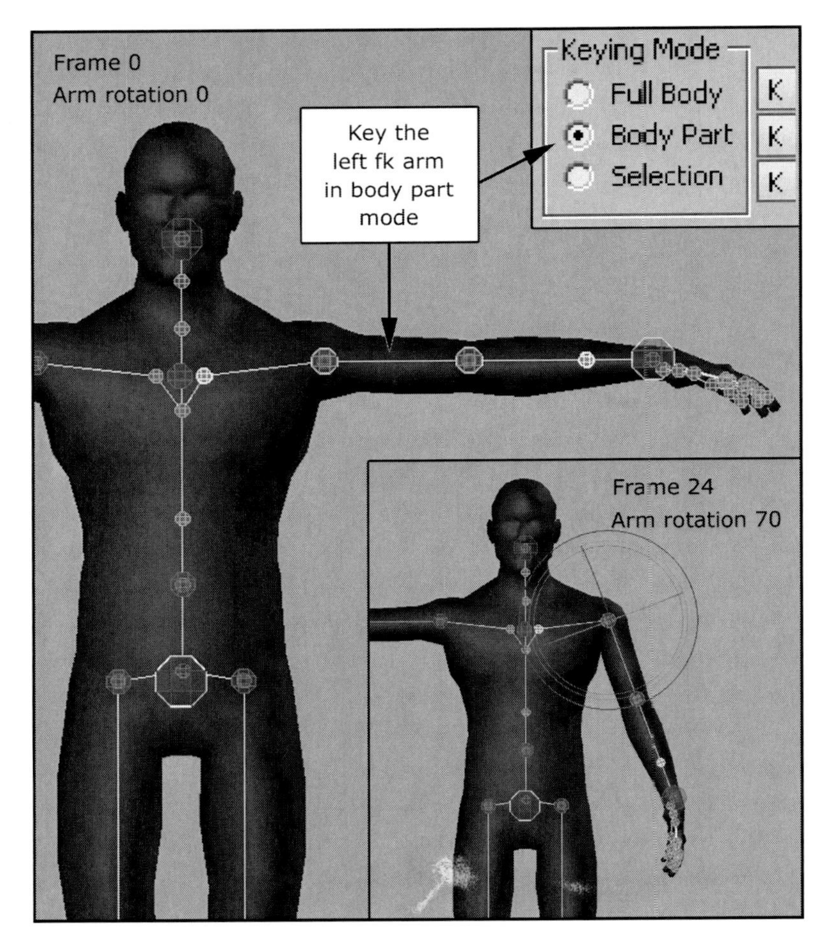

Figure (13_10).

Switch the *Keying Mode* to *Selection*. Go to frame 0, select the *Right Arm FK* controller and key it. Go to frame 24 rotate the controller 70 degrees on the local Y axis and key frame. When you play the animation you will see both arms go down at the same time (Figure 13_11).

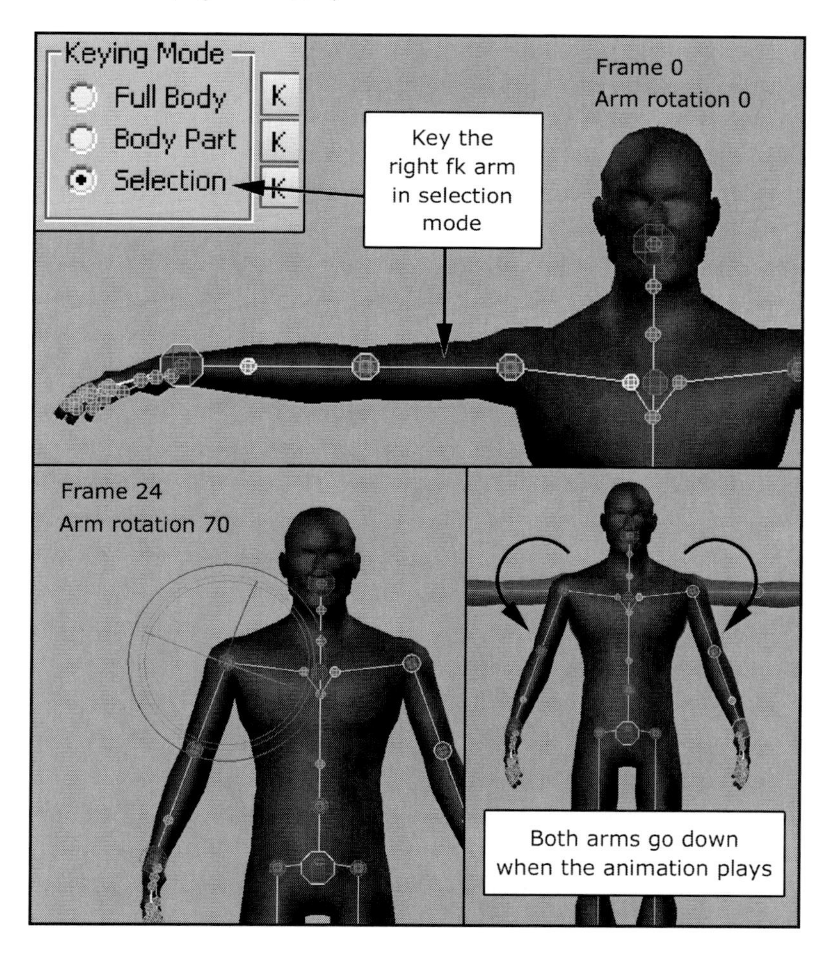

Figure 13_11.

Go to frame 24, select the *Left Wrist Effector* and the *Right Wrist Effector*; activate their *Reach T* and *Reach R* by setting them at a value of 100. You will see the right arm go back to the original T position while the left arm stays in place. When you play the animation the left arm still moves down but the right stays up (Figure 13_12).

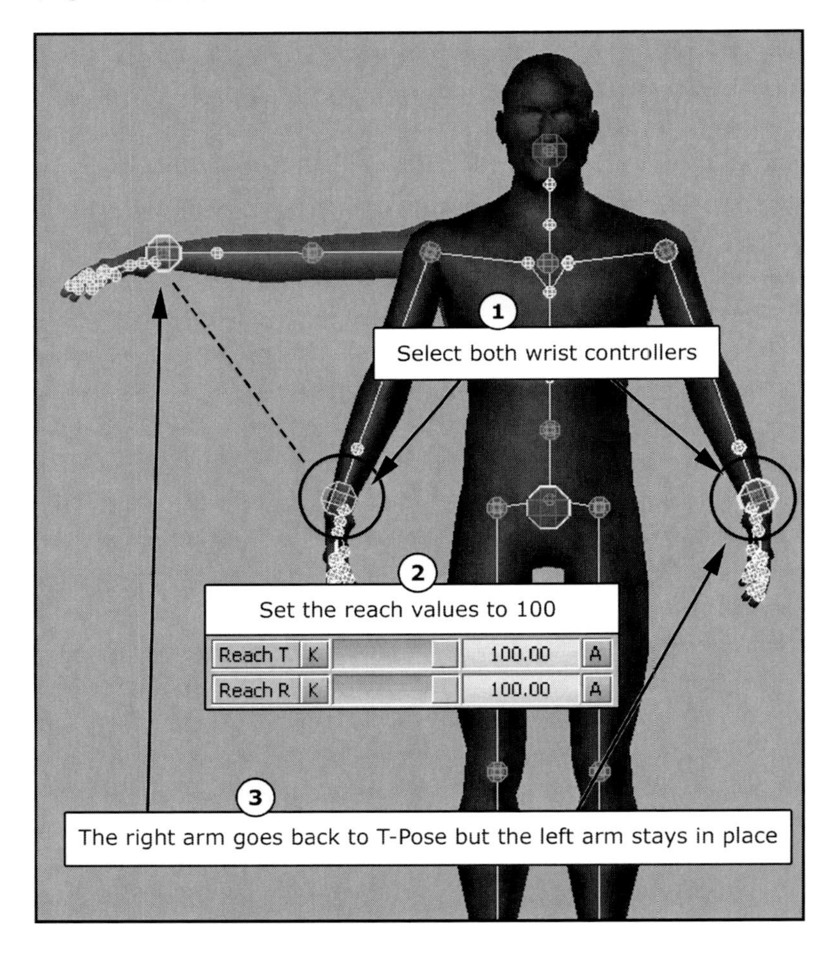

Figure 13_12.

By activating *Reach T* and the *Reach R* you are activating the *IK* controller. Since you keyframed the left arm in *Body Part* mode keys where placed in the *FK* components of the arm as well as the *IK* components keeping both controllers synchronized. In contrast, when you keyed the right arm in *Selection* mode you only affected the selected bone so the *IK* components were never selected or moved causing the *FK* controllers and the *IK* controllers to become un-synchronized (Figure 13_13).

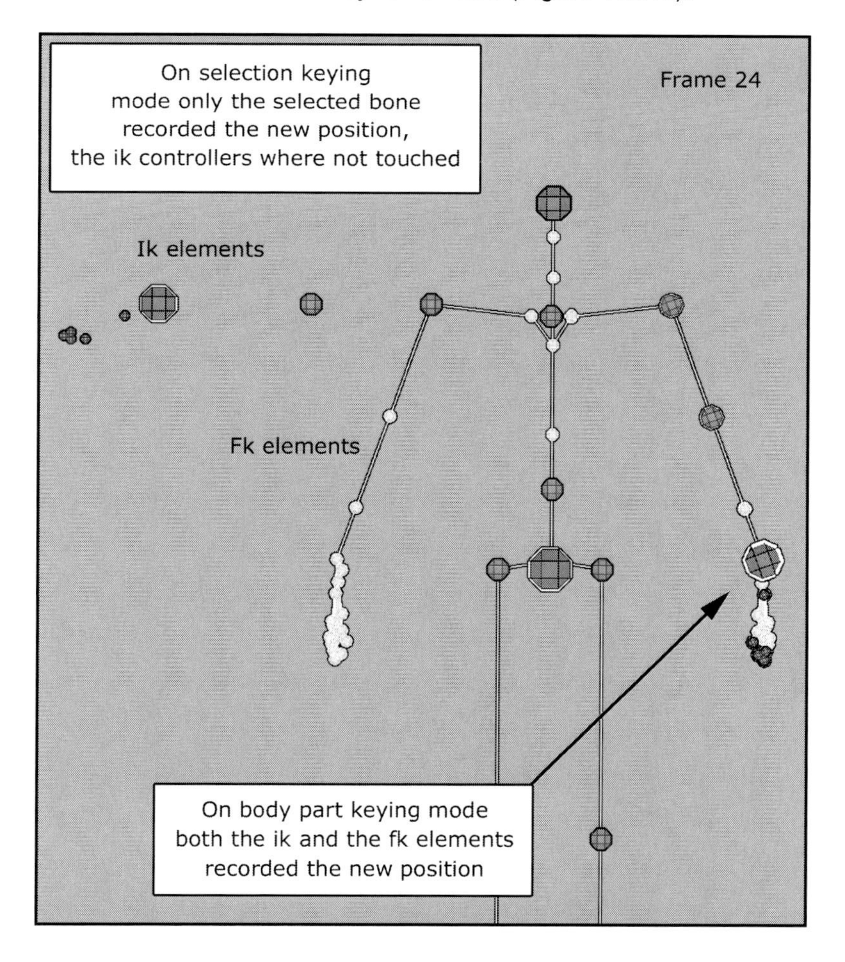

Figure 13_13.

Depending on which set of controllers are driving the rig (*FK* or *IK*) your character might move different between poses. Re-open the "Chapter 13" file to make sure that there is no previous animation on the rig and that we have the original settings on the controllers. Make sure that you are on *Body Part Keying Mode* on the *Character Controls* window. Go to frame 0 and key the *LeftWristEffector* and the *RightWristEffector*, go to frame 24 and pose and key the left and right arms so they look like figure 13_14 (Figure 13_14).

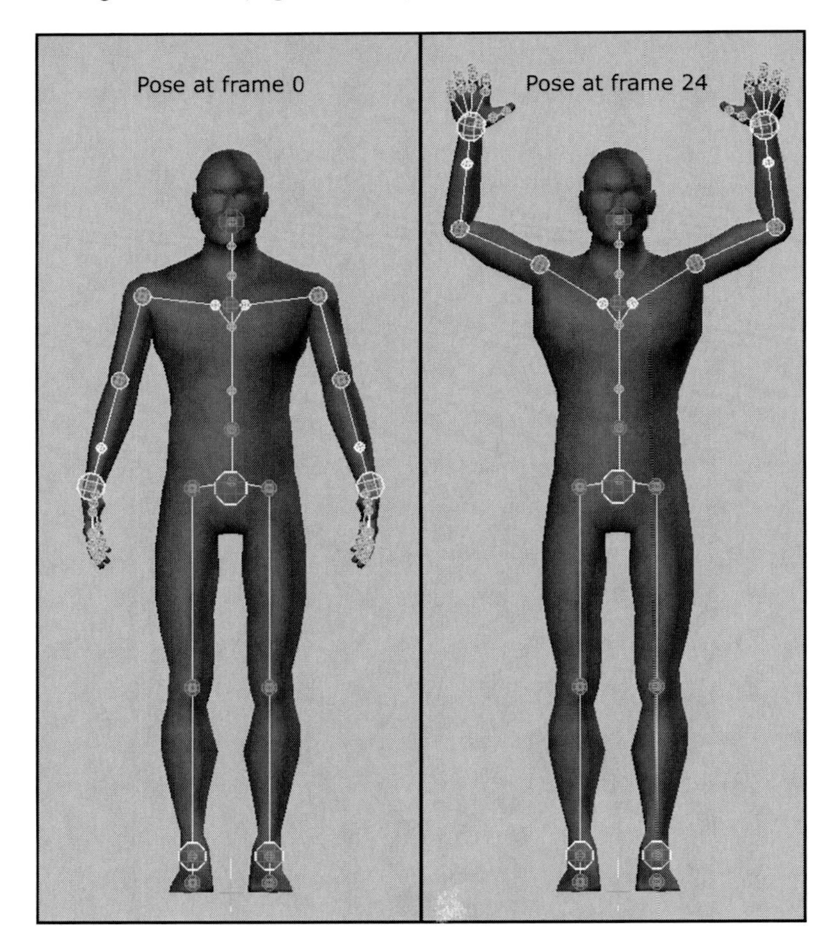

Figure 13_14.

Since we have been keying on *Body Part* mode, every element of both arms has been key-framed and the *FK* controllers as well as the *IK* controllers are synchronized. Even if the key-framed poses are similar the movement of a limb is very different when it is being controlled by *FK* or by *IK*.
Select the *RightWristEffector* and turn on the *Reach T* and the *Reach R* values by setting them to a 100. *Ctrl* select the *LeftWristEffector* and turn-on trajectories to see the difference in movement between the arms (Figure 13_15).

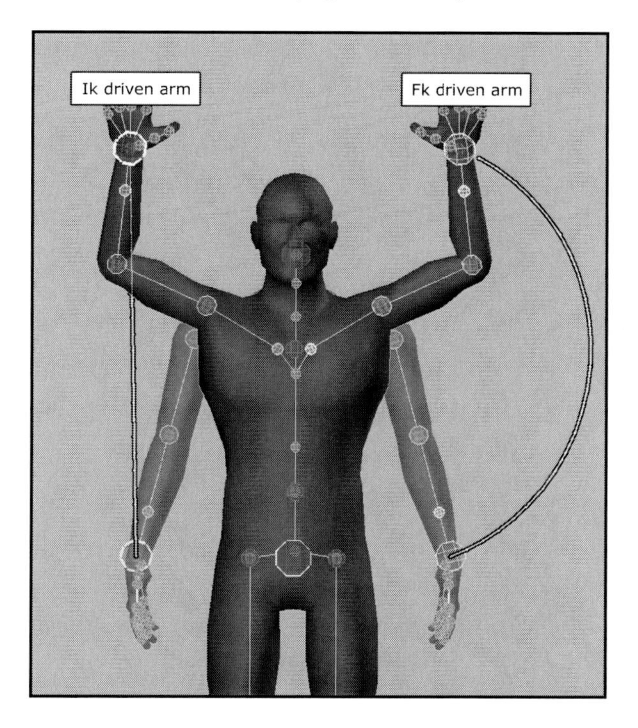

Figure 13_15.

Of course the left arm which is driven by *FK* controllers moves more natural because it moves in an arc which is one of the basic principles of animation.
When you are selecting elements in the *Character Controls* window you are basically posing your character through the use of *IK* controllers. But until you activate the *Reach T* and *Reach R* of a section, you will be driving its elements with the *FK* controllers.

When in *Full Body* keying mode it is possible to take advantage of the *Pinning* and *Releasing* options of the *Character Controls* window. *Pinning* is temporarily locking the translations and rotations of an *IK* effector for posing purposes. When you move and pose parts of your character around, the pinned sections will stay in place allowing you to achieve complex poses with ease. *Pinning* only affects the posing stage (Figure 13_16).

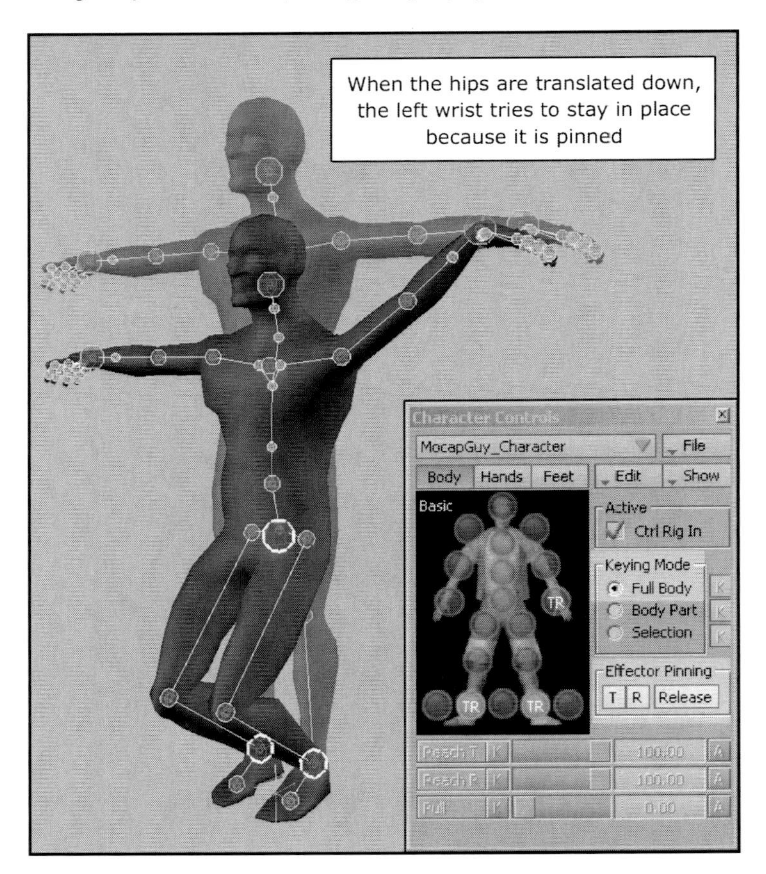

Figure 13_16.

You can pin any *effector* in the *Control Rig* by pressing the *"W"* key for translation pinning and the *"E"* key for rotation pinning while you have the *effector* selected. The *Release* button will allow you to temporary ignore the *pinning* of the *effectors* while in the posing stages.

This covers a basic introduction to **Motion Builder's** *Control Rig*. This rig is highly versatile and it takes a little bit of practice to fully understand its inner workings. The next chapter expands on the use of such rig to push and enhance already existing animation (Figure 13_17).

Figure 13_17.

Chapter 14
Enhancive Animation

There are two main schools of thought in motion capture production. One is to leave the motions exactly like the actor performed them. This philosophy calls only for minor corrections (if any) when the physiology of the digital character and its interactions with the environment impose changes upon the original motions (Figure 14_01).

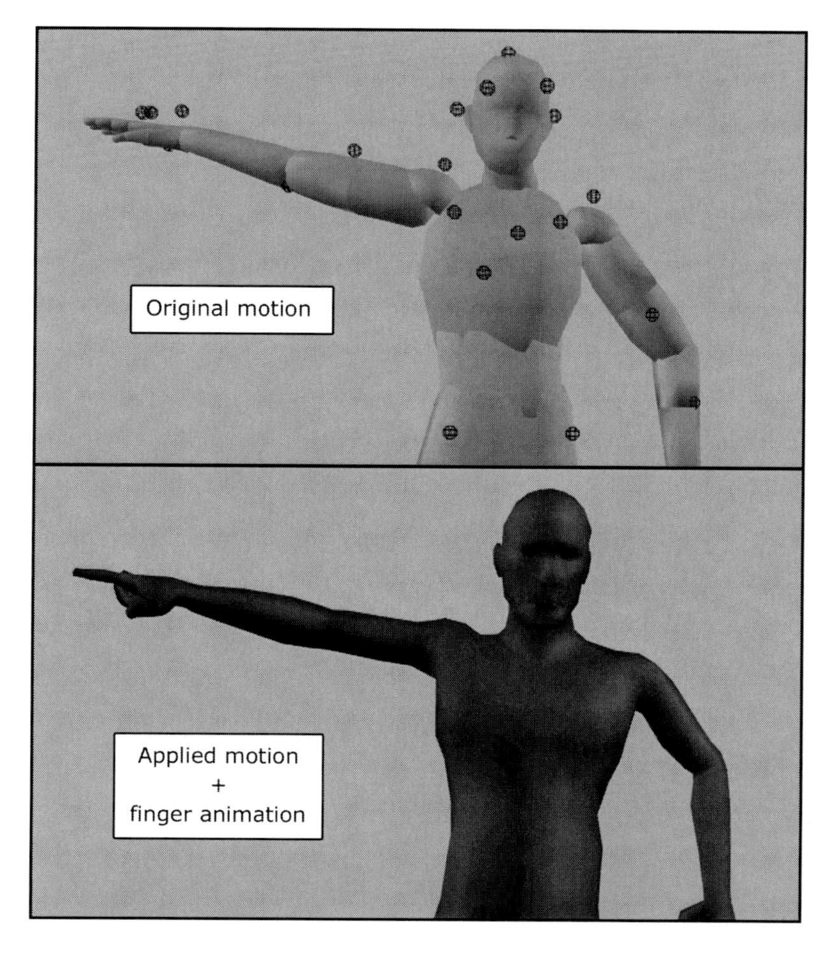

Figure 14_01.

There is a different philosophy that treats motion capture like a starting point not an end in itself. Very similar to the way a digital painter uses photographic manipulation to create an image that could not exist before just by the use of a camera. This second school of thought uses traditional animation techniques to augment motion capture by the application of the basic principles of animation (Figure 14_02).

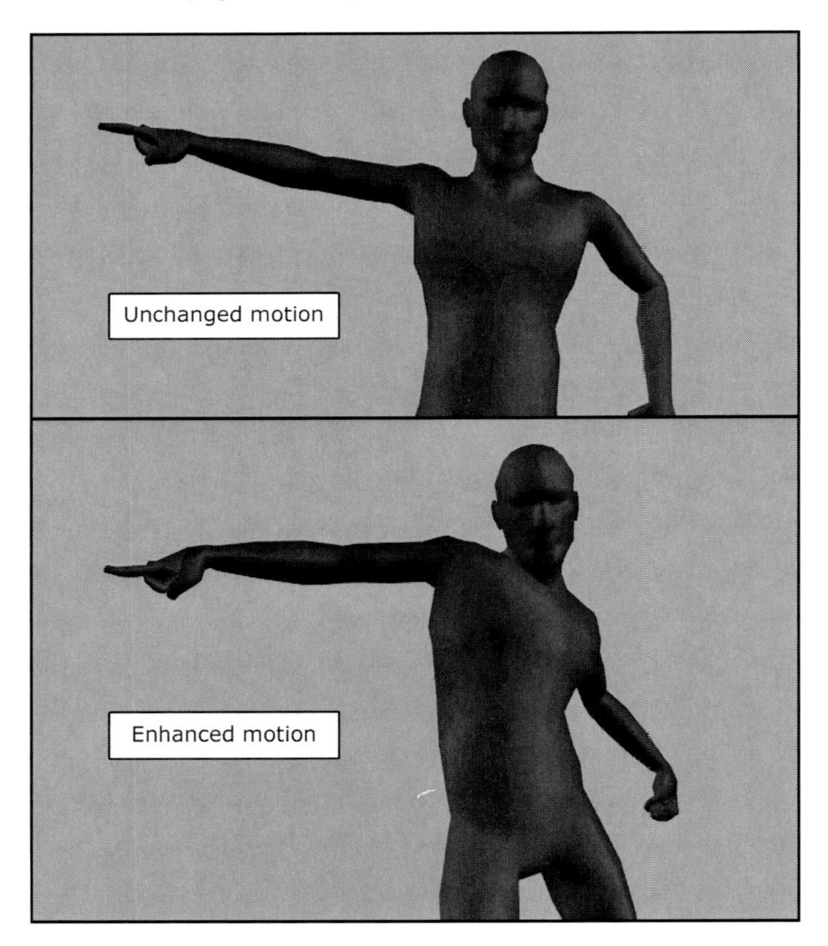

Figure 14_02.

This chapter will cover the process of enhancing motion capture with the use of traditional or hand key animation techniques. Figure 14_03 shows the comparison between the motion capture used in the tutorial as a starting point with the result after enhancing the mocap with traditional animation (Figure 14_03).

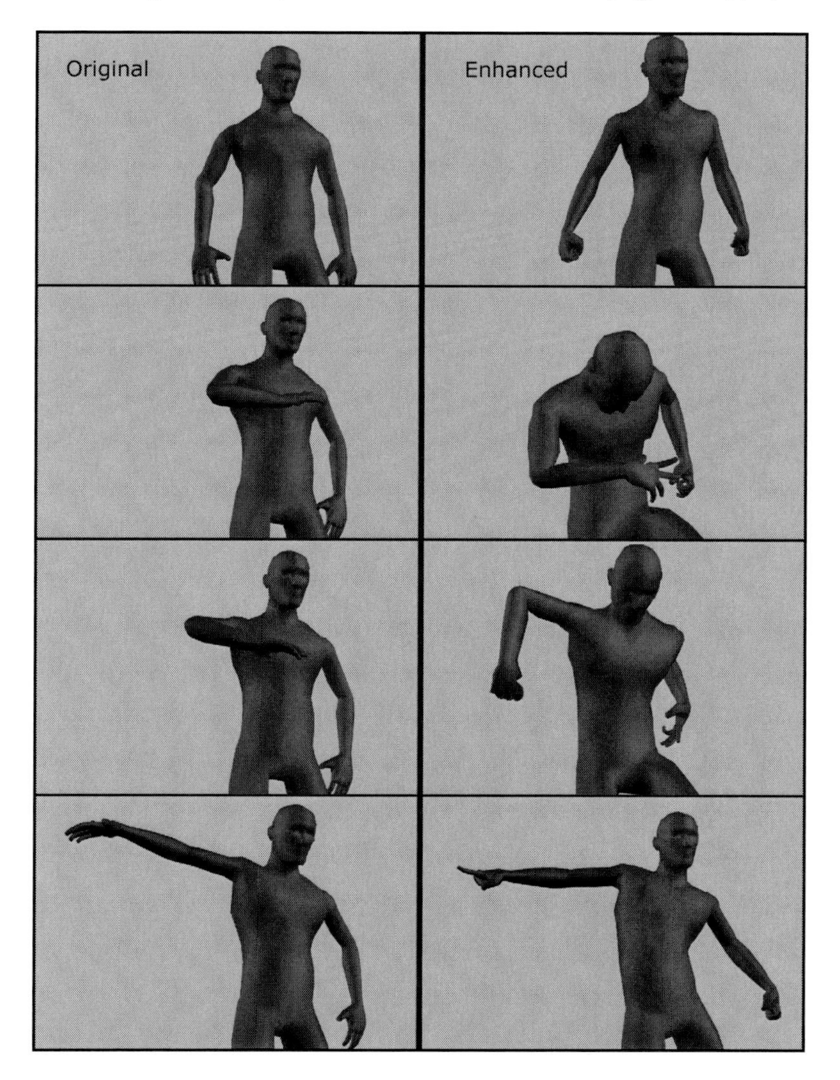

Figure 14_03.

We will first focus on fitting the motion capture to the digital character. Some production houses call this process *Cleanup*. It basically involves correcting any intersection and discrepancies that might occur because of the difference in physiology between the original actor and the digital counterpart (Figure 14_04).

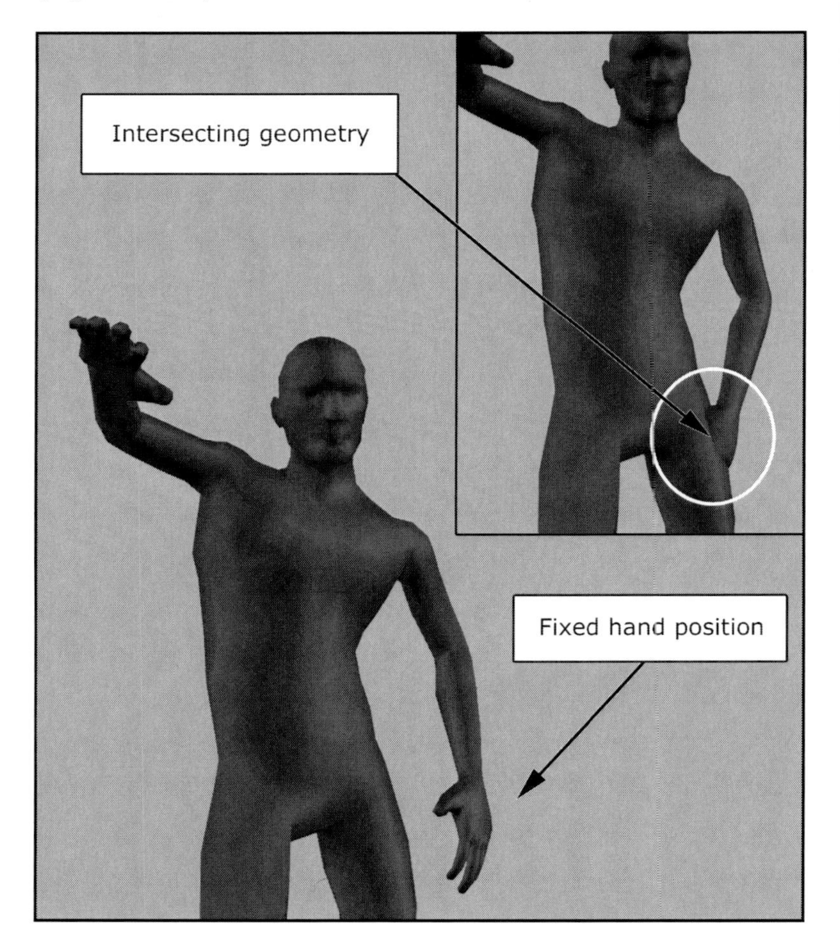

Figure 14_04.

Launch **Motion Builder** and open the "Chapter14"[11] file.

When dealing with motion capture for Film, TV and even Video Game Cinematics, it is very likely that the camera from which the action is going to be seen will already be setup for you. Working based on predetermined cameras helps in the sense that the artist's only worry is to make the performance work for a specific view. On the other hand, when working with in-game motions the artist is forced to make the actions pleasing from multiple points of view. For this chapter we will work with the predetermined camera approach. Click on the *View* drop down on the top left area of the *Viewer Window* and select "CameraShot01" from the *Perspective* options. This is the camera that we are going to be working from (Figure 14_05).

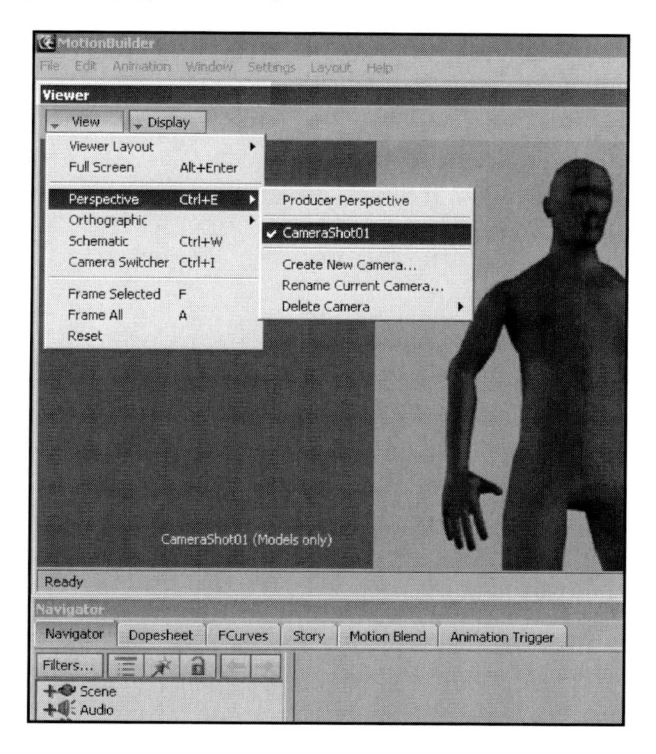

Figure 14_05.

[11] Download the files from www.MocapClub.com/TheMocapBook.htm

At the moment the animation has been retargeted to the character's skeleton. However, in order to make changes to the motion we need the animation to be transferred to the *Control Rig*. Under the *Character Controls* window go to the *Edit* drop-down menu and select *Plot Character*. Press the *Control Rig* button on the *Character* pop-up window that appears, click on the *Plot* button of the new *Character* window that pops (Figure 14_06).

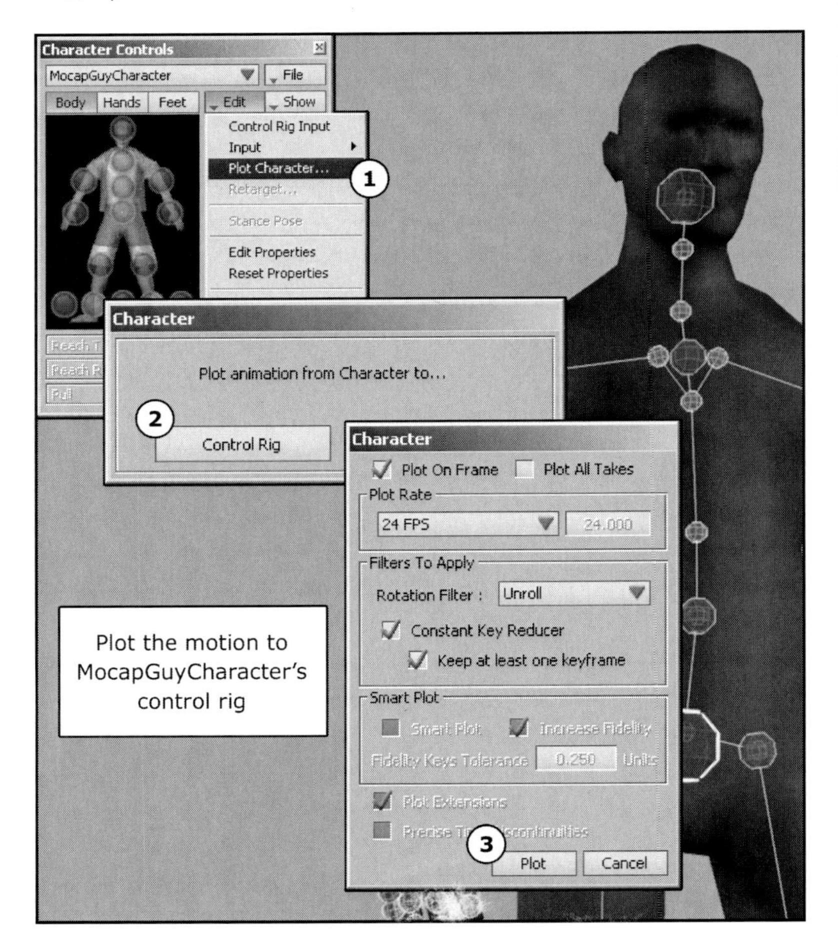

Figure 14_06.

First thing that we can see when we play the motion is that the clavicles are a little too low, specially the left one. To fix this lets create a new layer by selecting the *FCurves* tab from the *Navigator* window and pressing the add button from under the *Layer* tab on the right lower area of the window. Rename that layer "ClaviclesCorrections" (Figure 14_07).

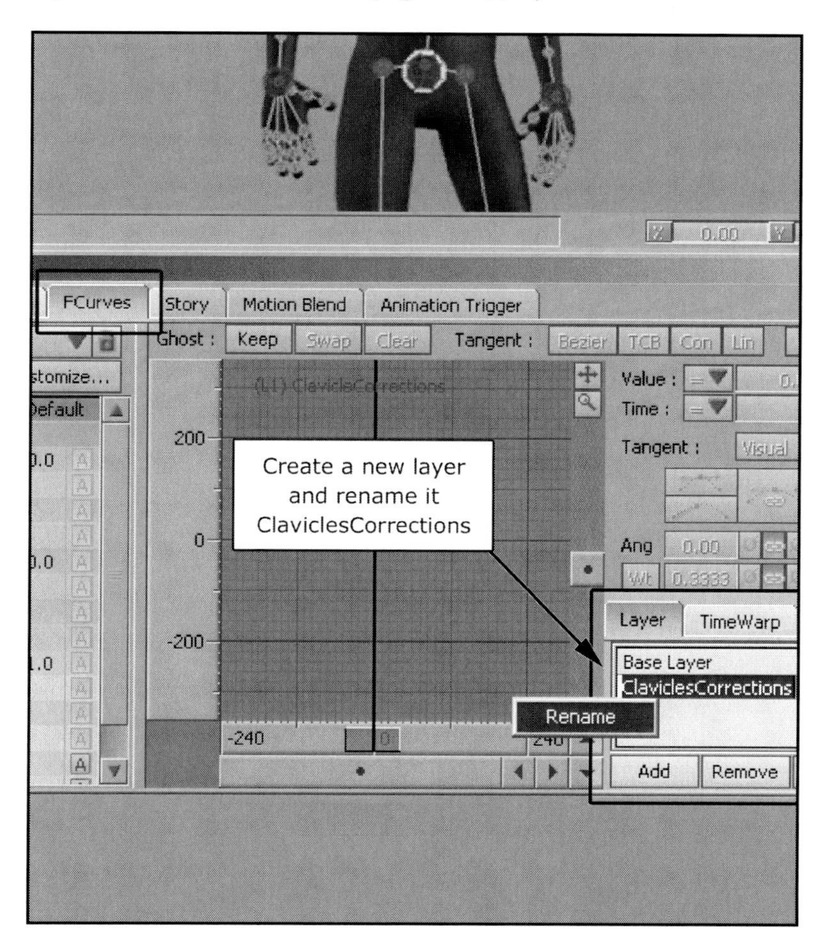

Figure 14_07.

We are going to correct the left clavicle (Left Shoulder) first because it just needs what I call a minor *Global Change*. We can basically just raise the height of the shoulder through the whole animation. To do this go to the beginning of the animation, select the *LeftShoulderEffector* and raise it until it looks more natural to the character. Make sure that *Body Part* is selected in the character controls window and set a key frame by pressing *"K"* on your keyboard. By doing this you have effectively raised the left shoulder for the duration of the entire animation (Figure 14_08).

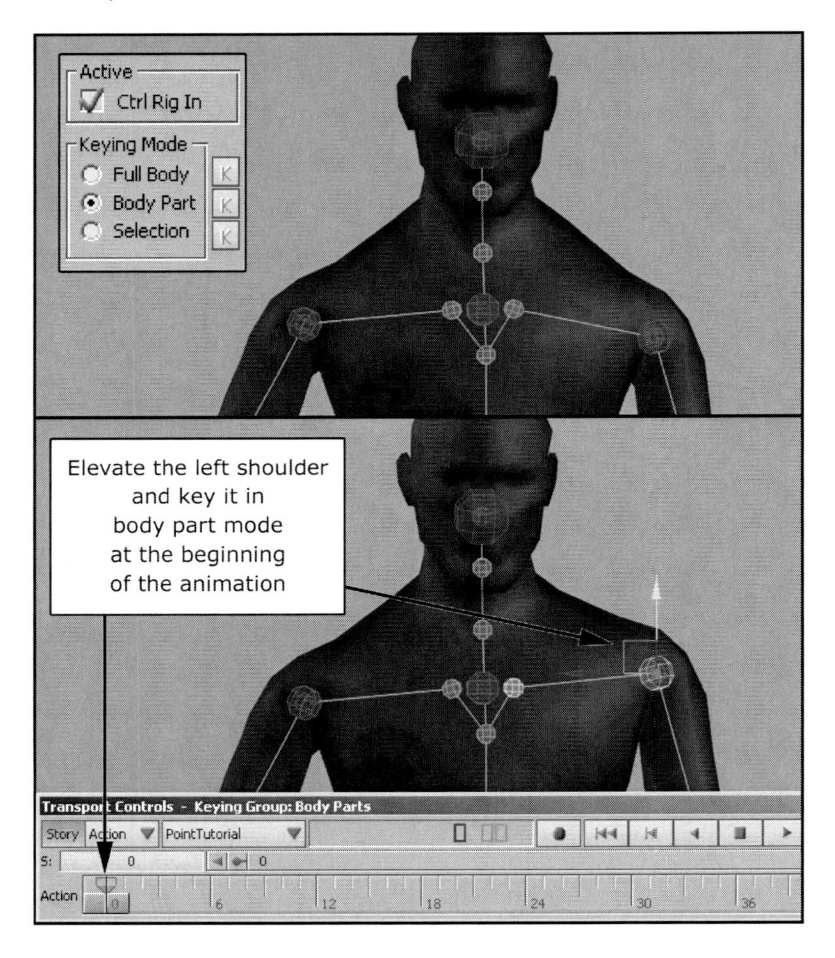

Figure 14_08.

Now we need to elevate the right shoulder in certain parts of the animation. The difference here is that we want to raise the shoulder when the arm is down but leave it at the same height after the character points. To do this we are going to use what I call *Placeholder Key Frames*. Go to the beginning of the animation and set a *Zero Key* Frame by pressing *Shift* + *"K"*. Set additional *Zero Keys* at frame 30 before the right shoulder starts to raise and at frame 40 when the shoulder finishes its elevation. By doing this we have set keys that guide us on the main points of the shoulder motion but do not affect its motion yet. Go to frame 0 and raise the right shoulder to a position that looks natural, set a regular key frame by pressing *"K"*. Elevate the shoulder to a comfortable position at frame 30 and set another regular key frame (Figure 14_09). By doing this we have fixed the shoulder where it looked a little low but left it untouched where it looked OK.

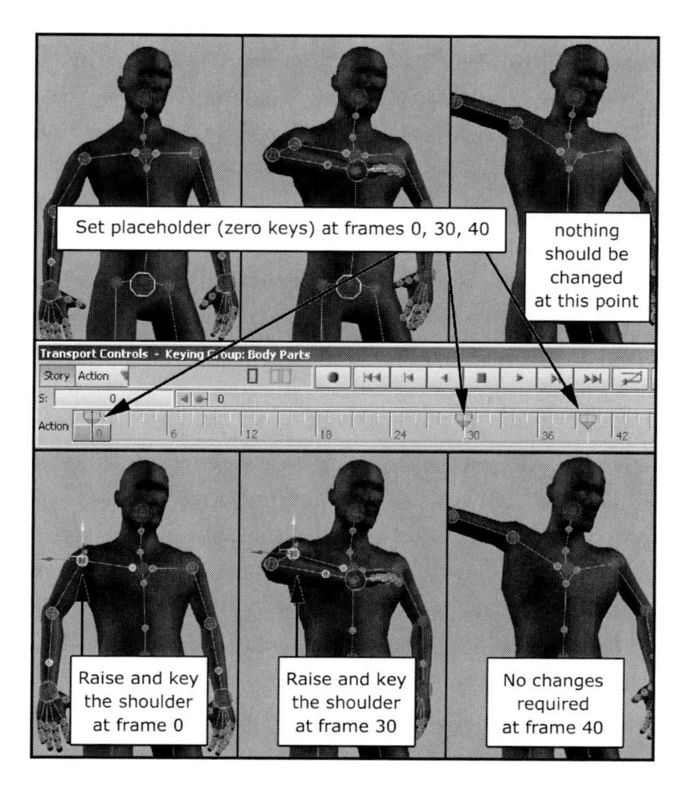

Figure 14_09.

When playing the animation from the *Perspective View* we can see that the right hand is going through the chest.
Since we like the corrections done to the shoulders we want to leave the "ClavicleCorrections" layer alone and create a new one for the hand adjustments. Create a new layer and rename it "RightArmCorrections". While in *Body Part Keying Mode*, select the *RightWristEffector* in the *Character Controls* window and set *Zero Keys* by pressing *Shift* + *"K"* at the beginning of the animation, at frame 17 before the performer starts to lift the arm, at frame 30 and at frame 38 when the arm motion stops.
At frame 30 move the arm so it does not intersect with the chest and press *"K"* (Figure 14_10).

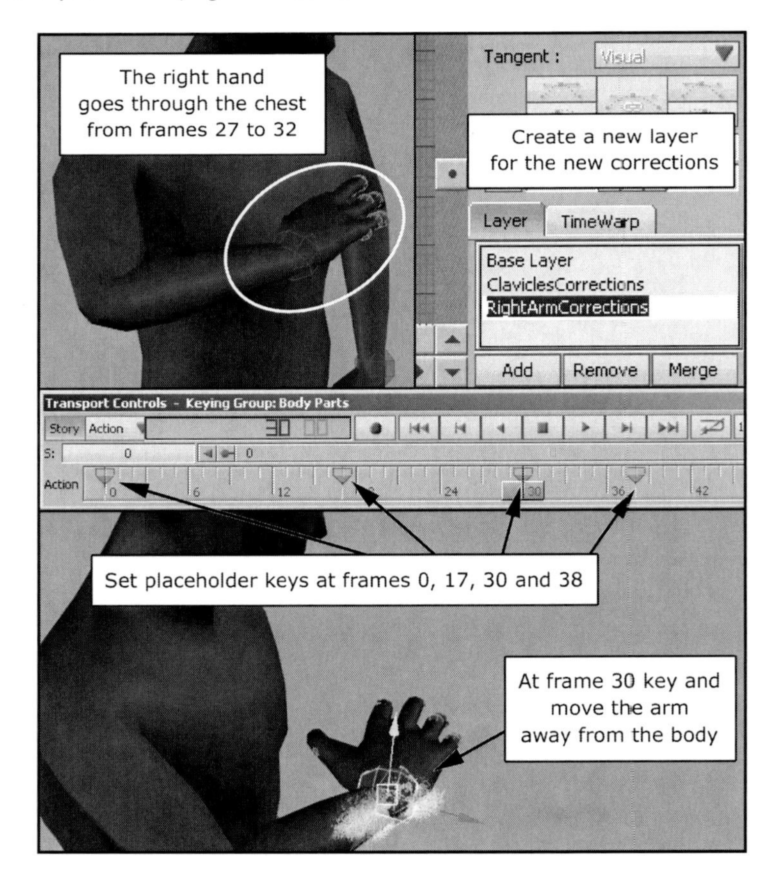

Figure 14_10.

Now the hand does not intersect the chest when the animation is played but the arm does not stay bent for the same amount of time before pointing outward.
Access the *Schematic View* by pressing *Crtl* + *"W"* while the cursor is inside the *Viewer* window. Select the *RightShoulder, RightArm, RightForeArm, RightForeArmRoll* and *RightHand* .
Like it was mentioned in "chapter 13", although we used the *IK Effectors* to pose the arm, the keys that control its motion are on the *FK* rig because the *effectors* are off (Figure 14_11).

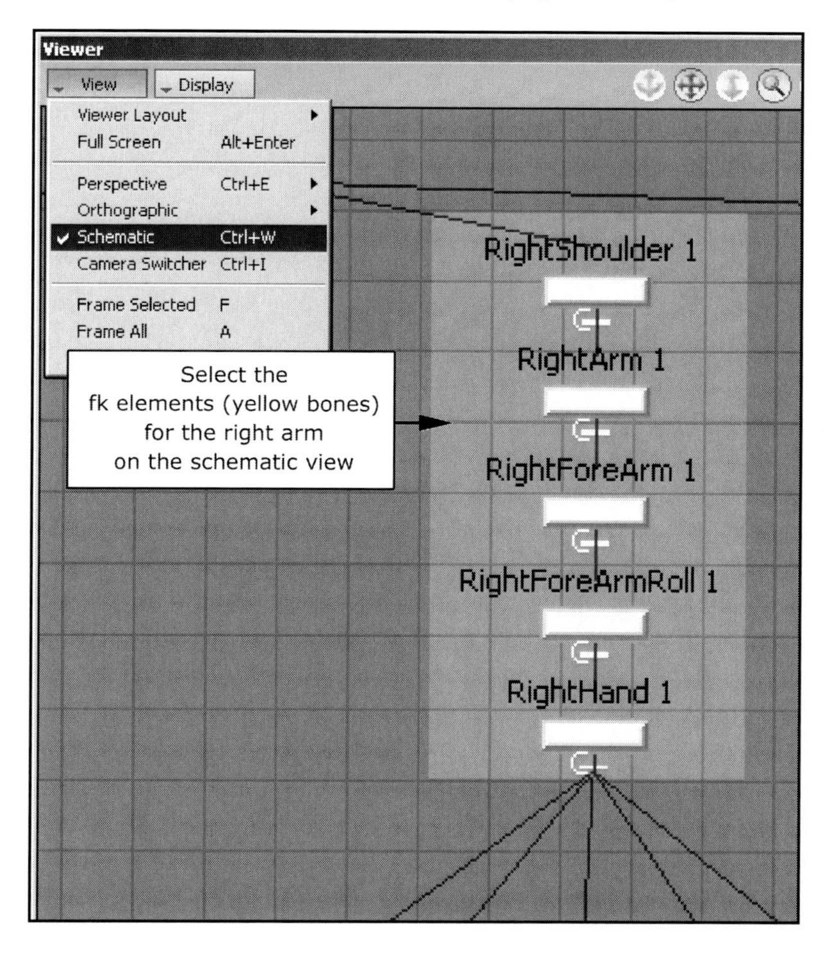

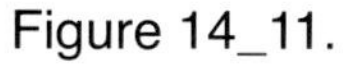
Figure 14_11.

Go to the *FCurves* tab, select the rotations and press *"F"* to frame the curves. The default interpolation for animation curves in **Motion Builder** is *Bezier*. With this interpolation the arm is not stopping at frame 30 but actually changing directions a few frames afterwards

To keep the pause of the arm while it is bent we need to change the tangents of the key at frame 30. Drag a box around the *FK* arm keys at frame 30 to select them and flatten the tangents by clicking on the *Set Flat* tangent button on the right area of the *FCurves* tab (Figure 14_12).

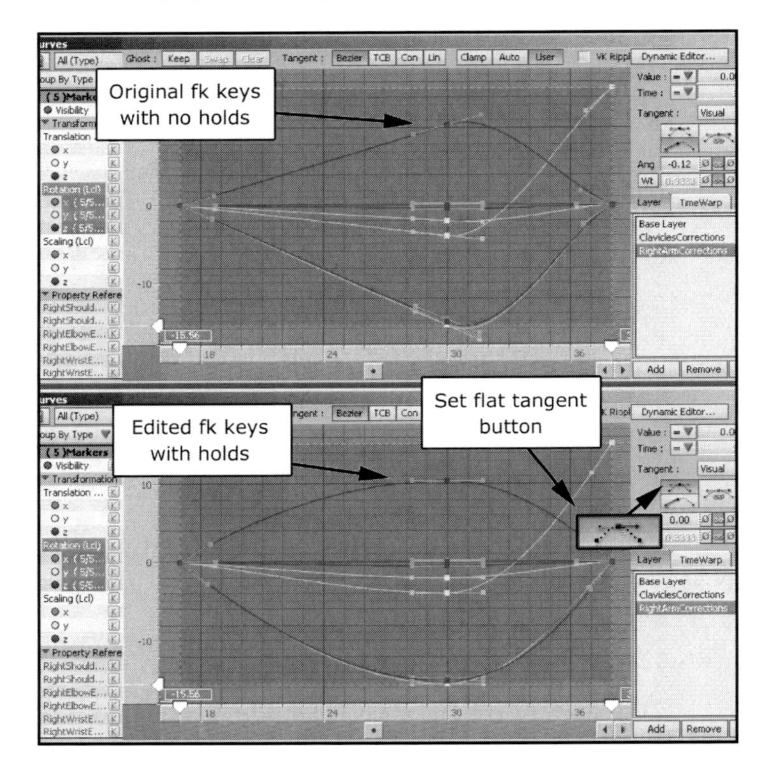

Figure 14_12.

The motion has been properly *cleaned* or fitted. By offsetting the shoulders, we have made them look a little more natural, and we have also fixed the intersection of the right arm with the chest. Now the motion fits our digital character a little better, we can concentrate on pushing the motion a little further. We will do this by the use of traditional animation techniques. One of the main strongholds of traditional animation is its reliance in strong poses. While the poses of the mocap that we are using right now are not bad, they can still have a little more kick to them. We can take a page from the artists of classical times and apply a little bit of *Contrapposto* to our standing poses so they become stronger. One of the key elements of the *Contrapposto* is that the orientation of the hips contrasts the orientation of the chest and shoulders providing a very interesting line of action (Figure 14_13).

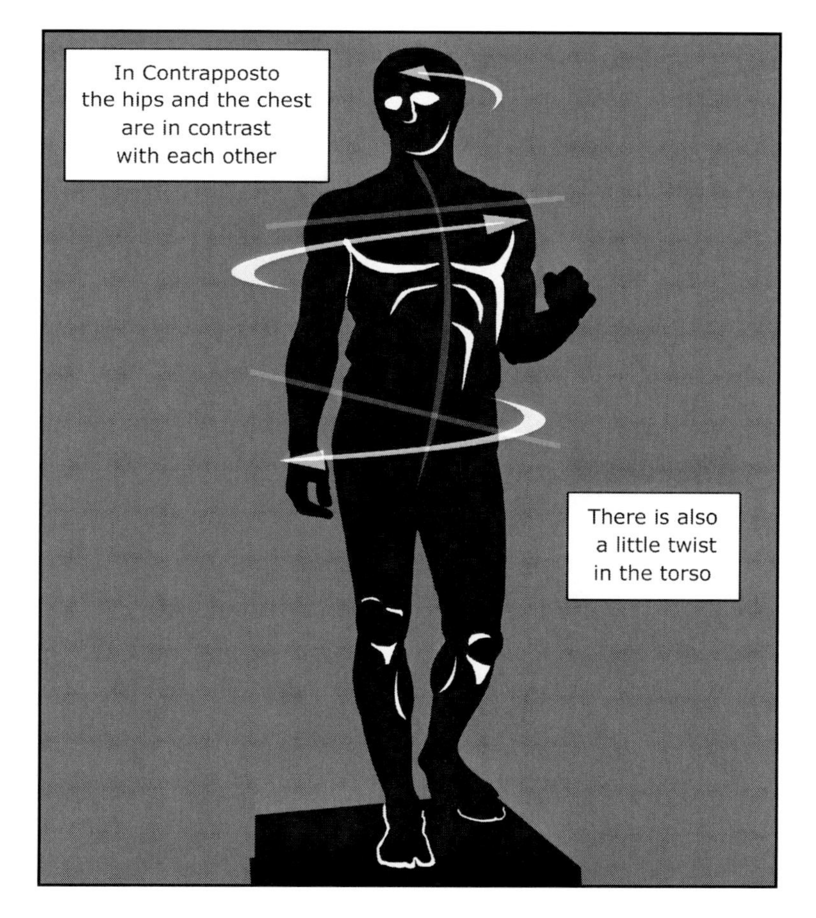

Figure 14_13.

Create a new layer and rename it “LineOfAction”. Activate *X-Ray* mode on the *Viewer* by pressing *Ctrl* + *“A”* until you can see the *Control Rig*. Go to the *Window* menu and select *Groups* from the list. On the window that appears uncheck the “LeftArm” and the “Right Arm”. This will allow us to concentrate on the spine (Figure 14_14).

Figure 14_14.

With *Body Part* checked on under the *Character Controls* window set *Zero Keys* for both the spine and the hips at frames 0, 18, at 28 when the right arm is fully bent, at 38 when the arm is fully extended, and at 46.
At frame 0 the spine is slightly curved to the right (stage right), exaggerate the curvature of the spine by rotating the *ChestEndEffector* and the *ChestOriginEffector* in the local Y axis to the right of the character. Rotate the hips of the performer in the opposite direction and set a key for both the chest and the hips effectors.
Right click on the key from frame 0 and choose copy for the list that appears then right click on frame 18 and choose paste from the list. Now both keys have a well-defined line of action (Figure 14_15).

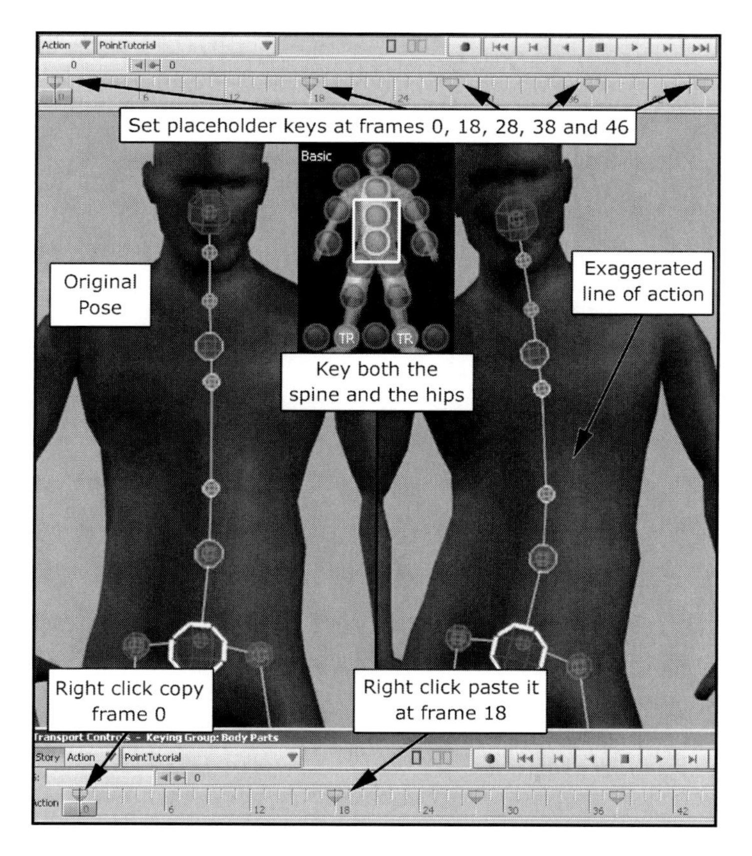

Figure 14_15.

At frame 28 rotate the chest *effectors* forwards on the local Z axis, also rotate the hips to keep the character's balance. Use the side views to make sure that the performer is properly balanced. Lower the hips to accentuate the pose squash, key frame both the chest and the hips *effectors* (Figure 14_16).

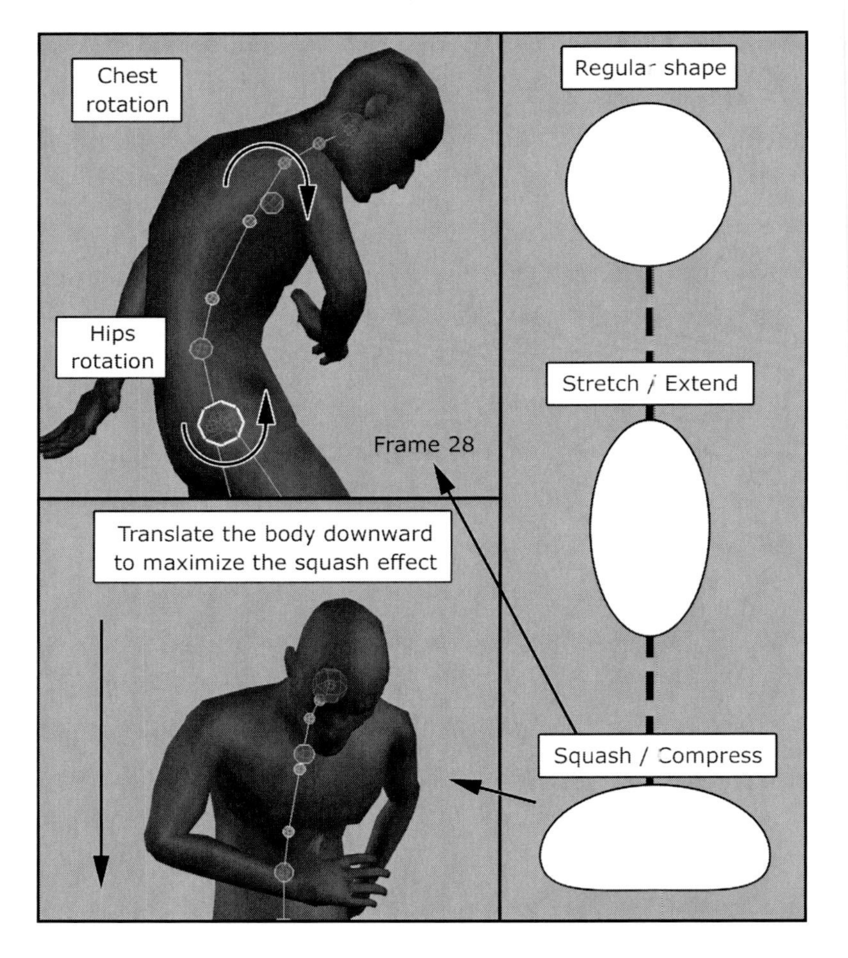

Figure 14_16.

At frame 38 rotate the chest *effector* backwards and rotate the hips to balance the pose. Right click and copy frame 38 to frame 46.
Go back to frame 38 and rotate the spine and hips a little further, also lift the hips a little higher to accentuate the stretch of the pose. This will create a little bit of an overshoot of the body at the frame that the pointing takes place (Figure 14_17).

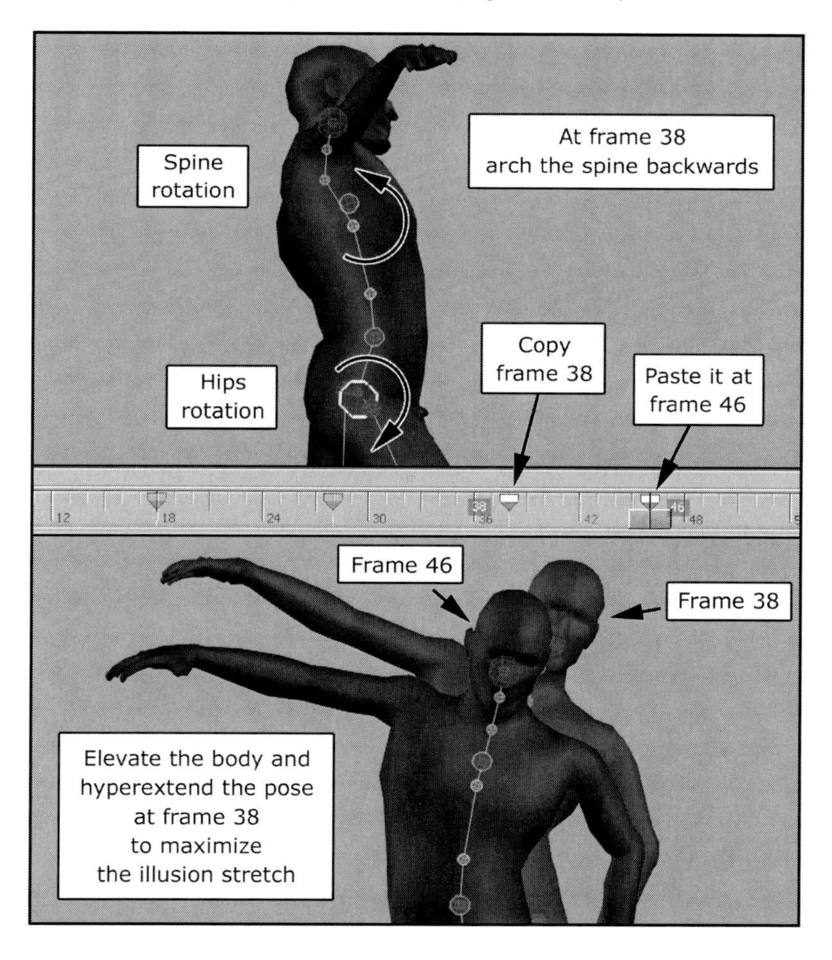

Figure 14_17.

Like we did on the "RightArmCorrections" layer, select all the *FK* bones for the spine as well as the hips *IK* control and edit the curves to your liking. How you edit the curves has a lot to do with how good or how bad the animation looks.
The last thing to do on the "LineOfAction" layer is offsetting the spine keys from the hip keys to create a little bit of overlapping action. Select the chest *effectors* and drag a box over its keys inside the *Action Timeline*. Click in the green area between key frames and move them towards your right one frame. Select the keys for the hips and shift them to your left one frame (Figure 14_18).

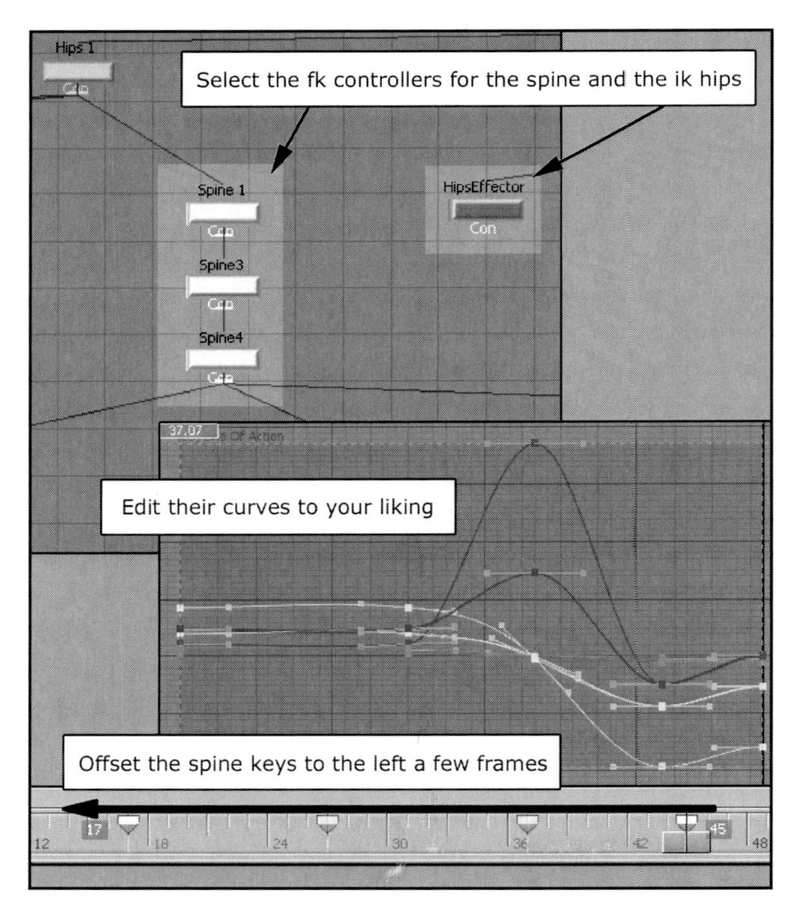

Figure 14_18.

When the animation is played we can see that although the line of action for the character looks much better, the right arm now points too high and the head looks a little stiff. Let's take care of the head stiffness first. Create a new layer and call it "HeadOffset". Set *Zero Keys* at frames at 0, 20, 30, 37 and 42. At frame 30 when the digital performer goes down rotate the head forward to increase the compression effect on the body (squash). Although we are not necessarily using *Squash and Stretch* in the cartoon sense for this animation, we can still take advantage from the realistic counterpart of the concept to push the animation which is *Compression and Extension.*
At Frame 37 rotate the head backwards and stage right to add a little bit more pop to the action and to compensate for the curve of the spine. At frame 42 readjust the position of the head so it is a little straighter and the neck extends a little forward to create a stronger pose. (Figure 14_19).

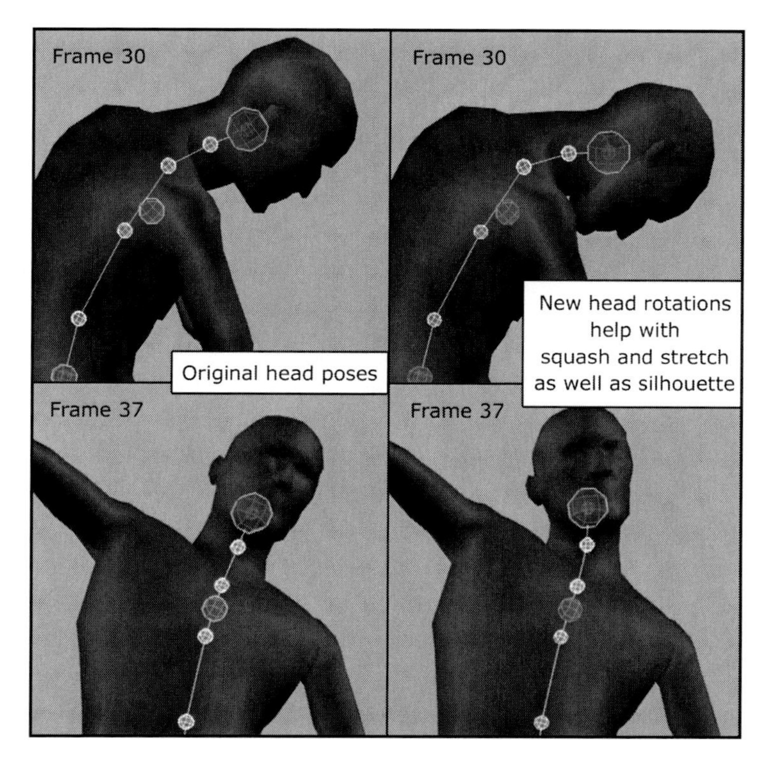

Figure 14_19.

As previously mentioned the arm is pointing a little high after the spine corrections. Let's fix this problem. Create a new layer and rename it "RightArmOffset". Set *Zero Keys* at frames 0, 21, 30, 37, 42, 52 and 67.
At frame 0 open the arm a little (Figure 14_20). Copy the key frame and paste it on frame 21.
Rotate the arm on the local Y axis on frame 37 so it is a little more leveled. Pose the arm even more parallel to the ground at frame 42 and copy this new key frame-to-frame 52. At frame 67 lower the arm a little bit to give the impression of the arm settling in to the pose (Figure 14_20).

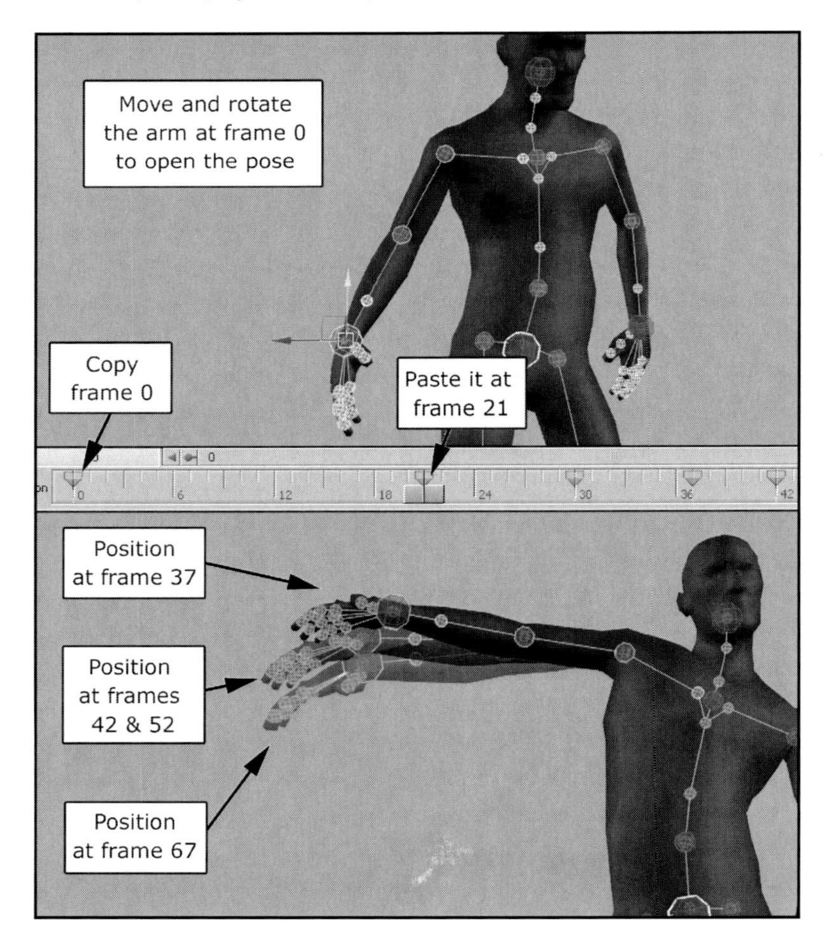

Figure 14_20.

Using the same logic that we applied to enhance the right arm we can improve the motion of the left arm. Set *Zero Keys* at frames 0, 27, 32, 39, 50 and 67.
Open up the left arm pose at frame 0 and copy the new key frame to frame 27. At frame 32 bend the arm to help with the pose compression (squash). At frame 39 open the left arm pose even more than before to help with the pose extension (stretch). At frame 50 keep the left arm pose open although a little less than before and take the elbow out a little to help with silhouette. At frame 67 lower the hand a little bit to sell the settling in of the pose (Figure 14_21).

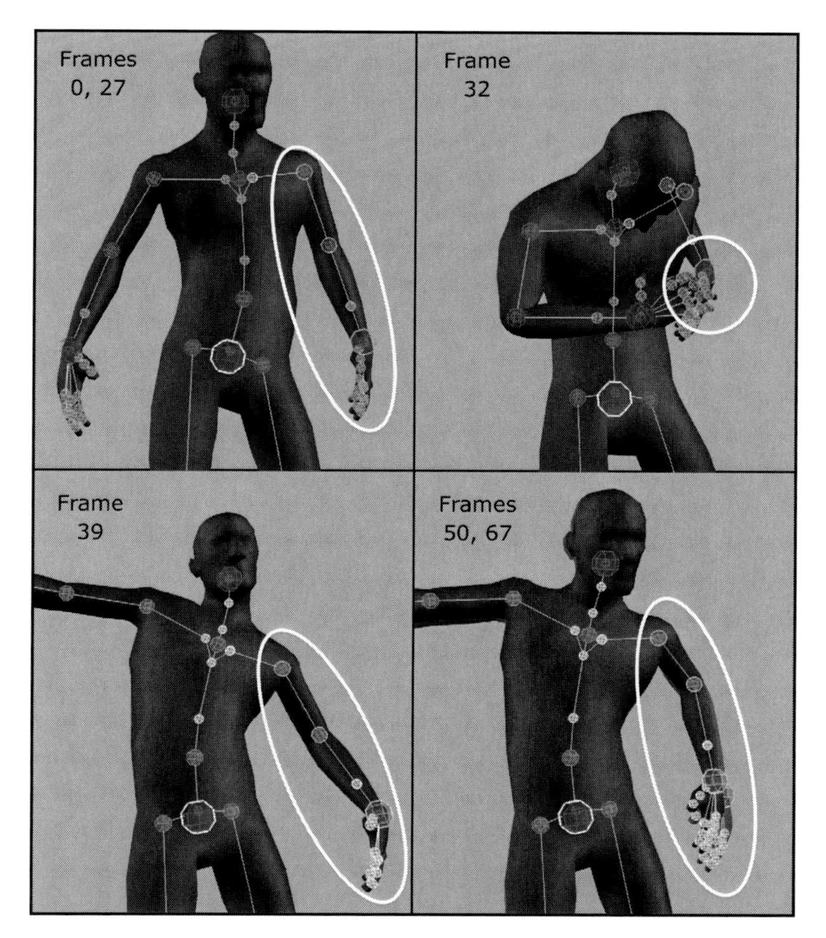

Fig 14_21.

The animation is already looking stronger than what we started with, but in order to ad a little more accent to the pointing motion we are going to add a step to it. Create a layer and rename it "Step". Making sure that you have *Body Part Keying Mode* selected under the *Character Controls* window *Zero Key* frame both the *HipsEffector* and the *LeftAnkleEffector* at frames 20, 30 and 38.
At frame 30 elevate the body by translating the hips upwards and rest the body over the right leg. Also lift the left leg from the ground and key both body parts (Figure 14_22).

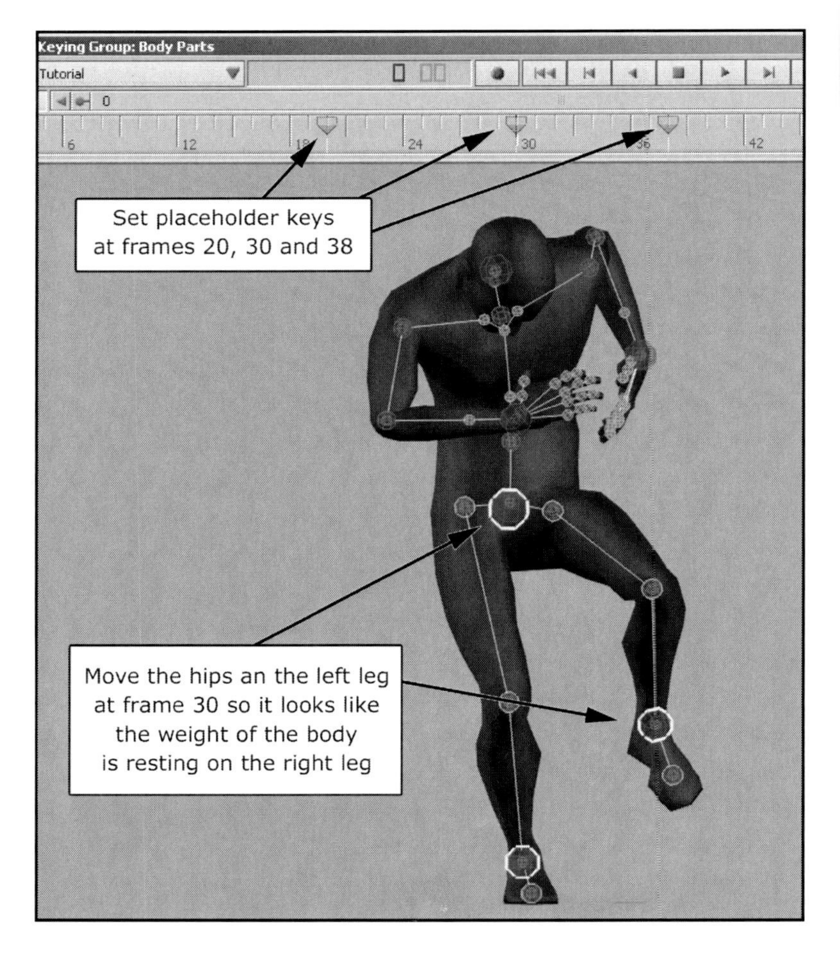

Figure 14_22.

Edit the curves from the hips and the leg so the character holds the elevated pose a little longer. Select the leg keys on the timeline and offset them 1 or 2 frames to the right to create a little bit of overlap between the body parts (Figure 14-23).

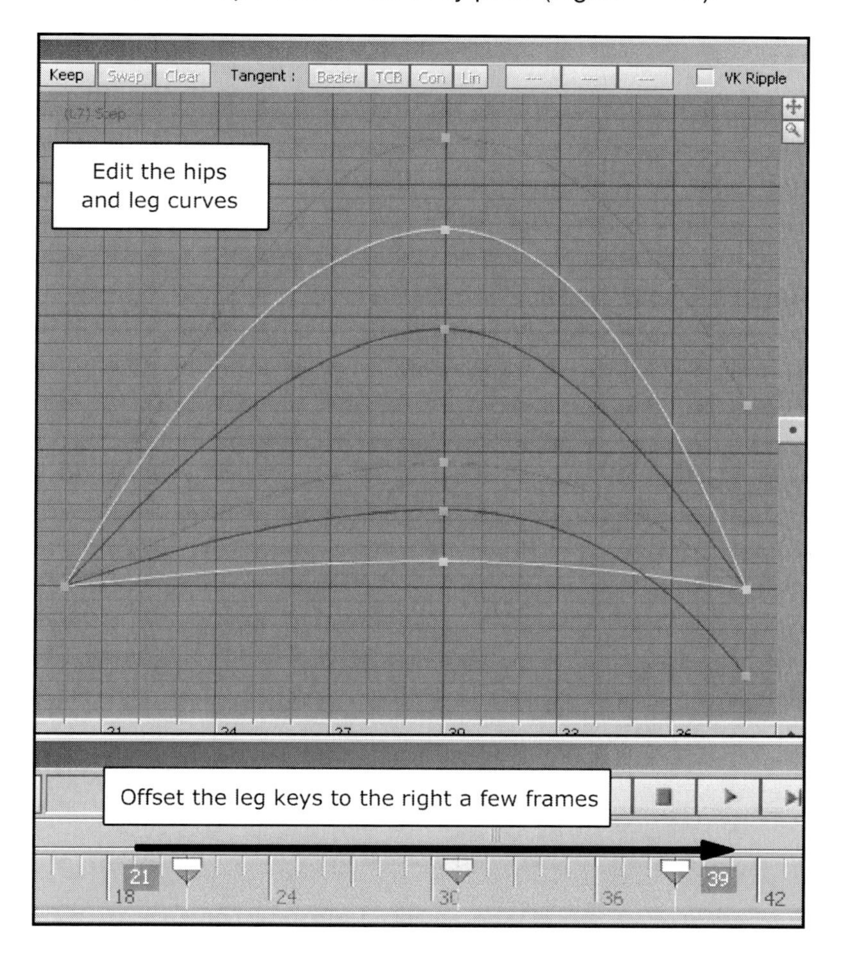

Figure 14-23.

To add little bit more punch to the right arm during the pointing motion we are going to give it a little bit more flexibility and apply somewhat of a whip-like kind of motion to it.
Create a new layer and rename it "ArmSnap". Set *Zero Keys* on *Body Part* mode for the right arm at frames 30, 34, 37 and 40. At frame 34 when the right arm is roughly at the highest point of its motion pose the limb with the elbow pointing upwards. At frame 37 pose the right arm bent with the elbow pointing downward (Figure 14_24).

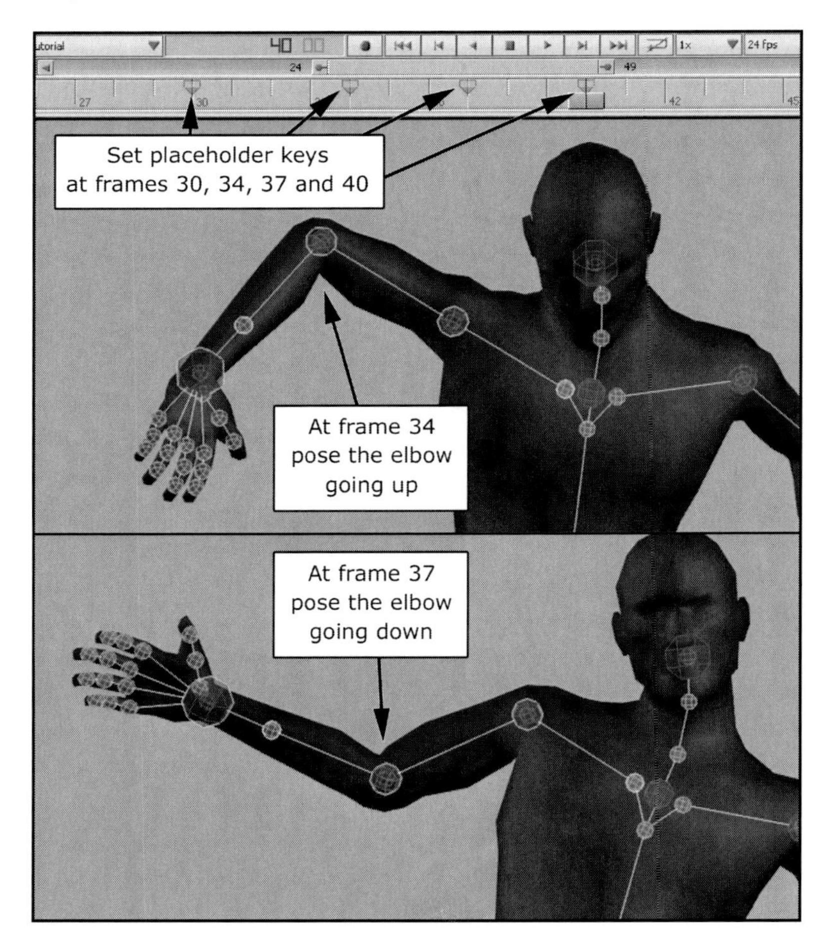

Figure 14_24.

It is time now to add a little bit of follow through to both wrists. Let's start with the right one. Create a new layer and rename it "RightWristFollow". Set *Zero Keys* on *Body Part* mode for the *RightWristEffector* at frames 20, 23, 28, 30, 34, 38 and 41. At frames 20 and 23 orient the right hand so it is perpendicular to the ground. At frame 28 rotate the hand downwards, at frame 30 stabilize the hand, at frame 34 rotate the hand downwards, at frame 38 rotate it upwards and at frame 40 stabilize the hand (Figure 14_25).

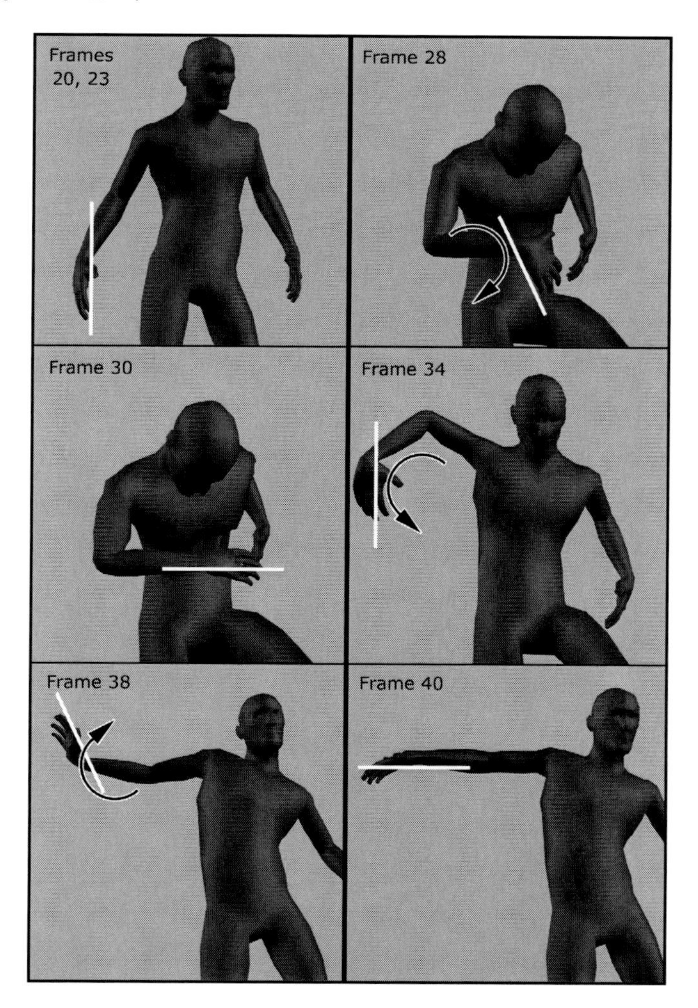

Figure 14_25.

Now let's take care of the left wrist. Create a new layer and rename it "LeftWristFollow". Set *Zero Keys* on *Body Part* mode for the *LeftWristEffector* at frames 20, 23, 26, 29, 32, 38, 41 and 45. At frame 23 level the hand with the forearm, at frame 26 set the hand perpendicular to the ground. At frame 29 level the hand with the forearm and at frame 32 rotate the hand towards the body so it helps with the pose compression. At frame 38 rotate the hand towards the body, at frame 41 level the hand with the forearm and at frame 45 rotate the hand downwards a little bit (Figure 14_26).

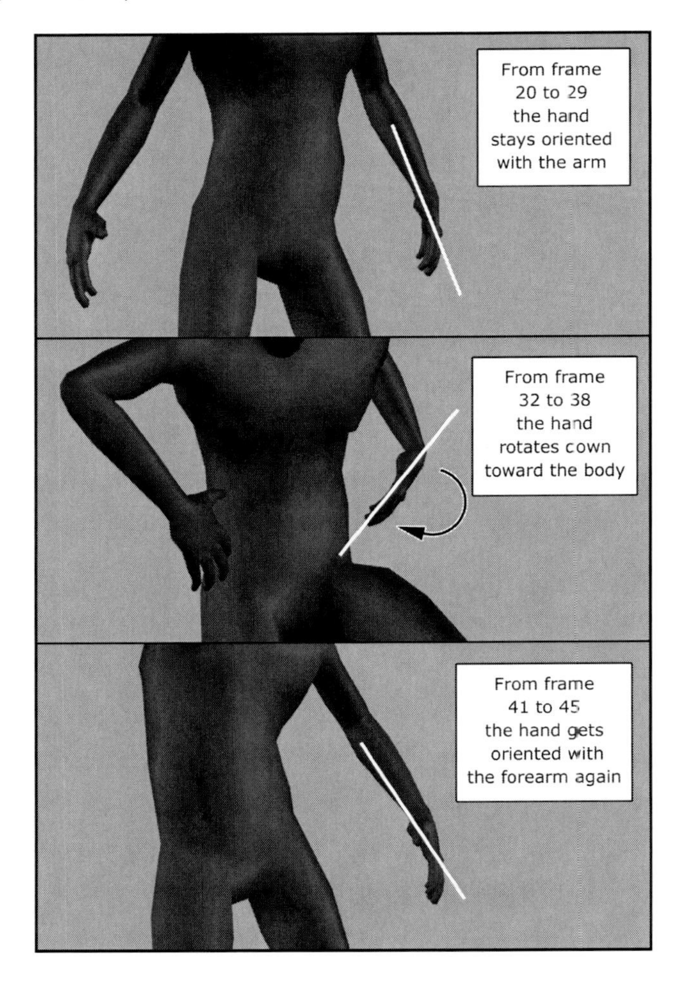

Figure 14_26.

To finalize the animation we need to add life to the fingers. Let's start by animating the fingers from the right hand. Create a new layer and rename it "RightHandFingers". You can select and key the fingers in the *Schematic View* (Figure 14_27).

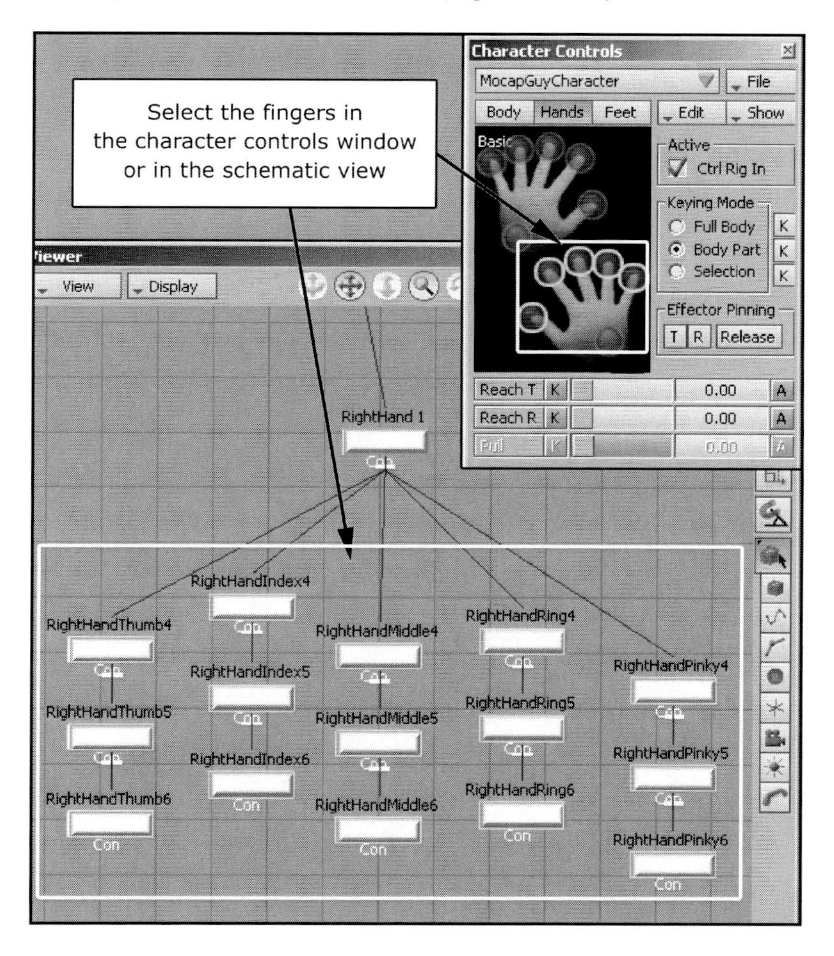

Figure 14_27.

Select the fingers for the right hand and set *Zero Keys* at frames 0, 21, 28, 31, 34, 37 40 and 48. At frame 0 pose the right fingers in a fist, at frame 21 keep the fingers in a fist to hold the pose. At frame 28 open the character's hand making sure that you are getting a strong pose, hold the pose at frame 31. At frame 34 pose the hand in a fist and hold the pose at frame 37. Set the hand in a pointing pose making sure to hyper-extend the index a little bit at frame 40 and relax the index finger a little at frame 48 (Figure 14_28).
Feel free to offset the fingers roughly one frame from each other to get a little bit of overlapping action.

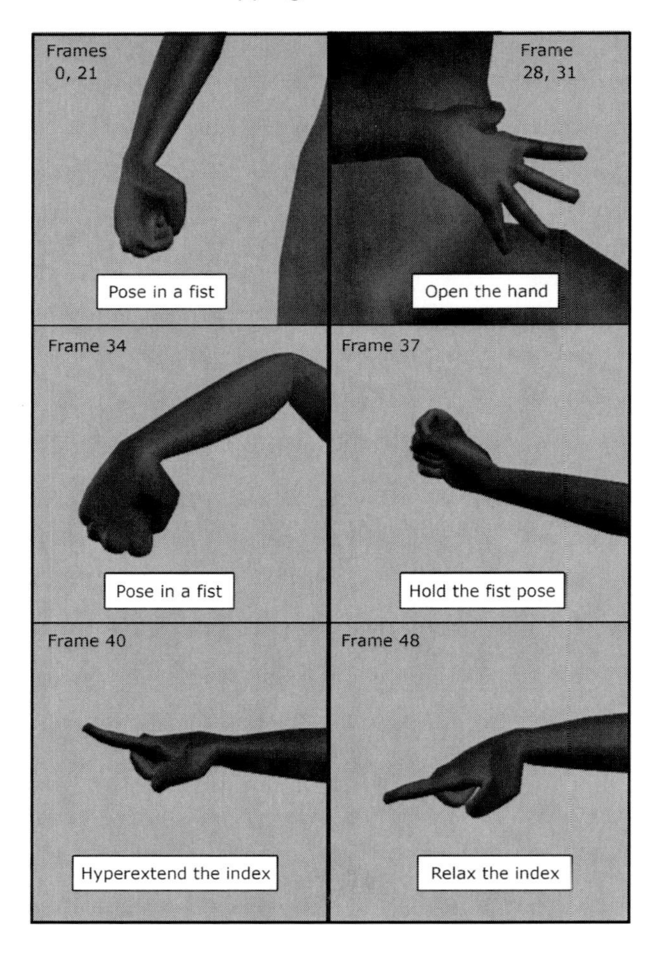

Figure 14_28.

Let's now apply the same process to the left fingers. Create a new layer and rename it "LeftHandFingers". Select the fingers for the left hand and set *Zero Keys* at frames 0, 24, 32, and 39. At frame 0 pose the left hand in a fist and hold the fist pose at frame 24. At frame 32 open the hand and at frame 39 close it in a fist again (Figure 14_29).
Note: Like you did with the right hand, feel free to offset the fingers in the left hand roughly one frame from each other to get a little bit of overlapping action.

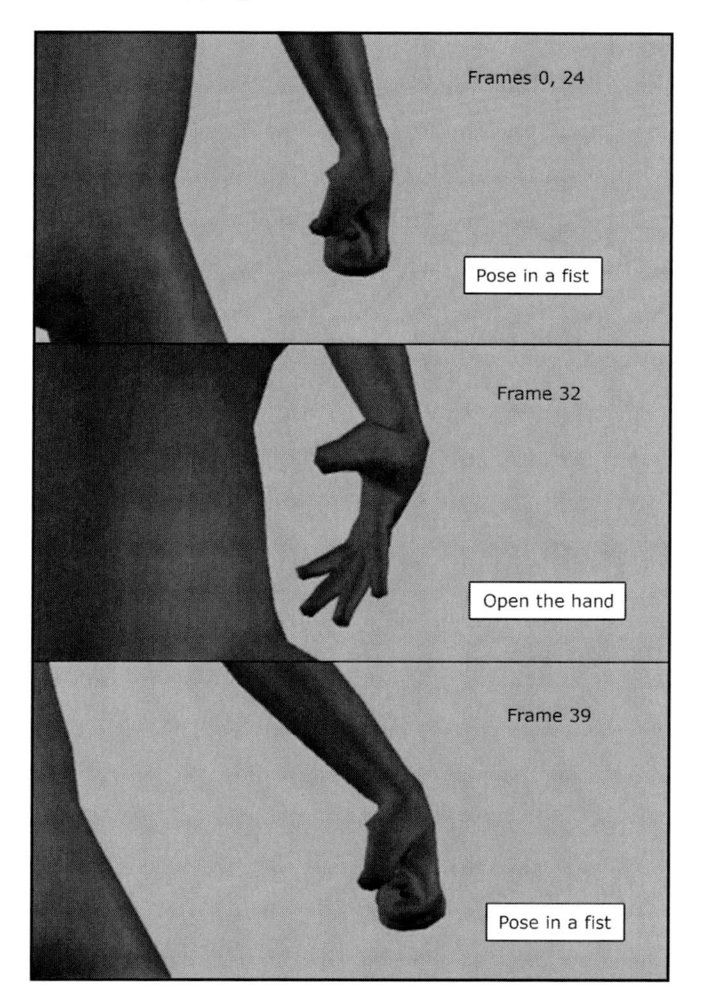

Figure 14_29.

We are done enhancing the motion. The last step will be to transfer the animation from the *Control Rig* onto the *Skeleton* so it can be later taken into **Maya**. Inside the *Character Controls* window, select *Plot Character* under the *Edit* drop-down. Select *Skeleton* from the pop-up window that appears (Figure 14_30)

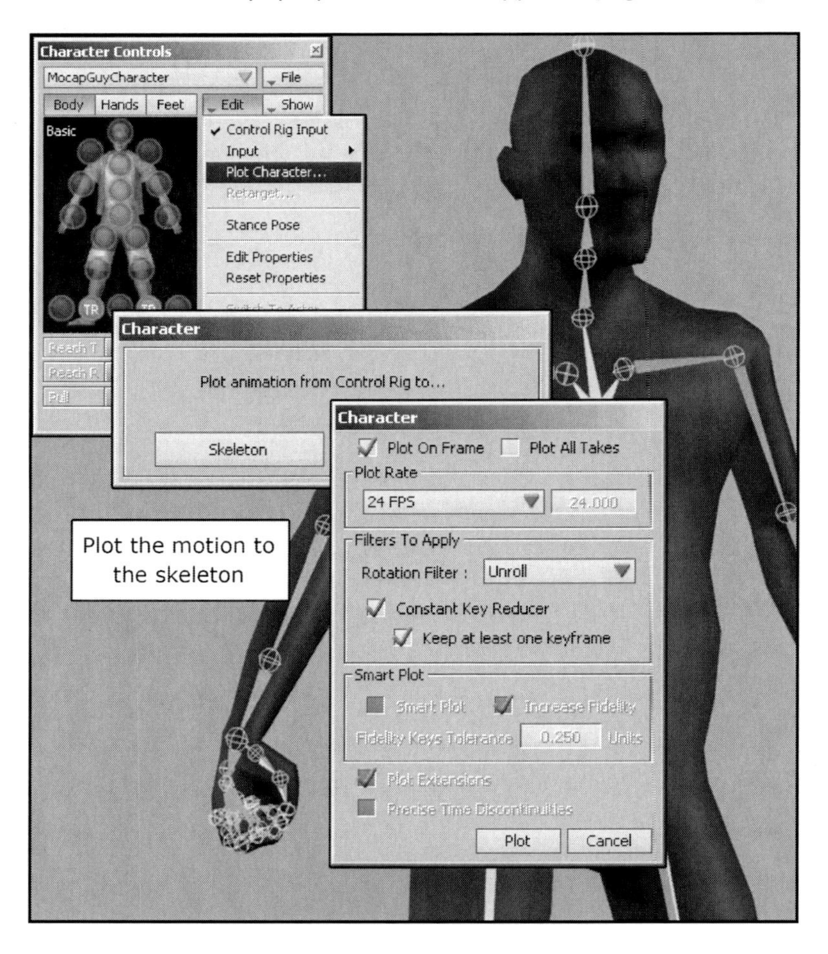

Figure 14_30.

The animation data has successfully been plotted to "MocapGuy's" original skeleton. "Chapter 16" will cover the process of how to get this animation into **Maya**.

Through the use of traditional animation techniques we went beyond fitting a motion from a physical performer to a digital counterpart. By combining motion capture and traditional animation we have obtained a performance that would have been extremely difficult to obtain otherwise. We pushed the given motion by the use of layer animation, the basic principles of animation and ancient practices like the Greek sculpting technique the *Contrapposto* (Figure 14_31).

Figure 14_31.

Chapter 15
Combining Motions

When dealing with motion capture it is sometimes necessary to string motions together to create a specific performance. This is particularly true in the video game realm where motions need to be combined with one another when responding to the player's commands. In this chapter we will concentrate on editing and combining motion with **Motion Builder's** *Story Tool* (Figure 15_01).

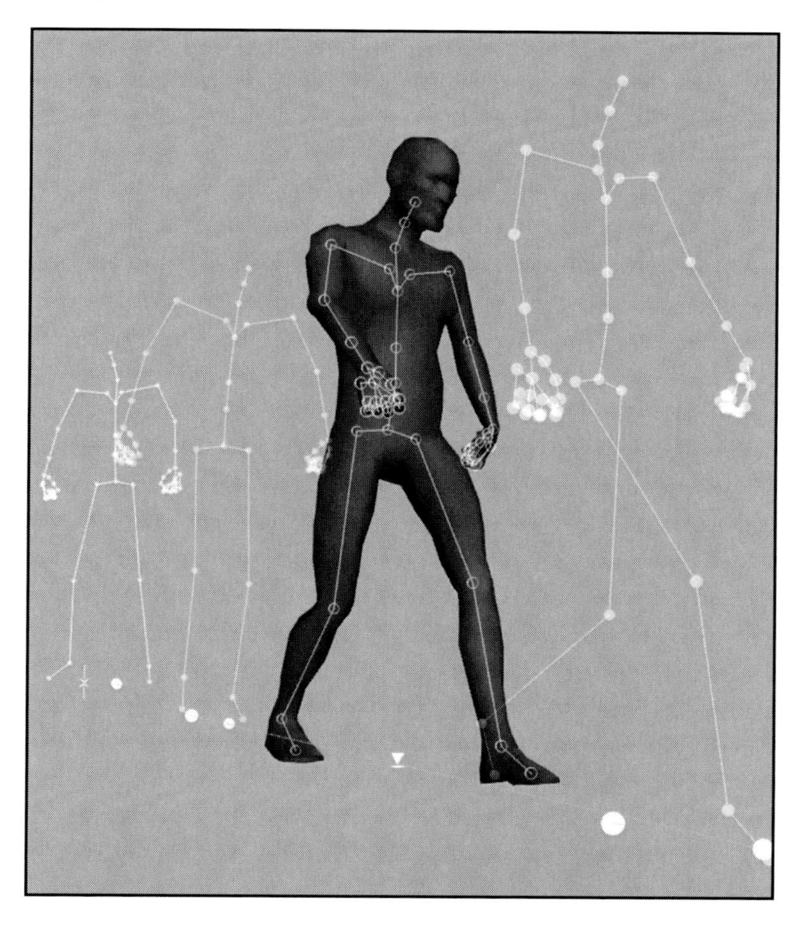

Figure 15_01.

Launch **Motion Builder** and open the "Chapter15"[12] file.

Select the *Story* tab in the *Navigator* window. Right click on the *Action Timeline* and select *Character Animation Track* from the *Insert* option in the pop-up menu. Pick the "MocapGuyCharacter" from the *Character* drop-down inside the track (Figure 15_02).

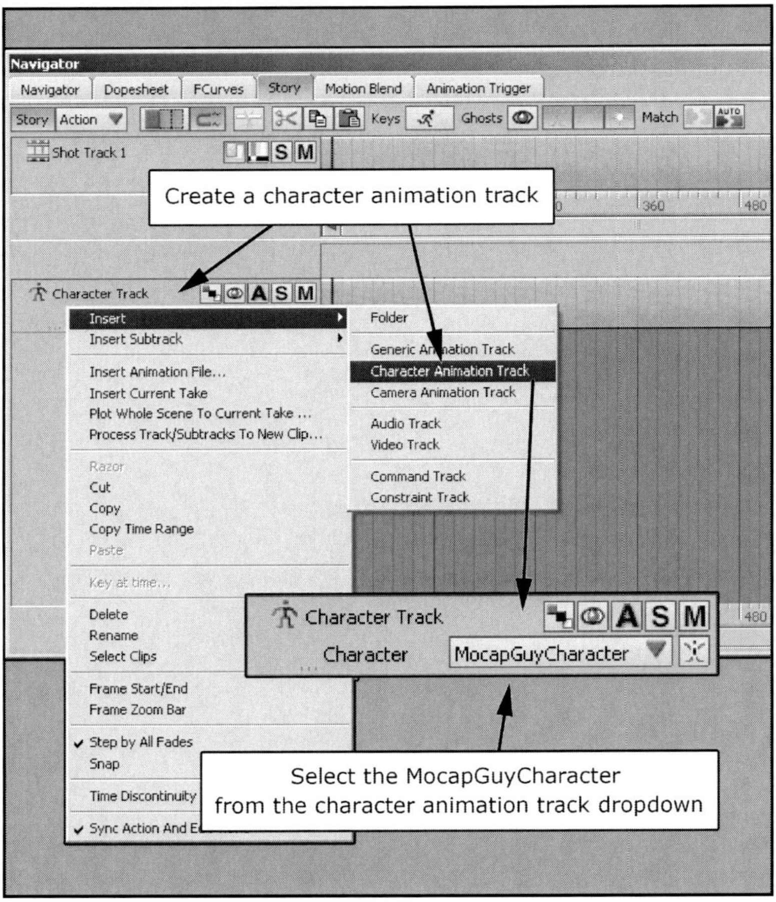

Figure 15_02.

[12] Download the files from www.MocapClub.com/TheMocapBook.htm

Right click on the left section of the *Assets Browser* window and select *Add favorite path* from the pop-up list. Navigate to the place where you downloaded the book files and select the "Chapter 15" folder (Figure 15_03).

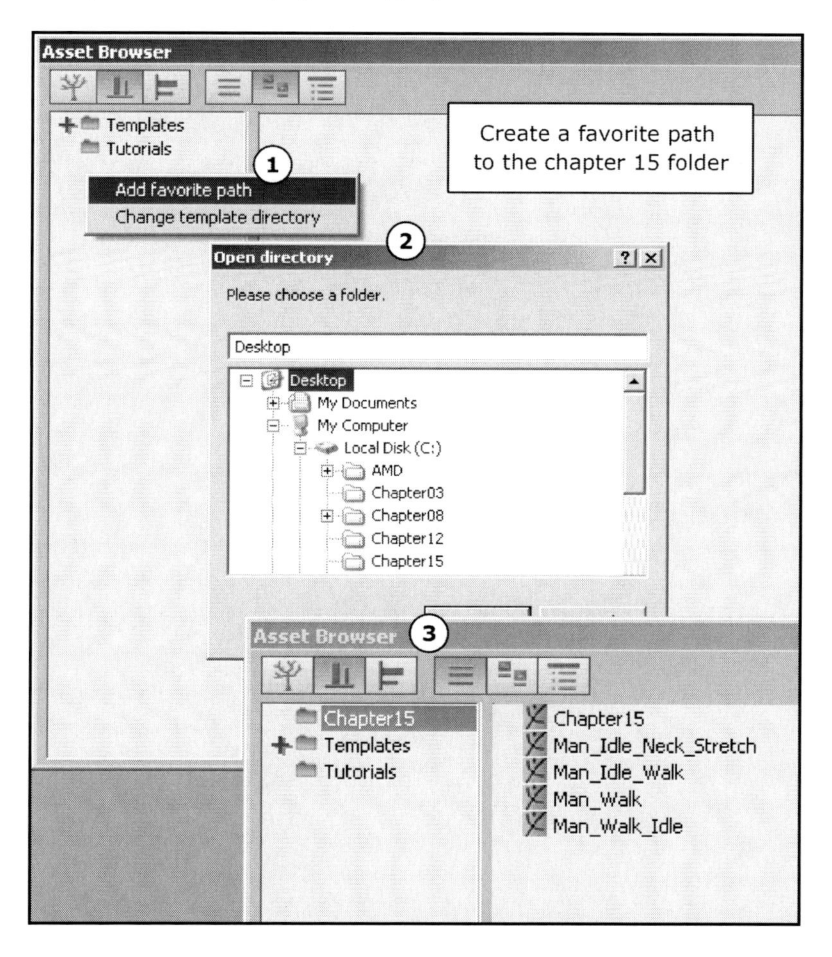

Figure 15_03.

Drag the “Man_Idle_Neck_Stretch” from the *Asset Browser* onto the *Character Track*. Place the *Timeline Indicator* at frame 0 and click + drag the clip until it snaps to the beginning of the timeline. Trim the tail of the clip by placing the cursor at the end of it and click + dragging to the left until frame 185 (Figure 15_04).

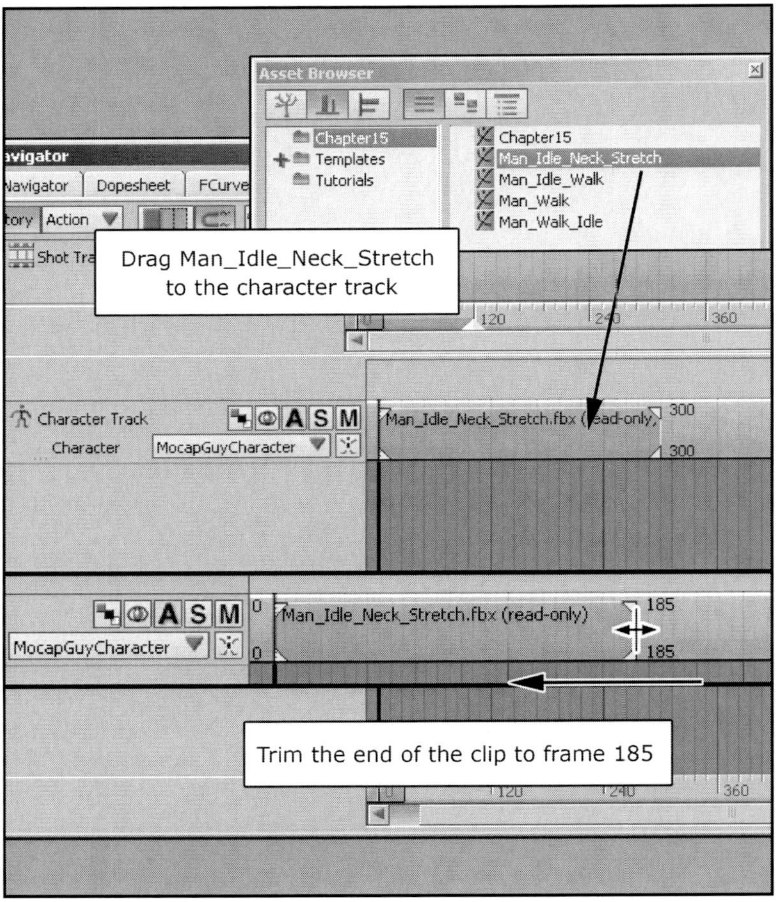

Figure 15_04.

Drag the "Man_Idle_Walk" file to the *Character Track*. Snap the clip to the end of the "Man_Idle_Neck_Stretch" motion. When the animation plays the "Man_Idle_Walk" motion starts a few grid units away from the previous clip. The spatial difference between the two clips is made more evident when you activate the *Show Ghost* button (Figure 15_05).

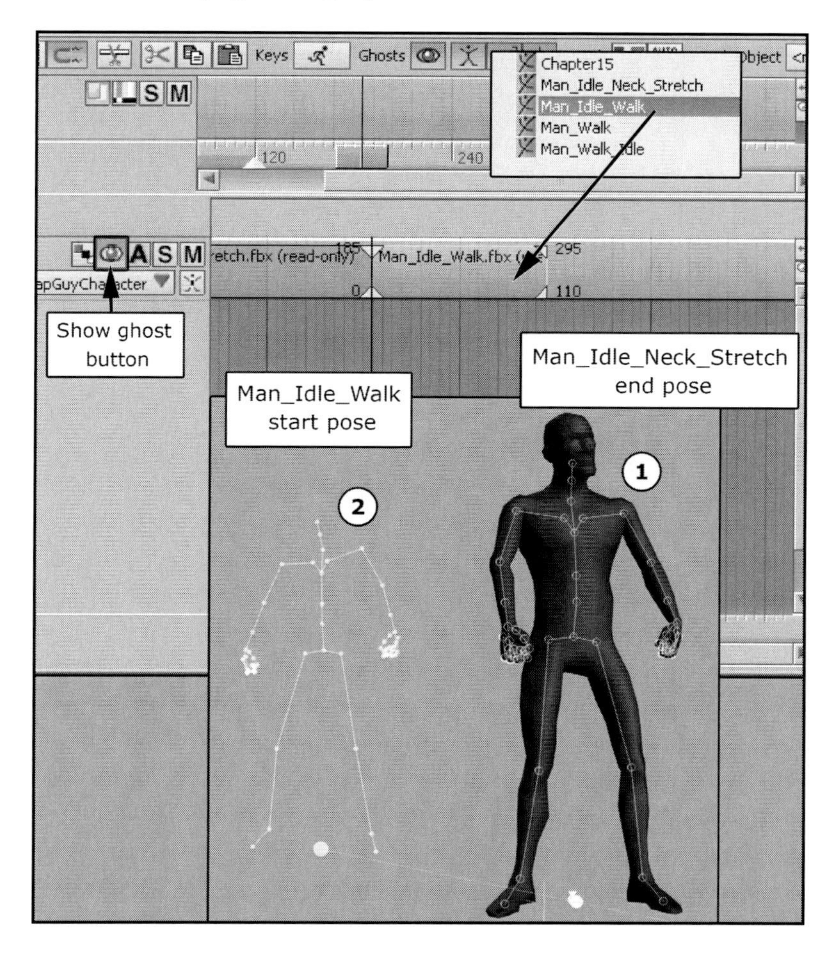

Figure 15_05.

With the “Man_Idle_Walk” clip selected, click the *Match* button in the *Story Controls*. Apply the options that figure 15_06 shows to the *Match Options* window and press the *OK button*. The “Man_Idle_Walk” clip starts its motion around the same area as the previous clip (Figure 15_06).

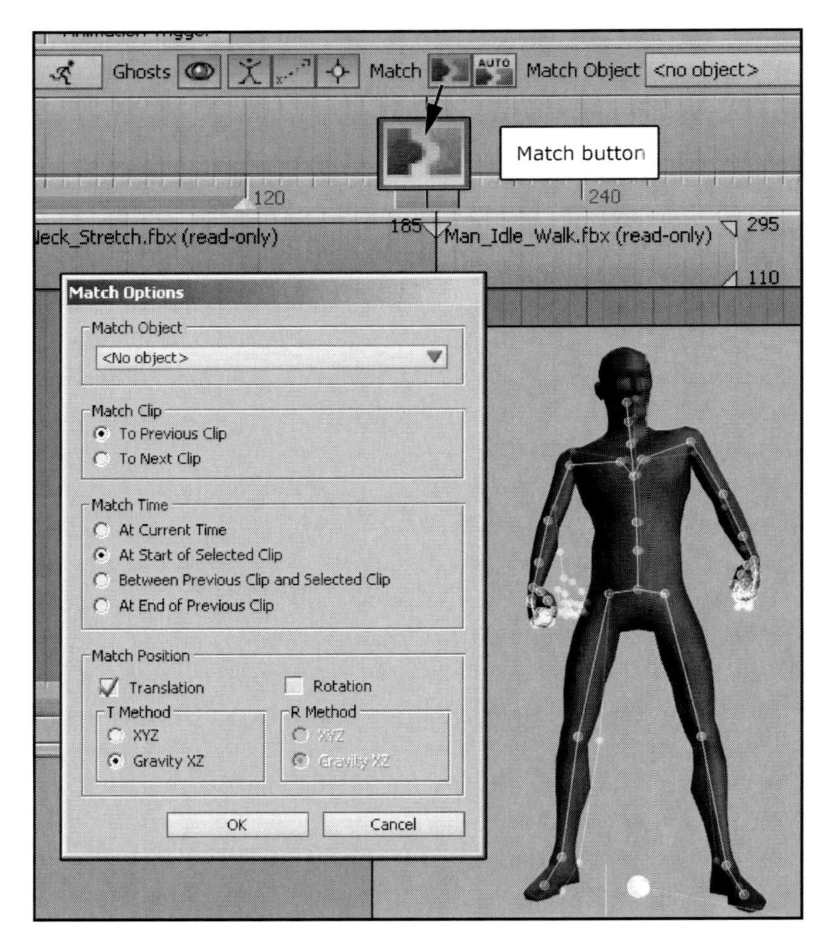

Figure 15_06.

Even though the poses from the end of the first clip and the beginning of the second one are fairly similar, there is still a jump between them.

Shorten the beginning of the clip by click + dragging the left edge until frame 17 appears in the left bottom of it. Overlap the "Man_Idle_Walk" motion onto "Man_Idle_Neck_Stretch" by 12 frames. Zoom-in to the clips if needed for increased accuracy by using the navigation buttons on the left section of the *Action Timeline* (Figure 15_07).

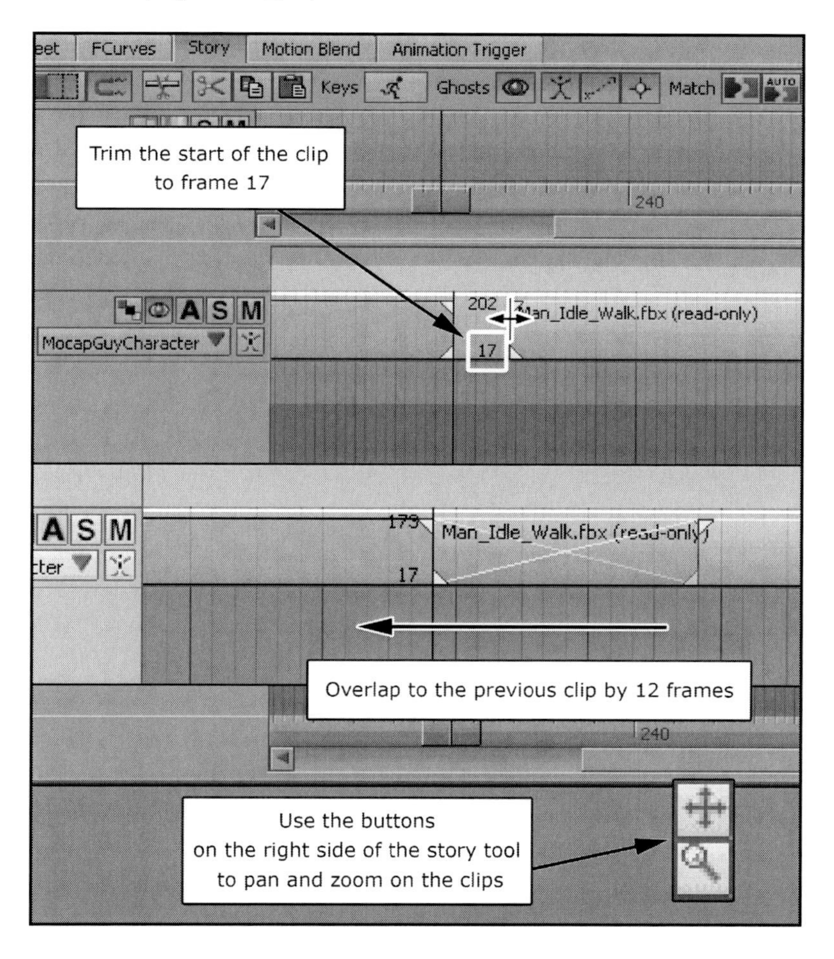

Figure 15_07.

Although the two motions are blending better, the feet are "*skating*" a little during the transition. Go to the middle of the blend select the "RightFoot" bone and press the *Match* button. Setup the settings of the *Match Options* window like figure 15_08 shows. The "RightFoot" now stays planted has the two motions blend (Figure 15_08).

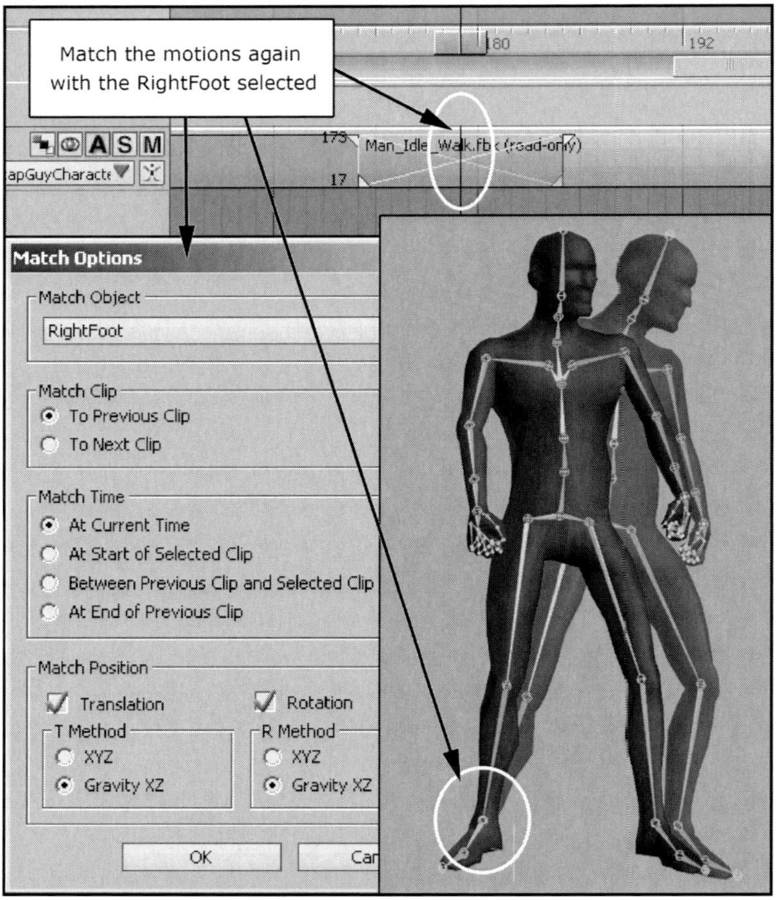

Figure 15_08.

We can still polish the transition a little bit. Double click the "Man_Idle_Walk" clip and go to the *Properties* tab. Change the *Fade In Interpolation* inside the *Fade In-Out* section from *Linear* to *Smooth*. Double click the "Man_Idle_Neck_Stretch" clip and change the *Fade Out Interpolation* to *Smooth*. This makes the transition between the two clips a little smoother than before (Figure 15_09).

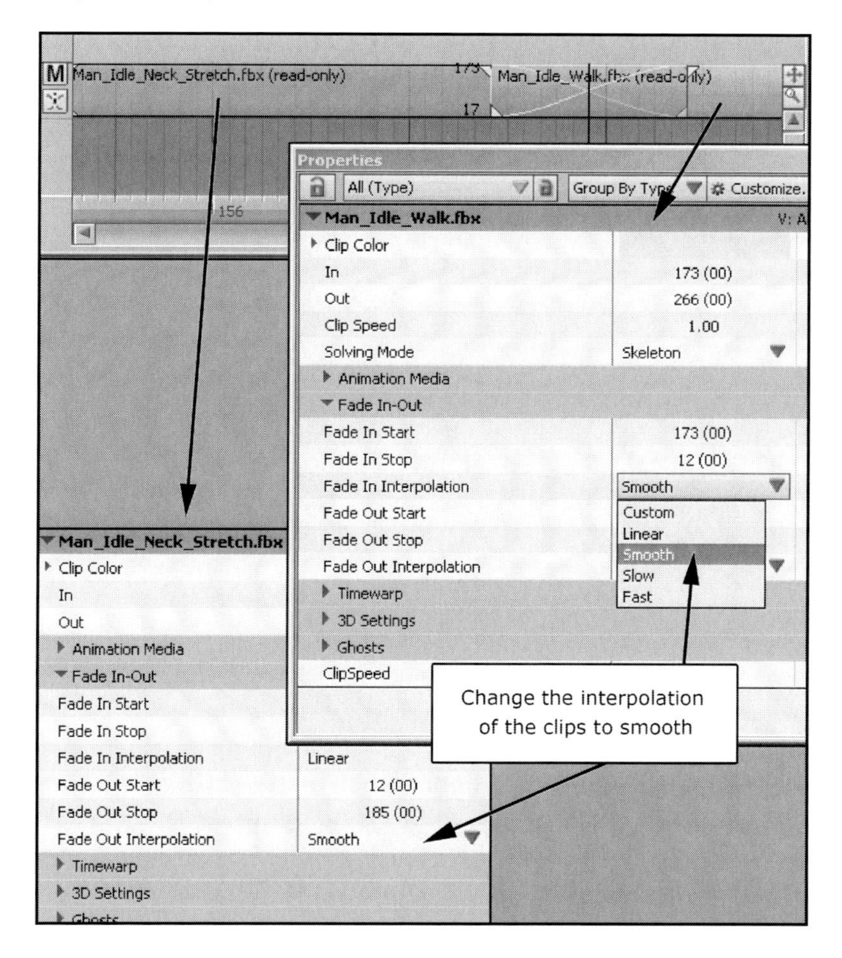

Figure 15_09.

Drag the “Man_Walk_Idle” file to the *Character Track* and snap it to the end of the previous clip. Match the motion to the “Man_Idle_Walk” clip using the settings that figure 15_10 indicates (Figure 15_10).

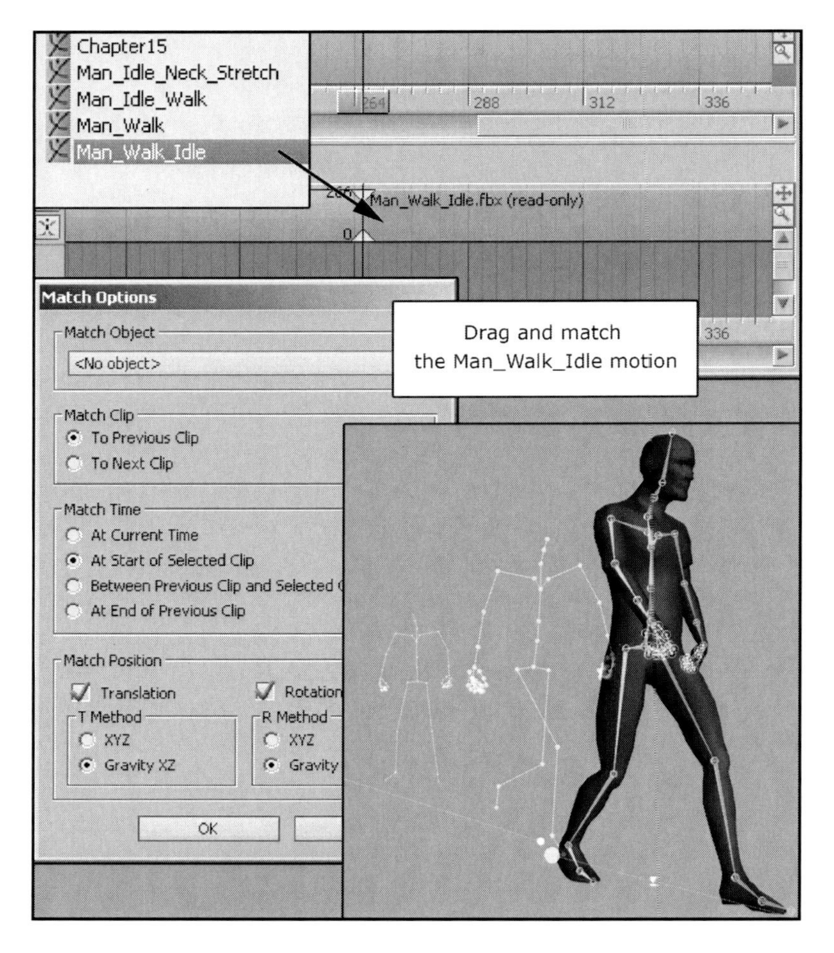

Figure 15_10.

When the animation plays the jump between the "Man_Idle_Walk" and "Man_Walk_Idle" clips becomes quite obvious. This is due to the difference between the leg positions between the end of one clip and the beginning of the next one. Trim the end of "Man_Idle_Walk" until the lower right number reads 98. Trim the beginning of "Man_Walk_Idle" clip until the lower left number reads 22. Snap the clips together and match them again (Figure 15_11).

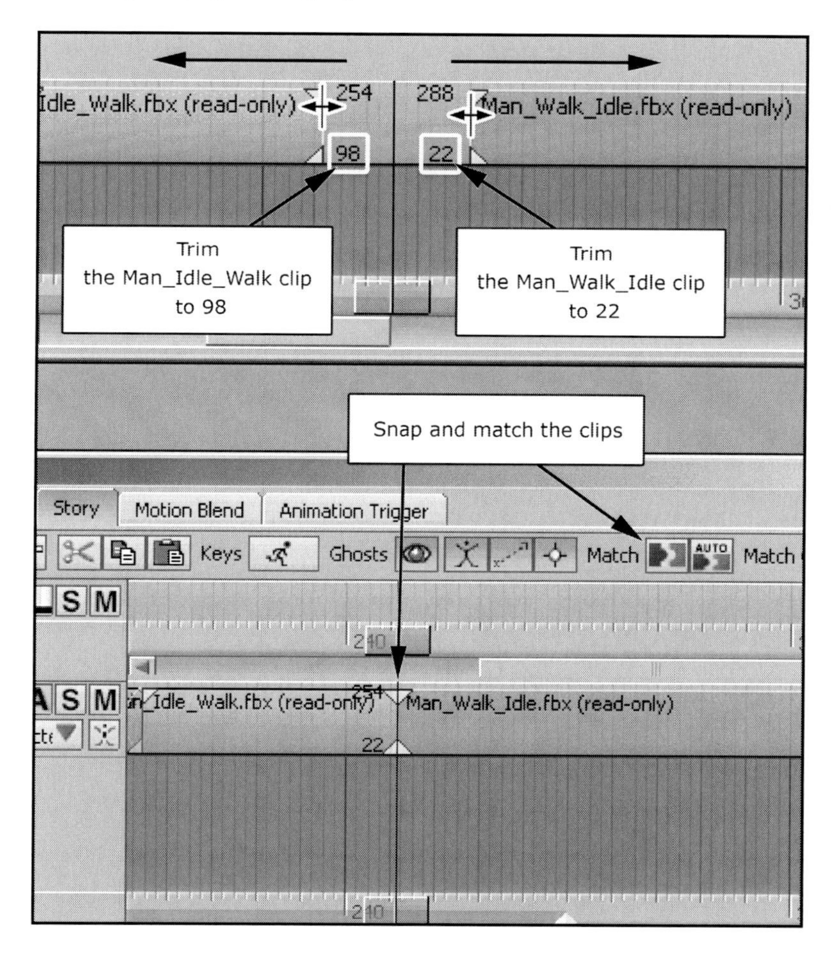

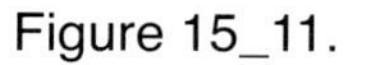
Figure 15_11.

Select the "Man_Walk_Idle" clip and extend the beginning of it so it overlaps with "Man_Idle_Walk" by 6 frames. Go to the middle of the blend and match the two motions based on the "LeftFoot". Change the *Fade Out Interpolation* of "Man_Idle_Walk" as well as the *Fade In Interpolation* of "Man_Walk_Idle" to smooth. Trim the end of the last clip to 168 according to the lower right corner of the track (Figure 15_12).

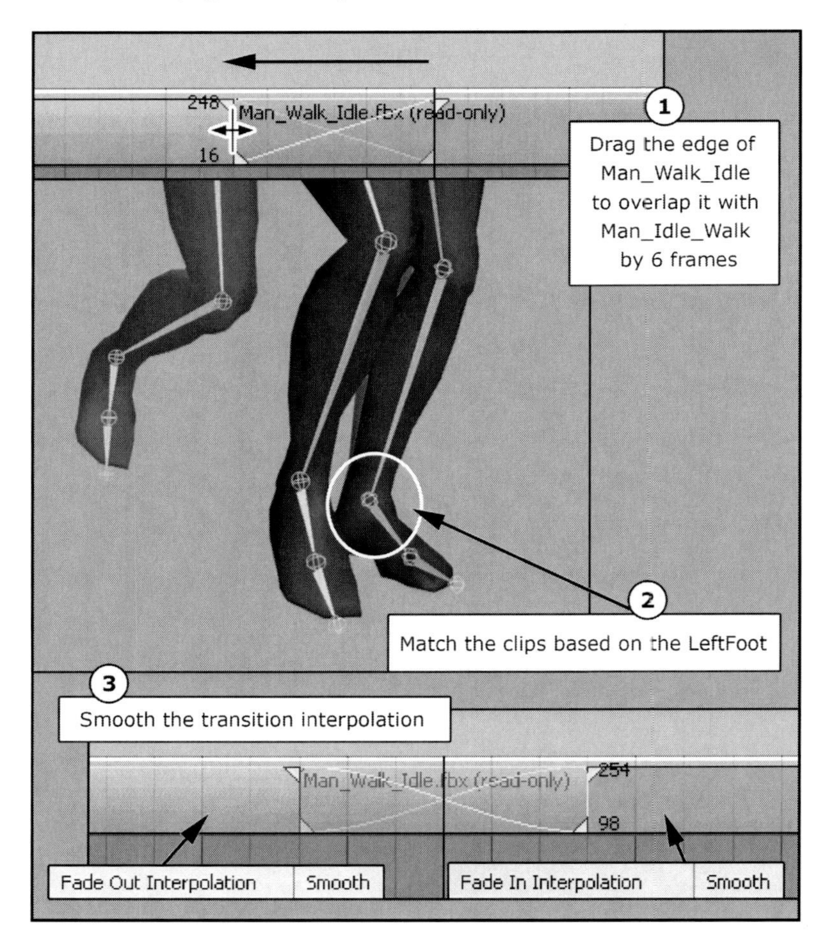

Figure 15_12.

The three independent motions that we started with are now transitioning properly from one to the next. The last step is to plot the animation onto the *Skeleton* so we no longer depend on the *Story Tool*. Right click inside the *Character Track* and select *Plot Whole Scene To Current Take*. This bakes the animation to the rig's bones. Right click in the *Character Track* and select *Delete*. Note how the "MocapGuyCharacter" moves exactly the same as it did before even though there are no clips in the story tool anymore (Figure 15_13).

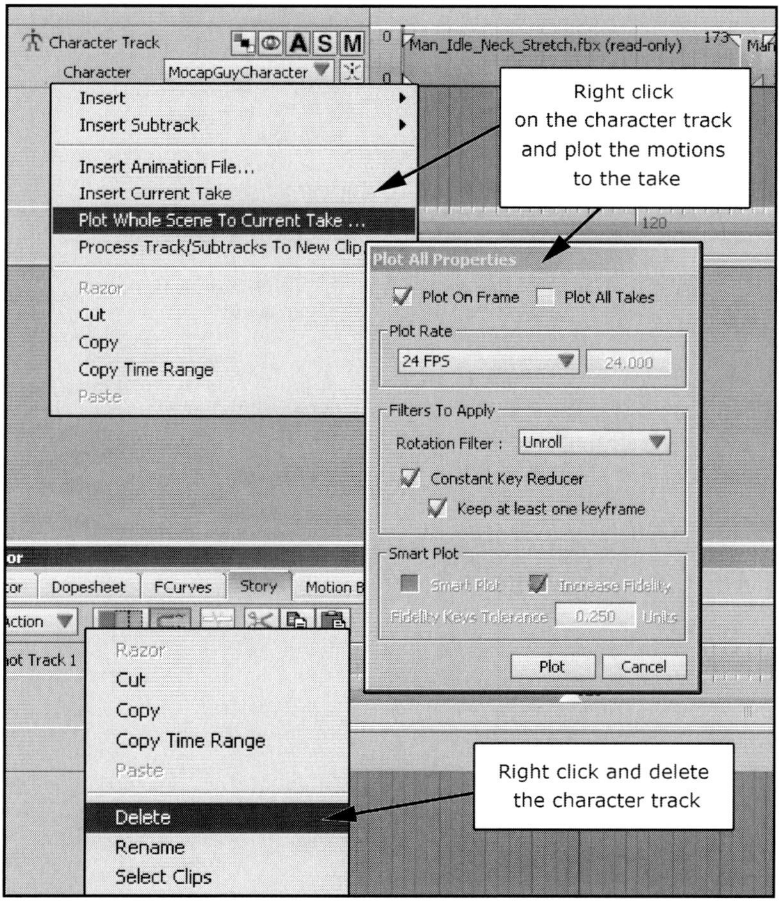

Figure 15_13.

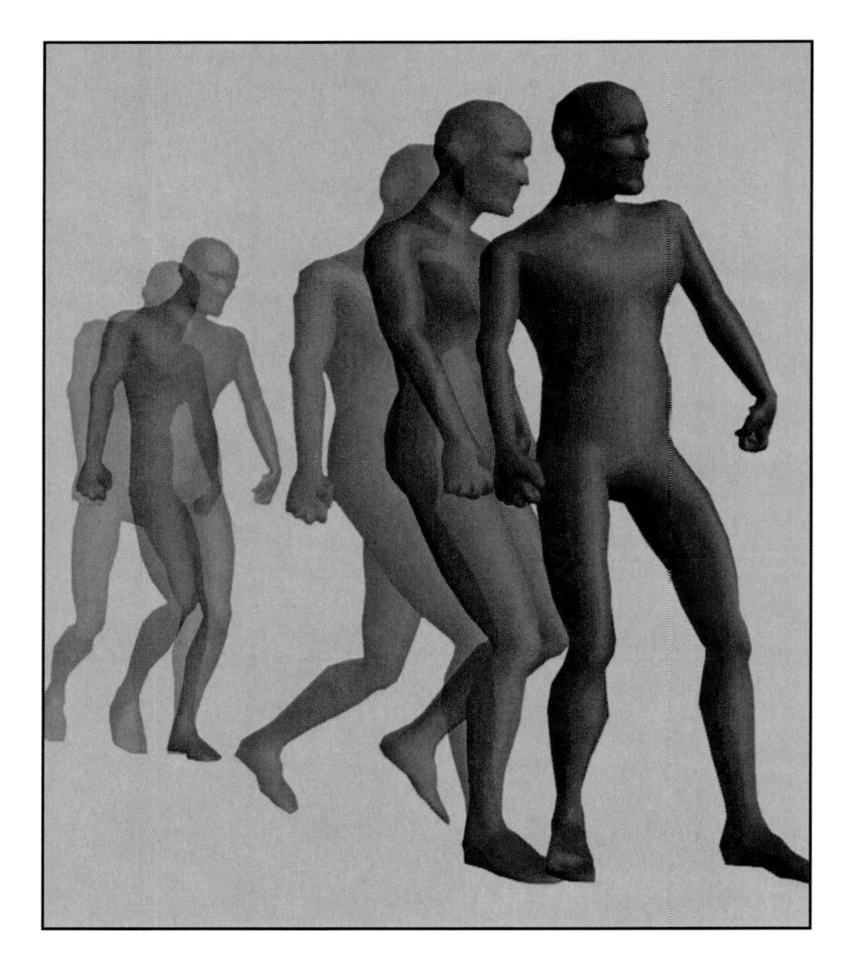

Figure 15_14

Part 06
Back In Maya

Chapter 16
File Integration

In this last section of the book we will concentrate on the steps needed to take the animation from **Motion Builder** to the rig that we created earlier in **Maya**. We are only interested in taking the back animation because we already have a **Maya** file with the character's geometry and a skeleton hierarchy. In fact, while the animator works in **Motion Builder**, the rigger can continue adding complexity to the rig in **Maya** by adding muscle effects and refining the deformations (Figure 16_01).

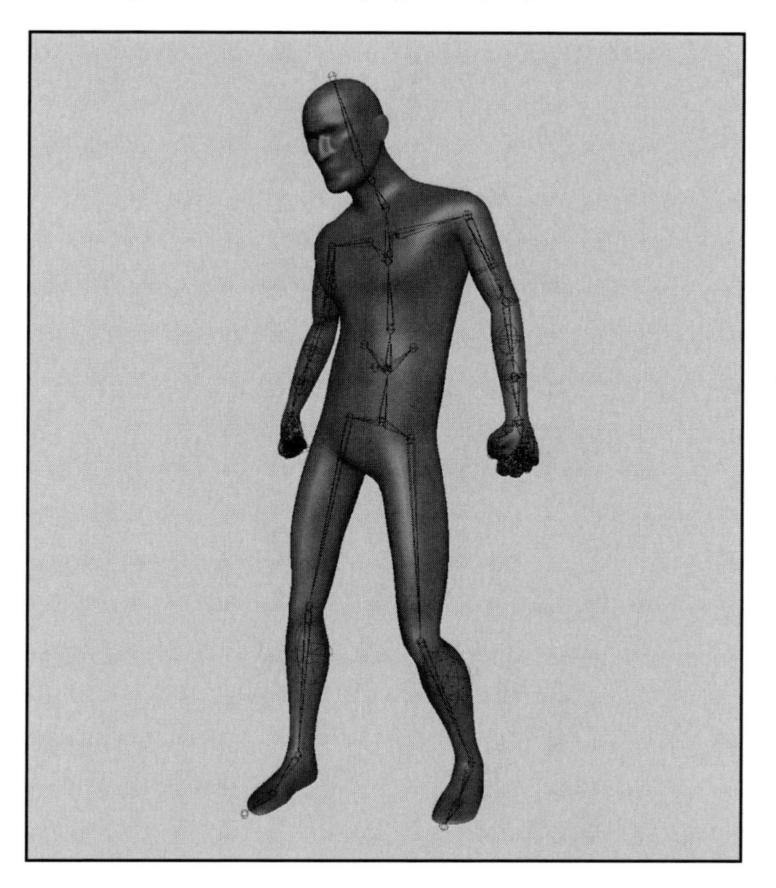

Figure 16_01.

Launch **Motion Builder** and open the "Chapter16MB"[13] file.

Let us start by cleaning this file. Since we are only interested in the animation of this file, we are going to delete everything but the original skeleton and the *Characterization*. Select "MocapGuyControlRig" in the *Navigator* window, right click and choose *Delete* from the drop down menu (Figure 16_02).

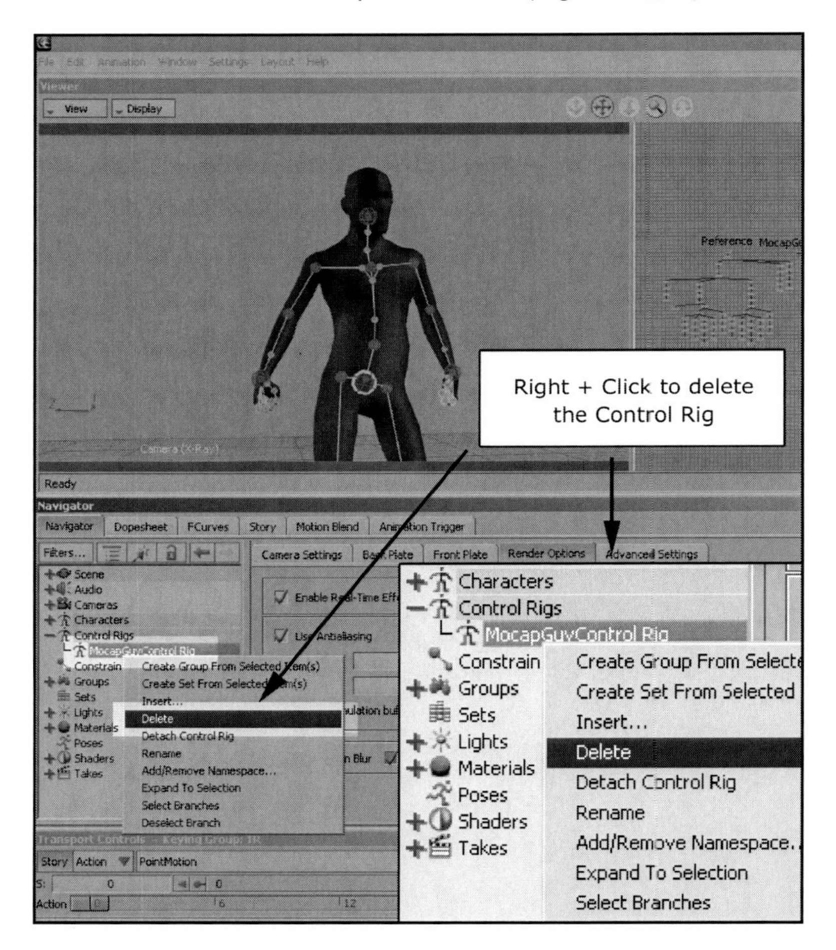

Figure 16_02.

[13] Download the files from www.MocapClub.com/TheMocapBook.htm

Note: Since the animation was plotted to the Skeleton in Chapter 14 we can delete the Control Rig and still keep the animation.

Call the *Schematic View* pressing *Ctrl* + *"W"* while the cursor is inside the *Viewer* window. Select "MocapGuy_Geo" and delete it. Also select *Camera* as well as *Camera Interest* and delete them (Figure 16_03).

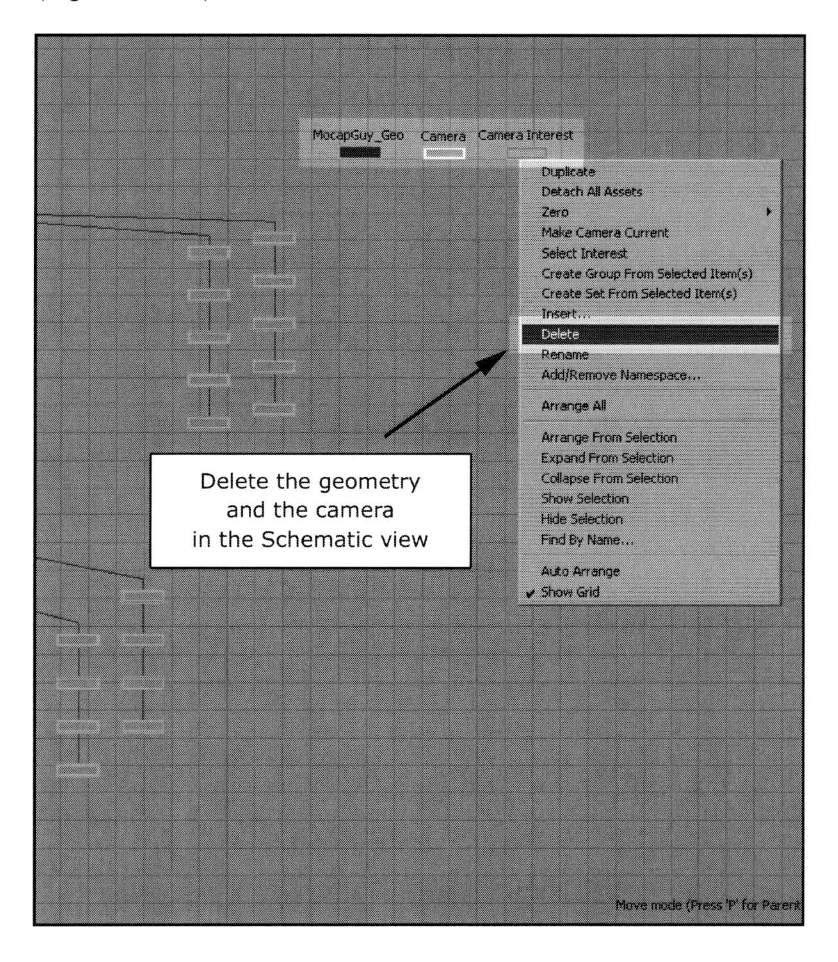

Figure 16_03.

We have successfully cleaned the file, only the original hierarchy and the *Characterization* remain (Figure 16_04).
Save the file as “Chapter16_Cleaned”.

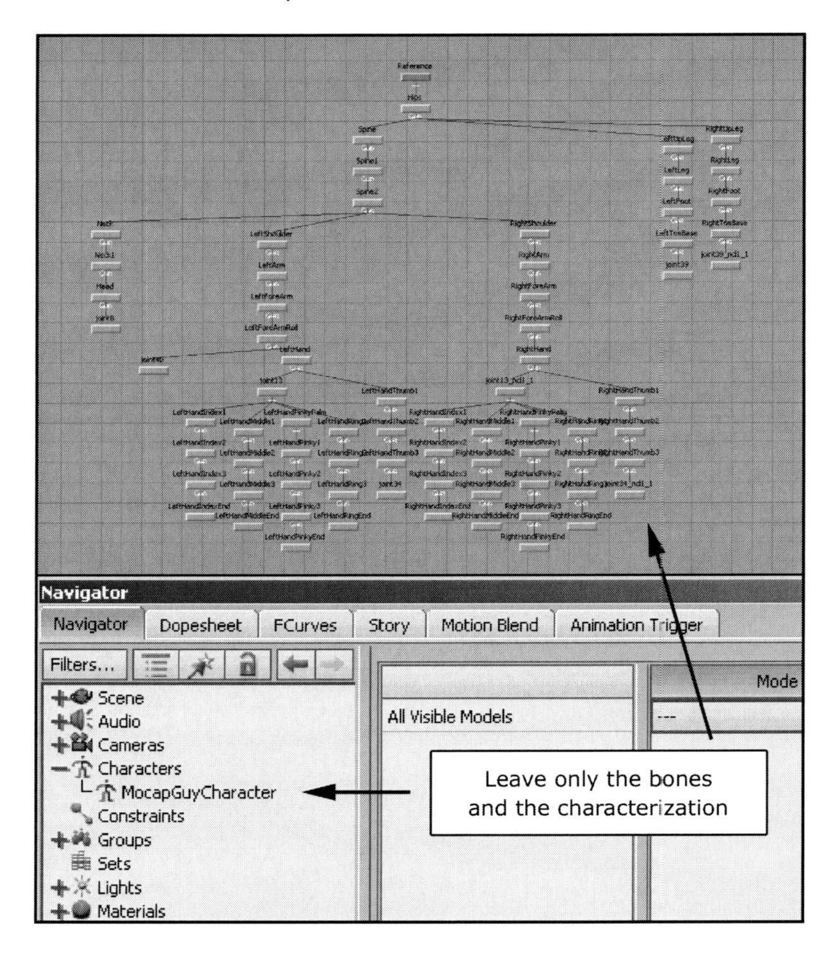

Figure 16_04.

Lunch **Maya** and open the “Chapter16MY” file.

If you rotate the elbows, wrists, feet and spine of the rig you can see how muscle effects have been added to “MocapGuy” by the use of corrective *Blend Shapes*, deformers and *Set Driven Keys* (Figure 16_05).

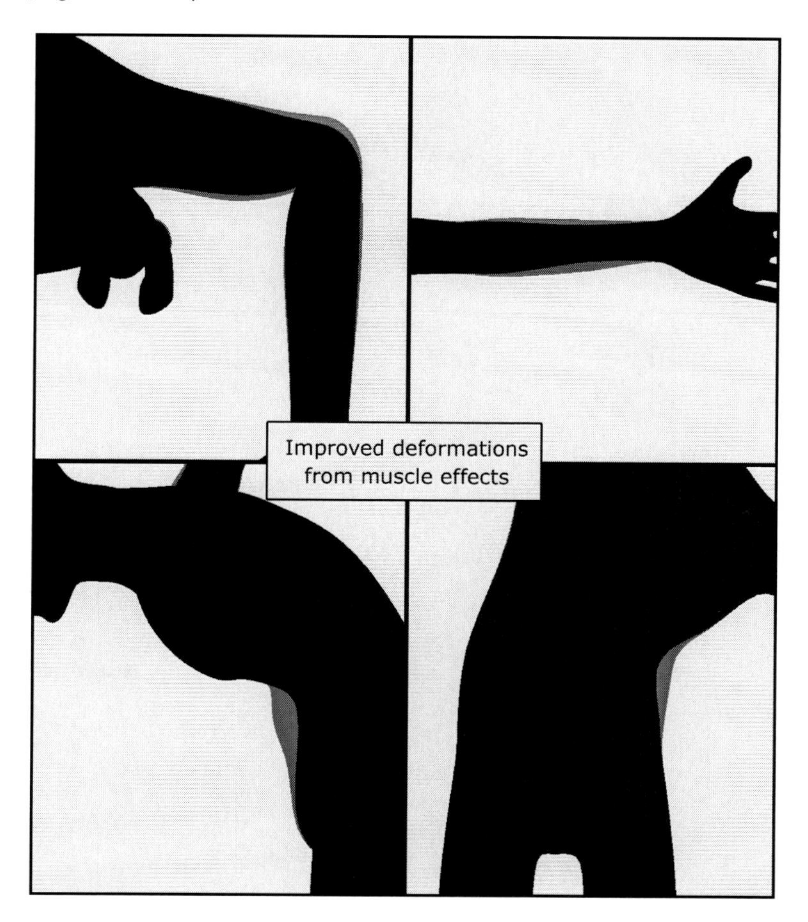

Figure 16_05.

Note: To learn more about Blend Shapes, deformers and Set Driven Keys for body rigs, refer to Chris Maraffi’s “Mel Scripting a Rig in Maya” book or the “Hyper-Real” series from Autodesk (formerly Alias).

Call the *Import* window from the *File* menu. Set the *Files of Type* option to *Fbx* and Navigate to the place where you saved the "Chapter16_Cleaned" file (You are also welcomed to use the "Chapter16MB_PlottedBook"). Press the *Import* button and populate the settings of the *FBX Importer* window as figure 6 illustrates (Figure 16_06).

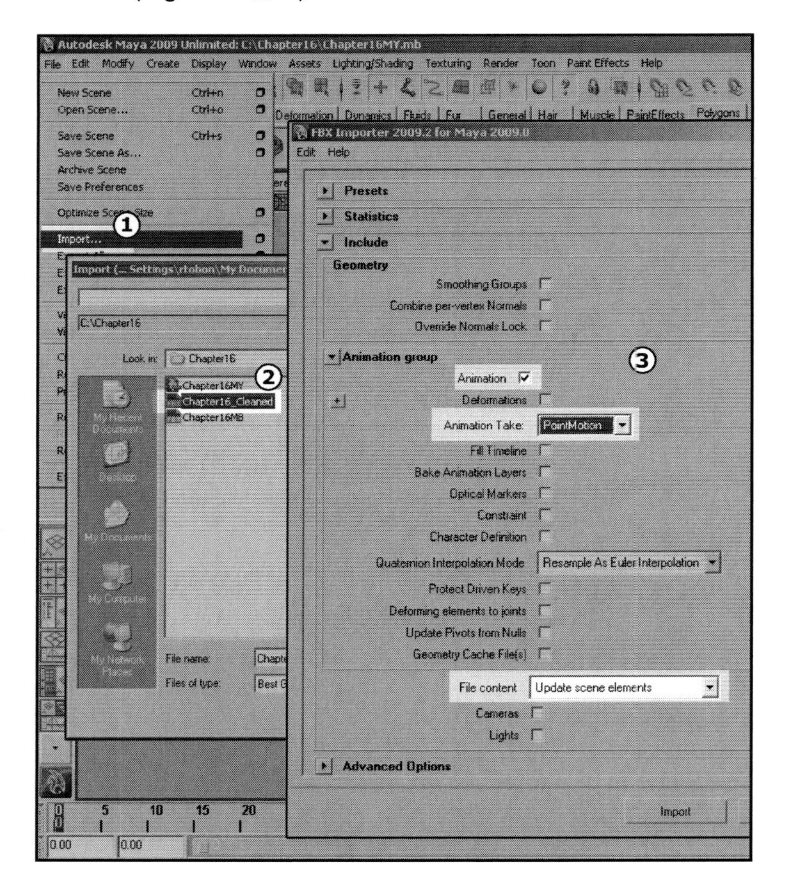

Figure 16_06.

Note: If the FBX option does not appear under the Files of Type drop-down of the Import window, activate the fbxmaya.mll option in the Plug-in Manager window. You will find the Plug-in Manager in the Settings/Preferences section of the Window menu.

It is easy to see that "MocapGuy" is a little big for the **Maya** scene. This can be fixed by restoring the "Reference" node to its original size. Select the *Hypergraph: Hyerarchy* option from the *Window* menu and click on the "Reference" node inside the window that pops-up. Type 0.5 in the X, Y and Z scale channels. You can see that "MocapGuy" returns to the size that it originally had when we first opened the file (Figure 16_07).

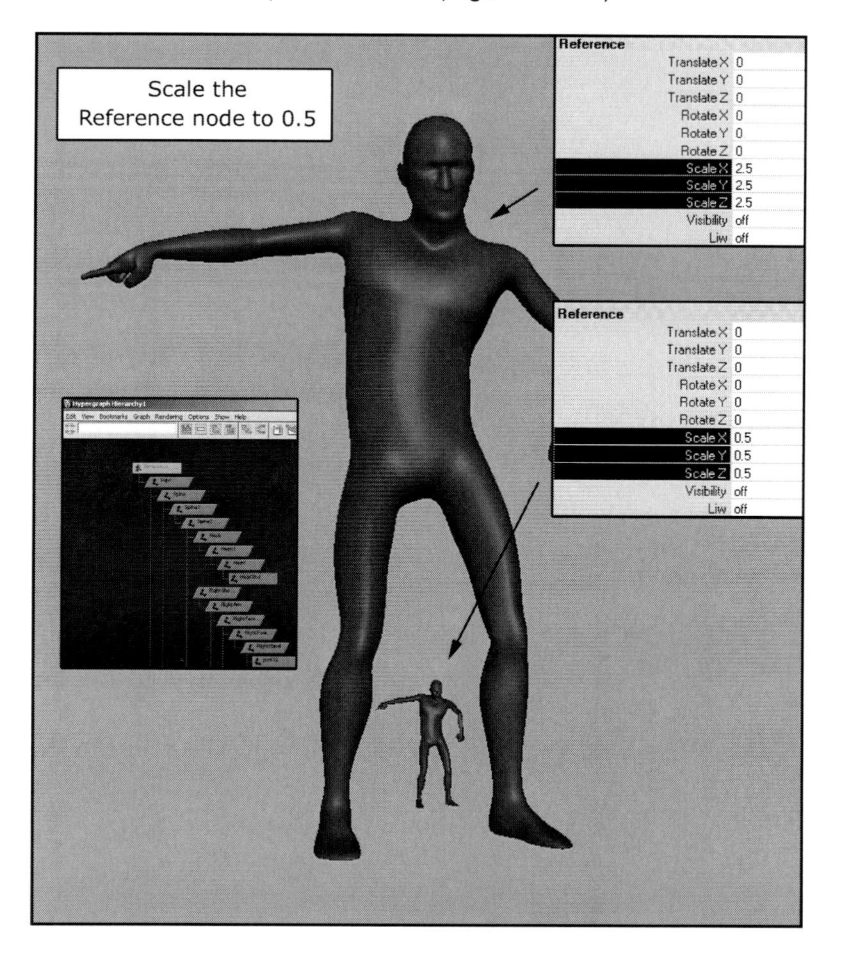

Figure 16_07.

Note: You can also use the Reference node to reposition "MocapGuy" around the scene.

The animation has been successfully retargeted to an upgraded version of "MocapGuy's" rig inside the **Maya** file. Cameras, lights and additional elements can be added to the scene so it can be rendered later (Figure 16_08).

Figure 16_08.

THE END

This book has guided you through the process of digitizing, motions and performances. It also covered how to enhance those performances and apply them to a CG character.

By going through the practical exercises in this book you have become more knowledgeable about capturing, tracking, solving, integrating, enhancing and combining motions.

This book also provided an example of a **Cortex** to **Motion Builder** to **Maya** motion capture pipeline

I hope you enjoyed the material!

To continue your motion capture learning and to download free motion capture data visit: www.mocapclub.com

Appendix A
Cortex Navigation Quick-Table

Command	Action
Alt + Left Mouse Button	Camera Orbit
Alt + Middle Mouse Button	Camera Pan
Alt + Right Mouse Button	Camera Zoom
I key	Zoom In XYZ Graphs
O key	Zoom out XYZ Graphs
F key	Next Frame
S key	Previous Frame
E key	Up the Marker List
D key	Down the Marker List
Y key	Make Markers Unnamed

Appendix B
Maya Navigation Quick-Table

Command	Action
Alt + Left Mouse Button	Camera Orbit
Alt + Middle Mouse Button	Camera Pan
Alt + Right Mouse Button	Camera Zoom
Q key	Selection Mode
W key	Move
E key	Rotate
R key	Scale
Ctrl + A	Attributes
F8 key	Object Mode
F9 key	Component Mode

Appendix C
Motion Builder Navigation Quick-Table

Command	Action
Ctrl + Left Mouse Button	Camera Orbit
Shift + Left Mouse Button	Camera Pan
Ctrl + Shift + Left Mouse Button	Camera Zoom
T key	Translate
R key	Rotate
S key	Scale
Ctrl + A	Display toggle
Ctrl + W	Schematic View
Ctrl + E	Perspective View
K	Set Key